BIOCHEMISTRY RESEARCH TRENDS

CHLORPYRIFOS

TOXICOLOGICAL PROPERTIES, USES AND EFFECTS ON HUMAN HEALTH AND THE ENVIRONMENT

BIOCHEMISTRY RESEARCH TRENDS

Additional books in this series can be found on Nova's website
under the Series tab.

Additional e-books in this series can be found on Nova's website
under the e-book tab.

BIOCHEMISTRY RESEARCH TRENDS

CHLORPYRIFOS

TOXICOLOGICAL PROPERTIES, USES AND EFFECTS ON HUMAN HEALTH AND THE ENVIRONMENT

CAITLIN MAYES
EDITOR

nova publishers
New York

Library of Congress Cataloging-in-Publication Data

ISBN: 978-1-63482-111-7

Library of Congress Control Number: 2015931178

Published by Nova Science Publishers, Inc. † New York

CONTENTS

PREFACE

Chlorpyrifos (CPF) is a broad spectrum organophosphate pesticide. Organophosphate pesticides comprise a group of substances used in agriculture for insect and plague control, infestations in buildings, man or domestic animals. The use of insecticides represents an environmental risk due to the high mass of product applied annually. This book discusses the toxicological properties, uses and effects on human health and environment of the organophosphate pesticide, chlorpyrifos.

Chapter 1 - In Colombia the organophosphate pesticide Attamix SB with active ingredient Chlorpyrifos in a concentration of 0.00125 g Kg^{-1} is commonly used as an insecticide of coffee, sugar cane, pine and eucalyptus crops; where the two firsts were the 85% of the total agricultural production in the country in 2013 according to the Agricultural Society of Colombia. This indicates that the use of this product is massive in this country and having in mind that the agricultural production is poorly technicized, it can cause a huge impact in the environment due to an incorrect use. As have been demonstrated that the Chlorpyrifos is toxic to fish, aquatic invertebrates, small mammals, and birds, it was decided to practice a toxicity characteristic leaching procedure (TCLP) of Chlorpyrifos from Attamix SB, because this test identifies if the waste is hazardous and should be handled as hazardous waste and try to replicate the conditions of a municipal landfill to detect a contaminant (metal or organic compound concentration) in the water that can leach from the soil. There were analyzed 10 samples by TCLP and it was identified the active ingredient by gas chromatography. The results showed that the Chlorpyrifos is retained mainly by the solid phase, indicating that in natural conditions, the soil is the reservoir of the contaminant with concentrations that permit classified it as hazardous waste.

Chapter 2 - Organophosphate pesticides comprise a group of substances used in agriculture for insect and plague control, infestations in buildings, man or domestic animals. The use of insecticides represents an environmental risk due to the high mass of product applied annually. Chlorpyrifos (CPF) is a broad-spectrum-organophosphate pesticide. The primary target of CPF toxicity is the central and peripheral nervous system, due to its ability to inhibit the acetylcholinesterase activity. However, other actions of CPF as the effects on synthesis of macromolecule like DNA, RNA and proteins, interactions with neurotransmitter receptors and other neurochemical affections were described. Breast cancer is the most frequent malignant disease in women. Exposure to estrogens throughout woman's life is a risk factor for this malignancy. Estrogen receptor alpha (ERα) is the major regulator of breast cancer tumor behavior. In the mammary gland, 17β-estradiol (E2) promotes cell proliferation in both normal and transformed epithelial cells by modifying the expression of hormone responsive involved genes. Endocrine disrupter (ED) is defined as a compound of industrial or natural origin that interferes with hormone biosynthesis, action and metabolism, resulting in a deviation from normal homeostatic control or reproduction. CPF was recognized as an ED since it has been demonstrated to possess the ability to interfere with the ERα responses. Moreover, CPF possesses antiandrogenic activity and significantly decreases testosterone biosynthesis. The authors demonstrated the ability of CPF environmental concentration to induce cell proliferation through ERα in hormone-dependent MCF-7 breast cancer cells. In contrast, higher doses of CPF promoted cell cycle arrest in S–phase modifying checkpoints proteins, through a mechanism that may involve changes in redox balance in MCF-7 cells. In hormone-non-dependent MDA-MB-231 cells, the authors demonstrated that CPF is not able to promote cell proliferation but induces G2/M cell cycle arrest. In breast cancer cells, CPF induces redox imbalance, which lead to the cell cycle arrest. In turn, redox imbalance was trigged by an increment of p-ERK1/2 levels which is a result of an increment of hydrogen peroxide production induced by CPF. *In vivo*, they have recently observed that virgin rats exposed to CPF present a proliferative mammary tissue similar to the pregnant state, lower estradiol circulating levels and a prolonged estrous cycle. These results point to CPF as an ED stimulating the mammary gland proliferation and altering the estrogenic balance at systemic levels. Taken all this into account, the use of CPF should be strictly controlled considering that this pesticide could affect human or animal health and environment.

Chapter 3 - Chlorpyrifos is an organophosphate insecticide widely used in a variety of food crops in Argentina and other countries. In a previous work, the authors have reported that subchronic exposure (14 days) to an environmental concentration of chlorpyrifos (5 $\mu g\ L^{-1}$) caused severe effects on reproduction of the freshwater hermaphrodite gastropod *Planorbarius corneus*. The aim of the present work was to study the effects of chlorpyrifos on the reproduction of this adult snail after an acute (2 days) exposure. Also B-esterase activities (cholinesterases and carboxylesterases) were measured in the gonads, to study the possible link of the biochemical responses to the reproduction effects. Snails were exposed during 2 days to 5 $\mu g\ L^{-1}$ of chlorpyrifos. Water control and solvent control (acetone) group were included in the bioassay. After 2 days of exposure, some snails per group were used to measured enzyme activities and other snails were transferred to water. Then, the bioassay was continued for 14 days. Snails were fed once a week. During the 14 days in water, the number of egg masses was registered and, for each egg mass, the number of egg masses without eggs, the number of eggs per mass, the time for hatching, the hatchings and the survival of the offspring were counted. After 2 days of exposure and after 2 and 14 days in water, snails were placed on ice, the shells were removed and gonad homogenates were realized. Supernatant fraction (6 for each time and for each group) was used to measure cholinesterase and carboxylesterase activities and protein content. After an acute exposure, 5 $\mu g\ L^{-1}$ of chlorpyrifos caused a decrease on cholinesterase activity using acetylthiocholine as substrate (45 %) and carboxylesterase activities (26 % using p-nitrophenyl acetate as substrate and 40 % using p-nitrophenyl butyrate) in gonads. After 14 days in water, no recovery of cholinesterase activity was observed. However, carboxylesterases measured with both substrates returned to control values. Regarding the reproduction effects, only an increase in the number of egg masses without eggs and an increase in the time for hatching were observed. Recovery of carboxylesterases in gonads could be related to the lower effects observed on reproduction in comparison with the subchronic exposure. After 14 days of exposure, 5 μg L^{-1} of chlorpyrifos had produced a strongly decrease in B-esterase activities and severe effects on reproduction, such as the decreases in the % of hatching and survival of the offspring, the increases in egg masses without eggs and in the time of hatching. Exposures to environmental concentrations of chlorpyrifos caused different toxic effects on the reproduction of *Planorbarius corneus* depending on the time of exposure.

Chapter 4 - Organophosphates are used frequently in Europe for pest control due to their low price and broad spectrum of activity. In 2003, they accounted for over 59% (4645 tonnes) of insecticide sales in the EU, with chlorpyrifos the top selling insecticide (15.6%, 1226 tonnes).

Chlorpyrifos was also one of the most widely used organophosphates in the US for pest control, but the US Environmental Protection Agency (EPA) imposed a ban on the sale of chlorpyrifos for residential use in December 2001. The primary route of exposure to the general population of children and adults is dietary. Many authors deal with the toxicity of chlorpyrifos today.

In the authors' experimental study different cell cultures, derived from kidney (RK13), liver (WBF344 and Hepa 1c1c7), and insect origin (Sf21) were used to examine the effect of the chlorpyrifos. The most sensitive cells to chlorpyrifos were Sf21 cells with significant suppression of their proliferative activity ranging from $10^{-1} – 10^{-5}$ M. However, the significant suppression of proliferative activity was also recorded in mammalian cell cultures: liver cells Hepa 1c1c7 ($10^{-1} – 10^{-3}$ M), WBF344 ($10^{-1} – 10^{-2}$ M), and kidney cells RK13 (10^{-1} M). Cytopathic effect and the LDH leakage into the medium were observed in RK 13 ($10^{-1} – 10^{-3}$ M) > WBF344, and Hepa 1c1c7 cells ($10^{-1} – 10^{-2}$ M) > the insect cell line Sf21 (10^{-1} M), compared to solvent control. The results indicate that chlorpyrifos exposure caused cell-type dependent decrease in cell proliferation and the cell membrane damage.

Chapter 5 - The aim of the paper is to give an overview of the mechanism of interaction between chlorpyrifos and the products of its environmental and metabolic transformations with some enzymes which regulate the key functions in many physiological processes. The metabolites of the parent compound formed due to the metabolic processes as well as chemical transformations after its application in the environment will be identified.

The enzymes from the group of ATPases, cholinesterases and peroxidases were chosen as the targets for evaluation of chlorpyrifos and the products of its metabolic/chemical transformations toxicity. The mechanism of inhibition of the selected enzymes induced by chlorpyrifos and the products of its transformations will be elucidated. The recovery effects on the inhibited activities of the selected enzymes will be analyzed.

The overview of the methods for the detection of the parent compound and the products of its transformations in the environment will be given. The toxic effects of chlorpyrifos and its metabolites towards the selected enzymes will be compared and discussed.

Chapter 6 - Over the last decade a considerable body of evidence has emerged indicating that chlorpyrifos (CPF) targets important processes during nervous system development causing developmental neurotoxicity. These processes include neuronal proliferation, differentiation, axonogenesis, synaptogenesis, myelination and apoptosis. This book chapter attempts to collate mechanistic information derived from in vivo and in vitro studies investigating the neurotoxic properties of CPF in developing nervous system. Furthermore, this existing mechanistic knowledge will be organized accordingly and linked to distinct adverse outcomes that CPF can cause in developing organisms such as motor, sensory and cognitive deficits. These neurodevelopmental effects vary in severity, duration and time of onset and can be reversible or can have permanent, life-long consequences. Thus, the developmental neurotoxicity induced by CPF is complex and here there will be also presented a chemical-specific predictive risk assessment approach that will encompass all the available mechanistic as well as complementary toxicokinetic information. Additionally, this book chapter will revise all the information available from several risk assessment reports performed by different international organizations regarding the neurodevelopmental effects of CPF. An analysis of the various outcomes and the differences in diverse aspects of these assessments will be provided. Finally, the implications of these evaluations across different regulatory frameworks will be also described.

Chapter 7 - Chlorpyrifos is a widely used broad-spectrum phosphorothioate insecticide. Chlorpyrifos fate in human is the process by which it is biotransformed through enzymatic catalysis to metabolites that are eliminated directly or after conjugation with endogenous cofactors. Biotransformation plays a major role in chlorpyrifos mediated toxicity and risk assessment. Chlorpyrifos is bioactivated by human cytochrome P450 (CYP) enzymes to the potent neurotoxic inhibitor chlorpyrifos-oxon and detoxified via dearylation to 3,5,6-trichloro-2-pyridinol. Human cytochrome P450 enzymes that mediate these reactions and the response mechanisms are under genetic and epigenetic control and influenced by other environmental stressors. Potential interactions of chlorpyrifos and its metabolites may occur in human during occupational and dietary exposure. This chapter provides a comprehensive knowledge on the fate, disposition and interactions of chlorpyrifos in human.

Chapter 8 - In this chapter, the role of biogeochemical processes in influencing the fate of chlorpyrifos in a lake environment is described using a case study. Lake Victoria being the second largest freshwater body in the world was selected for this purpose. Since chlorpyrifos is not readily soluble in water, sediments which serve as sink to strongly hydrophobic chemicals (SHOCs) were selected as a matrix of study. An understanding of sorption processes is key to describing the fate of chlorpyrifos in aquatic environments and was therefore investigated both on the walls of the containers used during laboratory analysis as well as on the lake sediments. In addition, biotic and abiotic degradation processes for chlorpyrifos in the lake sediments were studied. Surface sediment samples were collected using a core sampler, extracted for chlorpyrifos residues using solid dispersion methods and extracts cleaned up using gel permeation chromatograph (GPC). Analysis for Chlorpyrifos residues was done using a temperature programmed Gas Chromatograph equipped with an Electron Capture Detector. Validation of samples was done using a Gas Chromatograph equipped with a Mass Selective Detector. The average concen tration of chlorpyrifos in sediments was 17μg/kg. Sorption experiments were conducted in mixed solvents (methanol-water) and sorption coefficients in aqueous systems (using HPLC/UV) were calculated based on the Solvophobic model. Degradation of chlorpyrifos was defined as the disappearance of extractable concentrations with time. Experiments were conducted using both sterilized unsterilized samples. Although working in mixed solvents has been believed to eliminate sorption on container walls, results from this chapter indicate that even at volume fraction of 0.4 (40% methanol- 60% water) there was considerable sorption of chlorpyrifos on Teflon lined centrifuge tubes. Thus, sorption on containers can cause grave errors in the estimation of sorption coefficients, especially for SHOCs. To account for these effects for chlorpyrifos, sorption isotherms on sediments, were determined in methanol-water at volume fractions (f_c) greater than 0.5. Chlorpyrifos has been found to have a relatively high affinity to the lake sediment, with the sorption coefficients (K_m) decreasing with increasing f_c, while increasing with organic matter content at $f_c = 0$. Log K_{OC} values for chlorpyrifos were in the range of 3.29 – 3.81. Degradation studies indicate that chlorpyrifos degrades faster in unsterilized than in sterilized sediment samples of Lake Victoria, suggesting that microorganisms contribute to the degradation processes within the lake sediments. Thus, while many parameters are known to affect the degradation of organic pesticides in the environment, this chapter demonstrates that microbial presence and organic matter content play key roles in the degradation of chlorpyrifos in Lake Victoria sediments. The pesticide half-life ranged from 22-39 days.

Chapter 9 - Chlorpyrifos is used worldwide in agricultural and urban environments for insect control. The major environmental studies on the occurrence of chlorpyrifos in the air and water have taken place in North America. Major areas of study in the United States have included California, Midwestern United States and along the Mississippi River. In Canada the major usage region for chlorpyrifos is the prairie agriculture region which has the highest grain and oil seed production in Canada. Environmental measurements of chlorpyrifos have also taken place in National Park sites in both Canada and the United States. The focus of this chapter is a review of atmospheric, surface water, or precipitation measurements that have taken place in North America. New measurements in two agricultural regions of Canada (Okanagan Valley and Prairie Agricultural Region) are also presented. Atmospheric concentrations of chlorpyrifos have been measured at Bratt's Lake, SK since 2003. Large annual shifts in atmospheric concentrations of chlorpyrifos have been observed in the Canadian prairies with a seasonal maximum concentration of chlorpyrifos in late July-early August.

Evidence for potential for regional to long-range atmospheric transport of chlorpyrifos is presented from historical measurements at National Park or remote sites as well as new measurements in areas with little local/regional usage of chlorpyrifos. Both urban and agricultural sources have contributed to the occurrence of chlorpyrifos in the atmosphere or surface waters. In the United States environmental levels and usage of chlorpyrifos have declined since the 1990s. Usage and atmospheric concentrations of chlorpyrifos in the Canadian prairies has also declined since 2003 and attributed to less demand for chlorpyrifos for control of grasshoppers in the prairies.

Chapter 10 - Biosensors represent alternative electro-analytical techniques that can be used to complement or replace classical analytical methods for pesticide detection. Biosensors have also emerged as simple, rapid and sensitive tools that can be applied for environmental monitoring, food safety and quality control. This electrochemical technique further allows for simpler sample preparation and efficient on site analysis, thereby decreasing the cost per sample analysis.

This chapter reports the work done in the construction of a thick-film acetylcholinesterase biosensor for the determination of chlorpyrifos pesticide concentrations in selected aqueous organic solvent solutions. The constructed biosensor was designed by encapsulating acetylcholinesterase (AChE) as enzyme in a polyaniline (PANI) polymer composite, followed by the drop coating of poly(vinyl acetate) (PVAc) on top to secure the biosensor film from disintegration in organic solvents. Biosensor construction also involved the optimisation of various performance and stability parameters to determine its influence on the amperometric behaviour or the biosensor. These factors included optimal enzyme loading, effect of pH, long-term stability of the biosensor, and the effect of both polar and non-polar organic solvents on enzyme behaviour. The constructed AuE/MBT/PANI/AChE/PVAc thick-film biosensor offered a reasonable linear range between 0.24 – 0.70 μg/L for chlorpyrifos.

The biosensor also exhibited a detection limit of 0.012 μg/for chlorpyrifos. It was further determined that the detection limit was below the allowable concentration of 1 μg/L set by the WHO guideline value for drinking water.

In: Chlorpyrifos
Editor: Caitlin Mayes
ISBN: 978-1-63482-111-7

Chapter 1

TOXICITY CHARACTERISTIC LEACHING PROCEDURE OF THE ACTIVE INGREDIENT CHLORPYRIFOS IN THE COMMERCIAL ORGANOPHOSPHATE PESTICIDE ATTAMIX SB USED IN COLOMBIA

Rosalina Gonzalez**, *PhD.
La Salle University, Philadelphia, PA, US

ABSTRACT

In Colombia the organophosphate pesticide Attamix SB with active ingredient Chlorpyrifos in a concentration of 0.00125 g Kg^{-1} is commonly used as an insecticide of coffee, sugar cane, pine and eucalyptus crops; where the two firsts were the 85% of the total agricultural production in the country in 2013 according to the Agricultural Society of Colombia. This indicates that the use of this product is massive in this country and having in mind that the agricultural production is poorly technicized, it can cause a huge impact in the environment due to an incorrect use. As have been demonstrated that the Chlorpyrifos is toxic to fish, aquatic invertebrates, small mammals, and birds, it was decided to practice a toxicity characteristic leaching procedure (TCLP) of Chlorpyrifos from Attamix SB, because this test identifies if the waste is hazardous and should be handled as hazardous waste and try to replicate the conditions of a municipal landfill to detect a contaminant (metal or organic compound concentration) in the water that can leach from the soil. There were analyzed 10 samples by TCLP and it was identified the active ingredient by gas chromatography. The results showed that the Chlorpyrifos is retained mainly by the solid phase, indicating that in natural conditions, the soil is the reservoir of the contaminant with concentrations that permit classified it as hazardous waste.

---

* rogonzalez@unisalle.edu.co.

INTRODUCTION

Colombia has been characterized as a country with a big tradition of farming, in fact, the major source of job is the farm activities. Farming has undergone important transformation processes, which are represented in an increment on planted areas especially in human consuming forage and forestry products. This situation has led to an increment of releasing chemicals especially pesticides (natural or synthetic), essentials for the control of pests mainly in intensive production processes (MinAmbiente, 1998).

According to Food and Agriculture Organization of the United Nations (1996), a chemical pesticide is defined as any chemical agent, which is used for the prevention, repression, repelling or controlling insects, mites, pathogens, nematodes, weeds, rodents, or other harmful organisms to plants and their products and derivatives. Also includes defoliants, plant physiological regulators and adjuvants.

Organophosphate Pesticides

They are synthetic organic substances, formed by a phosphorus atom bonded to four oxygen atoms, or 3 oxygen and one sulfur. One of the phosphorus - oxygen bonds is quite labile and phosphorus released from this "free group" is associated with acetyl cholinesterase to inhibit the nerve transmission and causing death. (Martinez 2006). This compounds are soluble in organic solvents, this makes easy the absorption of them through the biological barriers as skin and mucosa, and they can penetrate faster to the central nervous system rather than other substances. Some of these products can be stored in fat tissue which can cause delayed toxicity due to the late release. They have median vapor pressure which makes them volatile, facilitating their absorption through the livers and they are degradable in alkaline medium under hydrolysis, reducing their persistence in the environment.

Attamix

Is also known as Dursban 23, it has an active ingredient named Chlorpyrifos, which belongs to the family of pesticides and insecticides (ATSDR, 1997). It is a white crystal with a strong odor, not very soluble in water and for that reason is usually mixed with oily liquids before its application to crops or animals. It can also be applied to crops in the form of capsules.

In the United States, Chlorpyrifos was permitted for residential use until 2000 when it was restricted by the United States Environmental Protection Agency (US EPA) according with Rigas, 2001. In 1999 in Colombia were sold around 2.1 million liters and 1.6 tons of Chlorpyrifos, being the organophosphate with the highest percentage share in total sales of this group of insecticides (ICA 1999).

Physicochemical Characteristics (MinAmbiente 1998)

The structural form of Chlorpyrifos is appreciated in the next figure, as well as the physical and chemical characteristics of the pesticide Attamix in Table 1.

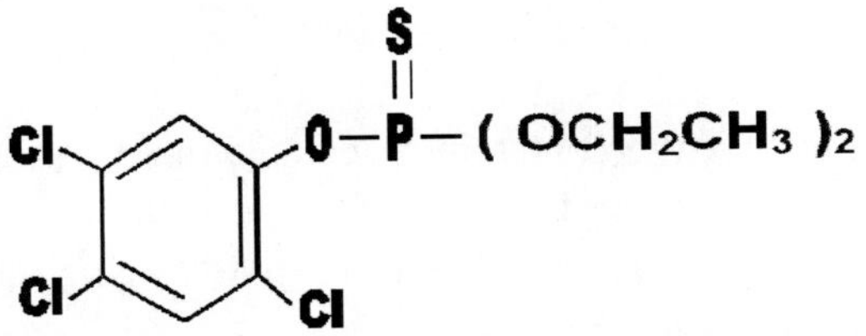

Source: Plaguicidas organofosforados (I): aspectos generales y toxicocinética. España. 2000.

Figure 1. Structural form of Chlorpyrifos.

Table 1. Physical and chemical characteristics of the pesticide Attamix SB

Trade Name	Attamix SB
Common Name	Clorpirifos
IUPAC Name	O,O-diethylO-3,5,6-trichloropyridin-2-yl phosphorothioate
Chemical Group	Organophosphate
Molecular Formula	$C_9H_{11}Cl_3NO_3PS$
Molecular Weight	350.62 g/mol
Use	Insecticide
Physical Appearance	Granulate Solid Brown Color
Melting Point	41-42 °C
Solubility in Water	2 mg/L at 25°C
Vapor Pressure	0.0000180 mmHg a 25°C
pH	5.5
Stability	Stable

Source: Resolución 2423 de 2006. Colombia. 2006.

TCLP (TOXICITY CHARACTERISTIC LEACHING PROCEDURE)

The test Leaching Toxicity Characteristic Leaching Procedure (TCLP) was developed in the late 80s in the United States by the US EPA to quantify the extractability of some hazardous chemicals from solid waste under a set of laboratory conditions. The chemicals are commonly constituents as metals, volatile and semi- volatile organic compounds and pesticides. The application of the technique is specific for each waste type and for each simulated situation. Although, the extraction is performed almost equally in all cases, in the technique is necessary to know the type of components to be analyzed, because there are procedures and calibration requirements that are specific for some substances as metals and volatile and non- volatile organic compounds. The Method is explained in the EPA 1311 ref. SW 846 section.

The TCLP test:

- Identifies whether the waste is hazardous and should be handled as hazardous waste.
- Try to replicate the conditions of a municipal landfill to detect a contaminant (metal or organic compound concentration) in water that can leach from the soil.
- Analyzes the leachate after an acidic solution passes through a sample of the waste

The TCLP test determines if the contaminant concentration is greater than the standard set for that substance. If the substance fails the TCLP test, the chemical is considered as a hazardous characteristic substance. In the TCLP test, the following scenarios for the disposal site and waste management are assumed:

- Mismanagement: location of sites without waterproofing background on vadose zone and aquifer
- Co- disposal of waste: disposal of industrial and domestic waste in the same place
- Percolation of leachate in steady state
- Transport of pollutants and natural attenuation

The TCLP uses fluids to simulate organic acids (acetate buffer) that should be formed by the decomposition of household waste in a landfill and by its design only predicts emissions by one route of contamination: the groundwater. Furthermore, it assumes a scenario of mismanagement and co – disposal and simulates a high degree of aggressiveness to the groundwater resources.

Is important to note that this method is designed to determine the toxicity characteristic leaching of a toxic compound that can be solid or liquid waste, or a mixture of these phases, simulating the conditions of a spill. From an additional analysis with the leachate sample or the extract as chromatography is identify the concentration of the compounds and the hazard potential of the waste leachate if the waste exceed the standard laws. (EPA, 1311 Method)

The method has an initial phase in which is determined the liquid and solid phases present in the sample and the percentage of solids are obtained, after that, the sample is stirred by 18 hours with the extract solution and the leachate obtained is quantified by chromatographic analysis.

The preparation of the extract is done by the EPA methods for semi and nonvolatile compounds, extraction liquid - liquid with separating funnel, and liquid -liquid extraction continued for liquid samples by 3510, 3520 3540 and 3541 Methods.

METHODOLOGY

TCLP System

The equipment and instruments used in the TCLP are in the next figure and the technical specifications in table 2.

Filtration Unit

Mixer and Extraction Bottles

Source: TCLP Project, 2009.

Figure 2. TCLP system.

Table 2. Specifications of the TCLP System

Equipment		
Equipment	**Function**	**Reference**
- Model		3750 LHWF
- Filtration Unit	Separate the phases of the sample	3750-LHWF 2.2 L
Filters		HWF-5044-0002, 142 mm, Glass Fiber, 0.7um
- Mixer	Mix the samples at 30 rpm	3740-BRE
Extraction Bottles	Contains the sample	3740-WGB

Source: Enviroment Sistem 2007.

Analytical Method to Identify Chlorpyrifos

Analysis of gas chromatography using certified standards of pesticides were used to quantify the Chlorpyrifos. The instrument used was a gas chromatograph Trace GC Ultra, Brand Thermo finnigan with flame ionization detector (GC- FID). An autosampler AI 3000 Thermo brand with a brand Hamilton syringe injection of 10 L capacity was used. The column was a general purpose capillary column Restek Rtx - 5 with a phase of 5 % diphenyl -

95% dimethylpolysiloxane 30 m long, 0.32 mm internal diameter and 0.50 μm of diameter of phase. The standard used was a mixture solution of 15 organophosphorus compounds (Dr. Ehrenstorfer reference standard) with a concentration of 200 mg/mL of each of the present substance, which was included in the Chlorpyrifos.

The following chromatographic conditions were determined for the analysis:

Injection volume: 2.0 uL (splitless mode)
Injector temperature: 250°C
Detector temperature: 280°C
Carrier gas: nitrogen at a constant flow of 0.8 mL/min
Fuel Gas: Air: 300 mL/min
Hydrogen: 30 ml/min
Make up Gas: Argon/Methane: 35 mL/min

Oven temperature: start 60°C for 1 min, Then a gradient of 7°C/min to 210°C and 210°C for 10 min, Then a gradient of 5°C/min to 250° C and 250°C for 10 min, with a total analysis time of 40 min.

Samples of Attamix – Chlorpyrifos

It was obtained from Minagro S.A. The pesticide was delivered in the commercial form of 24 bags of 500 gr. The following figure shows the product presentation:

Source: TCLP Project, 2009.

Figure 3. Attamix.

Procedure

Preliminary Evaluations

A preliminary evaluations were made, they involved the quality control method as cleaning tools and verification of the particle size of the sample and the selection of the fluid extractor. For quality control in the TCLP procedure was considered cleaning materials and equipment. The cleaning of equipment is necessary to remove any traces of substances that can contaminate samples analyzed. First a general washing with neutral soap and water is made, after, the material is dried to natural air and then a sulfuric-chromo solution is added, followed by rinsing the material with deionized water (Caballero, 2006). The sulfuric-chromo solution is a mixture of sulfuric acid and potassium dichromate used to oxidize and remove hard materials that not are remove by ordinary detergent. It is a very corrosive solution and is mandatory to use it with safety elements to protect eyes, hands and the inhalator human system.

A. Determination of Solids %

As Attamix is solid, the method suggests to perform a size reduction if the solid particles have a surface area per gram of equal or greater than 3.1 cm^2. Attamix presents a particle size between 0.3 - 0.5 cm^2 according the manufacturer, therefore, it was not necessary to reduce its size.

The method indicates that the percentage of solids defines the TCLP extract. As Attamix is solid, the extraction process involves the addition of the extraction reagent in amount equal to 20 times the weight of solids. The extraction reagent used is a function of the alkalinity of the solid phase and the type of residues analyzed.

B. Leaching Solution

The procedure described in Method 1113 of EPA indicates that a 5 gr of sample is transferred to an Erlenmeyer flask in which are added 96.5ml of deionized water, covered and stirred vigorously on a magnetic stirrer, after 5 minutes the pH is measured, if its value is closer to 6 units are added 3.5ml of 1N HCl, mix and cover again and heated in an oven for 10 minutes. Subsequently the pH is measured. According to the method if the pH is <5, the leaching solution # 1 is used, but if the pH is > 5 the Leaching solution # 2 is used 2. The extraction solutions are described below:

Leaching solution # 1: Add 5.7 mL of concentrated acetic acid in 500 mL of water, 64.3 mL of 1N NaOH and diluting to one liter. The solution should has a pH of 4.93 ± 0.05.

Leaching solution # 2: Dilute 5.7 ml of acetic acid in water to 1 liter. The solution should has a pH of 2.88 + 0.05.

The pH obtained by Attamix was 4.8 so, the Leaching solution #1 was used.

Test of the Study

The study involved the test to determine the amount of leaching solution, the stir of the sample, the extract preparation and the analytical method.

A. Amount of Leaching Solution

As was mentioned before the amount of leaching solution is equal to 20 times the weight of solids, in this case as Attamix is solid the amount used was 100 gr of solution.

B. Stir of the Sample

Stirring was done by adding both the liquid phase and the solid phase in the extraction bottles; with an angular speed of 30 ± 2 rpm over a period of 18 hours, as established in 1311 EPA Method. The following figure shows the stir system:

Source: TCLP Project, 2009.

Figure 4. Stir System.

C. Extract Preparation

After the stir of the sample, the content of the extraction bottles is filtered to separate the waste solution in a unit as the one shows in the next figure. Attamix had a period of filtration of 2 hr.

Source: TCLP Project, 2009.

Figure 5. Filtration System.

For Attamix the solid retained in the filter is collected and stored.

D. Analytical Method - Gas Chromatography

To perform the chromatographic analysis, it is necessary to prepare the samples according 3510C EPA method, because the samples cannot be injected directly into the chromatograph in high concentrations or very small, thus, necessary procedures are required.

Extraction: Liquid-liquid extraction was performed using a funnel, following the next steps:

- The pH was adjusted to 7 to a 90 mL of sample with sodium hydroxide (NaOH) 1M.
- 15 mL of dichloromethane (CH_2Cl_2) were added to the sample in a 500ml funnel.

The funnel was stir with in the solution, then, after 10 minutes the phase separation was done, after that, was removed the lower phase in an Erlenmeyer flask and this process was performed twice, adding a total of 45 ml of dichloromethane was necessary in order to separate the components present in the sample. The solution was filtered and the extractant was used to be quantified by gas chromatography. Solid extraction was performed in addition to determine the concentration of Chlorpyrifos using the TCLP extraction technique.

After the extraction described by the separation funnel, the sample is introduced in a roto-evaporator system to concentrate the sample to a 2 ml volume to GC of the liquid samples and a volume of 4 mL for the analysis of the solid samples.

Results

A. Chromatographic Analysis

The chromatographic runs were developed using the conditions showed in the methodology section and as was mentioned above, the analysis was performed to the TCLP liquid and solid extract. Once made the CG analysis of the TCLP liquid extract was done, none any peak was observed in the chromatograph at the corresponding retention time (27.6 min). As these negative results were obtained, the solid TCLP extracts were analyzed and in the next table the results obtained for the solids are presented:

Table 3. Solid concentrations of the TCLP extract

# ID test	Final Concentration (mg/L)
1	5275.82
2	14422.73
3	2642.30
4	8261.12
5	7462.20
6	4940.04
7	5607.30
8	6733.54
9	5569.86
10	6451.37

As is observed in the table the concentrations are very high indicating a potential impact of this chemical in the soil. These concentrations of the solid extract are high and exhibits toxicological impact to the environment and human health. In addition adverse effect on the soils includes the formation of surface crust, decrease in porosity, compaction and alteration processes of humus formation and decomposition of organic matter. The consequences for the adherence and persistence of the active ingredient in soil, also, can affect different environmental components due the generation of an ecological and biological imbalance, species extinction and ecosystem degradation. According to the PLAGSALUD project, 2000, Chlorpyrifos is listed in the 12 pesticides responsible of the morbidity and mortality increment from acute poisoning.

B. Environmental Impact Assessment

To determine the environmental impact assessment it was used the matrix importance of Conesa, 2003, where the level of importance of the impact caused by pesticides is determined by the next variables: -sign (+ o-): Intensity (I), area of influence (EX), time (MO), persistence (PE), reversibility (RV), accumulation (AC), effect (EF), frequency (PR) and Recoverability (MC). Each variable is explained below:

- Sign: Refers to the charitable nature (+) or detrimental (-) effect or impact.
- Intensity (I): It determines the impact of the affected item. The evaluation 12 corresponds to the total impact of the affected item, 8 to very high impact, 4 high impact, 2 to average impact and 1 to low impact.
- Area of influence (EX): Refers to the area of influence or impact in relation to the project environment. If the situation produces a localized effect, it is considered that the impact is very precise and the evaluation is (1). But, if the effect is not in a precise location, the impact will be (8), and considering an intermediate situation, a partial impact is evaluated (2) and large impact (4).
- Time (MO): The deadline refers to the time between the onset of action and the beginning of the effect on environmental factor considered. If the time is zero, the time is immediate, and if less than one, short-term year in both cases by assigning a value of (4). If a period of time is ranging from 1-5 years old, medium term (2), and if the effect takes to manifest more than five years, long-term value is (1).
- Persistence (PE) refers to the time that supposedly remains the effect from its appearance and from which the affected factor would return to initial conditions by natural situation or by introducing corrective measures. If persistence of effect occurs for less than a year, the assign value is (1). If it is between 1 and 10 years, is temporal, and the assign value is (2); and if the effect is more than 10 years, the effect is permanent and the assign value is (4).
- Reversibility (RV): This variable refers to the possibility of rebuilding the factor affected by the project, ie, the possibility of returning to the initial conditions by natural ways. If is a short term, the value assign is (1), if is a medium term the assign value is (2) and if the effect is irreversible the assigned value is (4).
- Recoverability (MC): This variable refers to the possibility of reconstruction, partial or total of the factor affected as a result of the project, ie, the possibility of returning to the initial conditions, through human intervention. If the effect is fully recoverable, is assigned a value of (1) or (2) if is in immediate or medium term. If the effect it is partially mitigated, and it takes a value of (4). When the effect is unrecoverable the assign value is (8). In the case of unrecoverable conditions, but the possibility of introducing countervailing measures, the adopted value is (4).
- Accumulation: This attribute gives an idea of a progressive increase of the effect, when it persists continuously or repeated actions generate it. When the action does not produce cumulative effects, the effect is evaluated as (1), but if the effect is cumulative the value assign is (4).
- Effect: Refers to the cause - effect relationship, ie the form of manifestation of the effect on a factor as a result of an action. The effect may be direct (4), in the effect is indirect (1),

- Frequency: Refers to the regularity of the manifestation of the effect, either constant in time (continuous effect), cyclical or recurrent (periodic effect) and unpredictably over time (irregular effect). A continuous effects are assigned with a value of (4), irregular and discontinuous (1).

For an overall rating of the importance, the equation proposed by Conesa, 2003 is:

$$I = \pm[\,3I + 2EX + MO + PE + RV + SI + AC + EF + PR + MC]$$

The importance of the impact values are between 13 and 88, the impacts values below 20 have irrelevant importance and are in accordance with the regulations. The moderate impacts have an importance values between 20 and 45. The severe impacts are when the importance values are between 45 and 70 and critical impacts when the value exceeds 70.

The evaluation for Attamix is:

Nature: is negative because the pesticide attacks directly pests, that are the target of its implementation, this has a negative impact on the environment by changing their initial conditions.

Intensity: It has a value of 4 corresponding to a very high degree of involvement, since its use is not for total destruction.

Extension: The area of influence value is (2), partial because it directly affects the point where it is applied or spilled.

Time: It was determined that the period of manifestation is 4, because the time in which the effects generated by the pesticide is immediately manifest due to the concentration at which it is spilled.

Persistence: Is evaluated in 2, temporal, because with corrective actions, the environment can return to the initial conditions

Reversibility: It is medium term (2) because the pesticide is slightly persistent.

Synergy: synergism has a value of 2 because the pesticide being an organic compound can be degraded in other substances that affect the environment due to chemical or microbiological reactions.

Accumulation: This value is 1, because no progressive increase by accumulation in a spill is observed, however its concentration increases the damage generated on the ground.

Effect: The effect is 4, directly, because it can change the natural conditions of the soil.

Frequency: is irregular with a value of 1, because in the case of a crop, the site of application and formulation can vary.

Recoverability: is medium term because the environmental recovery methods can vary by the type of soil and specific treatment is needed. Its value is 2.

Importance: -34

This value is within the range represented before is a Moderate Impact as proposed by Conesa, 2003, where the main variables causing the impact are the time and the effect, this due the concentrations that were generated in the study.

CONCLUSION

According to the TCLP test the active ingredient Chlorpyrifos present in the organophosphate pesticide Attamix Sb, not leach to the conditions studiedin the TCLP test, but the waste should be considered hazardous waste and treated as that.

Solid test concentrations ranging from 2000 to 14000 ppm. The results are consistent with the chemical properties because the active ingredient has a low solubility in water and showed that at acidic conditions that are supposed to have a place of disposal in a landfill or an open dump due the compound not leach.

The importance value obtained by the matrix of Conesa, 2003 showed a number of - 34, indicating that is a moderate impact, being the time and the effect the principal variables showing the greatest impacts caused due the high concentrations obtained in the solid TCLP extraction.

REFERENCES

ATSDR (Agency for Toxic Substances and Disease Resgistry). Clorpirifos. 1997. [en línea], Disponible en la World Wide Web: http://www.atsdr.cdc.gov/es/toxfaqs/es_tfacts84.pdf.

CABALLERO, LS. Análisis y determinación de la concentración de Hidrocarburos Aromáticos Policíclicos (hap´s) contenidos en el material particulado respirable (pm10) en la localidad de puente aranda de Bogotá en la zona de alta actividad industrial y alto flujo vehicular. p. 76. Tesis de Grado. Universidad de La Salle. Bogota. 2006.

CONESA, FV. Guía Metodológica Para La Evaluación Del Impacto Ambiental. Capítulo II. Pag. 76 – 130. 3 ed. Mundi-Prensa, Madrid. 2003.

FAO Programa De Las Naciones Unidas Para El Medio Ambiente Organizacion Mundial De La Salud Organizacion De Las Naciones Unidas Para La Agricultura Y La Alimentacion Roma, 1996 Eliminación de Grandes Cantidades de Plaguicidas en Desuso en los Países en Desarrollo - Colección FAO: Eliminación de Plaguicidas.

INSTITUTO COLOMBIANO AGROPECUARIO. Comercialización de Plaguicidas 1999. [en línea], Ministerio de Agricultura y Desarrollo Rural. División de Insumos Agrícolas. Bogotá, 2001. Disponible en la World Wide Web: http://www.dspac.uniandes.edu. co:5050/dspase /bitstream/1992/669/1/mi_1058.pdf.

MARTINEZ, A. Organofosforados. Cricyt - Conicet - Universidad Nacional de Cuyo - Gobierno de Mendoza – Argentina. 2006. Disponible en la World Wide Web: http://www.cricyt.edu.ar/enciclopedia/terminos /Organofosf.htm.

MINISTERIO DE MEDIO AMBIENTE MinAmbiente. Lineamientos De Política Ambiental Para El Subsector De Los Plaguicidas En Colombia. Dirección General De Desarrollo Sostenible. 1998. Disponible en la World Wide Web: http://www.ideam.gov.co/apc-aa/img_upload/467567 db4678d7b443628f8bc215f32d/Lineamientos_Plaguicidas.pdf.

PLAGSALUD. Proyecto Aspectos Ocupacionales y Ambientales de la Exposición a Plaguicidas en el Istmo Centroamericano. Proyecto financiado por la Agencia Danesa para el Desarrollo Internacional (DANIDA). 2000.

RIGAS ML, OKINO MS, QUACKENBOSS JJ. Use of a pharmacokinetic model to assess chlorpyrifos exposure and dose in children, based on urinary biomarker measurement. [en línea], Toxicol Sci. 2001. Disponible en la World Wide Web: http://kogi.udea.edu.co/ Revista/17/17-3-6.pdf.

TCLP Project. Rosalina González Forero. Comparación de la característica de toxicidad por lixiviación (TCLP) para los resultados obtenidos con los plaguicidas obsoletos thiodán, clorpirifos y malathion a partir de muestras obtenidas en Colombia. 2009.

VIRONMENTAL PROTECTION AGENCY – EPA. Method 1311 Toxicity Characteristic Leaching Procedure. SW846.

_____ . Method 3510c Separatory Funnel Liquid-Liquid Extraction.

_____ . Method 8141ª Organophosphorus Compounds by Gas Chromatography: Capillary Column Technique.

_____ . Method 8000 Determinative Chromatographic Separations.

In: Chlorpyrifos
Editor: Caitlin Mayes

ISBN: 978-1-63482-111-7

Chapter 2

THE ORGANOPHOSPHORUS CHLORPYRIFOS AS A BREAST CANCER RISK FACTOR

***Clara Ventura*[1], *Mariel Núñez*[1], *Andrea Randi*[2], *Andrés Venturino*[3] and *Claudia Cocca*[1]**

[1]Laboratorio de Radioisótopos, Facultad de Farmacia y Bioquímica, Universidad de Buenos Aires, Argentina

[2]Laboratorio de Efectos Biológicos de Contaminantes Ambientales, Departamento de Bioquímica Humana, Facultad de Medicina, Universidad de Buenos Aires, Argentina

[3]Laboratorio de Investigaciones Bioquímicas y Químicas del Ambiente (LIBIQUIMA), CITAAC, CONICET-Universidad Nacional del Comahue, Neuquén, Argentina

ABSTRACT

Organophosphate pesticides comprise a group of substances used in agriculture for insect and plague control, infestations in buildings, man or domestic animals. The use of insecticides represents an environmental risk due to the high mass of product applied annually. Chlorpyrifos (CPF) is a broad-spectrum-organophosphate pesticide. The primary target of CPF toxicity is the central and peripheral nervous system, due to its ability to inhibit the acetylcholinesterase activity. However, other actions of CPF as the effects on synthesis of macromolecule like DNA, RNA and proteins, interactions with neurotransmitter receptors and other neurochemical affections were described. Breast cancer is the most frequent malignant disease in women. Exposure to estrogens throughout woman's life is a risk factor for this malignancy. Estrogen receptor alpha (ERα) is the major regulator of breast cancer tumor behavior. In the mammary gland, 17β-estradiol (E2) promotes cell proliferation in both normal and transformed epithelial cells by modifying the expression of hormone responsive involved genes. Endocrine disrupter (ED) is defined as a compound of industrial or natural origin that interferes with hormone biosynthesis, action and metabolism, resulting in a deviation from normal homeostatic control or reproduction. CPF was recognized as an ED since it has been demonstrated to possess the ability to interfere with the ERα responses. Moreover, CPF possesses antiandrogenic activity and significantly decreases testosterone biosynthesis. We demonstrated the ability of CPF environmental concentration to induce cell proliferation through ERα in hormone-dependent MCF-7 breast cancer cells. In contrast, higher doses of CPF promoted cell cycle arrest in S–phase modifying checkpoints proteins, through a mechanism that may involve changes in redox balance in MCF-7 cells. In hormone-non-dependent MDA-MB-231 cells, we demonstrated that CPF is not able to promote cell proliferation but induces G2/M cell cycle arrest. In breast cancer cells, CPF induces redox imbalance, which lead to the cell cycle arrest. In turn, redox imbalance was trigged by an increment of p-ERK1/2 levels which is a result of an increment of hydrogen peroxide production induced by CPF. *In vivo*, we have recently observed that virgin rats exposed to CPF present a proliferative mammary tissue similar to the pregnant state, lower estradiol circulating levels and a prolonged estrous cycle. These results point to CPF as an ED stimulating the mammary gland proliferation and altering the estrogenic balance at systemic levels. Taken all this into account, the use of CPF should be strictly controlled considering that this pesticide could affect human or animal health and environment.

INTRODUCTION

Agriculture is one of the most important economic activities in development countries as Argentina and others where phytosanitary control and profitability are supported by the use of pesticides. However, it represents an important risk factor for workers, population health, plants and animals of the wild life and environment.

Organophosphorus compounds comprise a group of highly toxic chemicals for pest control. These substances represent the 36 percent of the insecticides used around the world. The application of insecticides, particularly organophosphates, constitute one of the main cause of environmental pollution in surface and groundwater affecting aquatic species populations and endangering human health [1].

The major mechanism of action of those organophosphorus compounds is the acetylcholinesterase (AChE) inhibition that conducts to an accumulation of acetylcholine within cholinergic synapses in the exposed organisms. Many unspecific signs and symptoms have been informed to appear within minutes to hours after exposure and can last for days or even weeks. Symptoms could disappear after organisms replace the inhibited enzymes in the nervous system. Exposure to low concentrations can cause runny nose, tears, and increased saliva or drooling. People may sweat, and develop headache, nausea, and dizziness. It has been reported that serious exposures can cause vomiting, abdominal muscle cramps, muscle twitching, tremors and weakness, and loss of coordination. Sometimes, people have informed to feel blurred and darkened vision. In severe poisoning cases, exposure can lead to unconsciousness, loss of bladder and bowel control, convulsions, difficulty in breathing, and paralysis [2, 3].

Chorpyrifos (CPF) is the main oranophosphorus pesticide used in our country where is used to control pests in field crops such as soybean. It is also used in vegetables, fruit and ornamentals crops. In many other countries, products containing CPF are used to control mosquitoes, ants, cockroaches, crickets, and termites. They are also used to treat wood fences and utility poles. The application of CPF is commonly made by pulverization or directly over the ground [4].

In spite of knowing that CPF produces embryotoxicity and neurotoxicity, and alters levels of enzymes involved in neurotransmitter biosynthesis, this pesticide is the most commonly used one in and rural farms [5, 6]. It is also demonstrated that CPF present neurotoxic actions because it modulates DNA synthesis [7], transcription factor activity [8], protein synthesis [9] and neurite outgrowth [10] in cell lines. Experimental studies have shown that CPF is a teratogen in aquatic species [11].

There are scarce studies performed to evaluate reproductive effects of CPF. In spite of the poor understanding of this action, CPF was postulated for the first time as an endocrine disruptor (ED) because it was found that this pesticide directly regulates gonadotropin-releasing hormone gene expression and biosynthesis in the GT1-7 hypothalamic cell line [12].

Endocrine disruptors are defined as compounds which may be of industrial or natural origin and interfere with hormone biosynthesis, action and metabolism, resulting in a deviation from normal homeostatic control or reproduction [13]. EDs may act via nuclear receptors such as ERα, non-nuclear steroid hormone receptors, non-steroid receptors, enzymatic pathways involved in steroid biosynthesis and/or metabolism, and several other mechanisms [14].

Residential use of CPF was forbidden in United States in 2001 and the application is now restricted to rural sectors. In our country, the use of CPF was also forbidden to domestic uses from 2008 [15]. It is known that CPF incorporation involves multiple routes of entry into the human body. Approximately 70 % is absorbed orally while a low dose may be incorporated through the skin. The pesticide can also be absorbed by inhalation and this route causes the fastest appearance of symptoms. After oral administration, CPF and its metabolite chlorpyrifos-oxon, are detected in blood, liver and tissues with high content of lipids including brain. Because of its hydrophobic nature, CPF is found accumulated in breast milk. An article published in 2003 reports about the presence of CPF in breast milk of women in lactancy [16].

In the last time, it has been reported that CPF interferes on the synthesis of different macromolecules [17], interacts with cellular receptors [18], modulates many cell signaling pathways [19] and modifies intracellular redox balance [20]. Lately, it has gained great interest the study of the pollutant effects in relation to cancer risk, with special attention to breast cancer because many common pesticides may act as ED affecting the normal physiology of hormone dependent tissues.

A great number of investigations inform about the critical role of external estrogenic stimulus on mammary carcinogenesis [21]. In this sense, the estrogenic action of different pesticides leads us to believe that these compounds could impact on the etiology of that malignancy. Additionaly, recent epidemiology studies inform about the relation between human occupational exposure to pesticides and glioma or colon and lung cancer [22-24].

The aim of the present publication is to inform about the new insights on CPF mechanism of action and its effect as ED in breast cancer development.

CPF Actions on Cell Proliferation and Intracellular Redox Balance

We first evaluated the effects of CPF on human breast cancer cell lines proliferation. For our studies, we used MCF-7 estrogen-dependent and MDA-MB-231 estrogen-independent human breast cancer cell lines. In both lines, we have studied viability, proliferation, doubling time and cell death when the cells were exposed to CPF. Doses of CPF have been selected taking into account the levels of the pesticide in water and soil found near the area of application [25-29].

In view of our results, CPF did not affect cell viability in the first 24 hours of exposure. Although after the first day of CPF at 50 μM exposure no changes were detected in cell viability, cells maintained arrested at different phases of cell cycle indicating that CPF could affect the control of cell growth. In this respect, we observed that CPF modulates D1 and E cyclins expression. So, we assumed that the alterations in cell cycle regulation would lead to death after longer periods of intoxication. Indeed, after two days of exposure, CPF at 50 μM induced an augment of necrotic cells [30].

Apart from the cell growth inhibition and cell cycle modifications, CPF at 50 μM increased intracellular reactive oxygen species (ROS) *in vitro*. These chemical species are normally produced by the metabolism of aerobic organism and participate in the strict control of diverse functions of the cells. The increment of ROS induced by CPF resulted entirely impeded by catalase (CAT) addition to the MCF-7 cells culture, so the hydrogen peroxide (H_2O_2) is probably the main specie involved in the redox unbalance induced by the pesticide. The total reversion provoked by CAT was not observed in MDA-MB-231 cells revealing that another species than H_2O_2 are produced by CPF exposure. In MDA-MB-231 cells, an increment in nitrogen reactive species (RNS) was also perceived [30].

With the aim of preserve the cellular homeostasis, aerobic organisms use a wide number of defense mechanisms when they have to face up to an increment of ROS. Among the strategies that cells use to recover redox balance, it has been included preventive mechanisms and antioxidant defenses. Antioxidant defenses comprise the antioxidant enzymes as superoxide dismutase (SOD) and CAT and non-enzymatic compounds as reduced glutathione (GSH) [31].

MCF-7 and MDA-MB-231 cell lines differ in their basal antioxidant activity. We have found that CAT and SOD enzyme activities are significantly higher in MDA-MB-231 than MCF-7 cells. These marked differences could be a reason for the different responses of those lines to the pesticide. As our results showed, 50 μM CPF induces an enhancement in CAT activity but SOD enzyme activity was found diminished only in MCF-7 cells. The inhibition of this activity could be due to an oxidation of protein active site because of ROS increment or due to an inhibition by the increment of the enzyme products. This inhibition was not detected in MDA-MB-231 cells, so it could be related to higher levels of SOD activity in these cells. Finally, levels of thiol groups levels were not found modified by CPF in any of the cell types [32].

Oxidation of macromolecules is one of the main consequences of redox unbalance. Lipid peroxidation was observed in MCF-7 cells when they were exposed to 50 μM CPF for 24 hours but no alterations were generated in MDA-MB-231 cells. It could be due to the less antioxidant capacity is present in MCF-7 cells in comparison with MDA-MB-231 [32].

A large number of investigations point to oxidative stress as a key in cell signaling that modulates various cell processes as cell death, proliferation, differentiation or malignant transformation. Among signals trigged by oxidative stress, MAPKs signaling is one of the most relevant. These proteins constitute a family of kinases proteins that regulate many important signals from the cell environment to the cellular inside. In mammary cells, it has been described a MAPKs family proteins conformed by p38, JNK and ERK1/2 which may be modulated by redox [33-35]. We lately reported that 50 μM CPF enhanced ERK1/2 phosphorylation after 15 minutes of exposure [32]. Recent publications have reported that this effect is induced by an intracellular increment of ROS levels [33]. In concordance with these reports, we have found a rapid increment in ROS content what could be related to the increment of ERK1/2 phosphorylation. This association was confirmed by the addition of catalase to the cell culture, which blocked this action of CPF [32].

Additionally, CAT and PD98059 MEK inhibitor impeded the clonogenicity reduction provoked by CPF bringing the evidence that hydroxide peroxide is the main specie involved in this action mediated by ERK1/2 phosphorylation [32]. In conclusion, our group demonstrated that 50 μM CPF induces an increment in intracellular oxidant species levels which enhances ERK1/2 protein kinases phosphorylation and conducts to cell death (Figure 1).

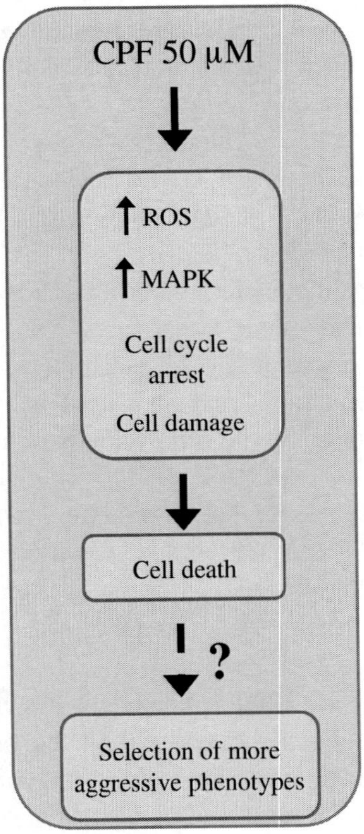

Figure 1.

CPF May Act As an Endocrine Disruptor

We have also studied the consequences of the exposure of the mammary cells to 0.05 μM CPF. As different groups of researchers have reported, these concentrations of the pesticide are coincident with those found in water from rural zones of our country [27-29]. In contrast with higher concentrations, 0.05 μM CPF promoted cell proliferation. It is important to highlight that this effect was only produced in MCF-7 cells that express estrogen receptor alpha (ERα) and depend on estrogens for growing. This action was not observed in MDA-MB-231 cells which do not express ERα and constitute an estrogen independent cell line.

In the last years, many chemical compounds were described as EDs when they are found at low concentrations in the environment. CPF and methoxychlor were found to slightly stimulate GnRH peptide levels, and this effect was blocked by the irreversible inhibitor of ERα, ICI 182,780, suggesting that this receptor mediates the effects of the pesticides on GnRH post-translational biosynthetic mechanisms [12].

In our experiments, ICI 182,780 abolished the action of CPF on cell proliferation what is directly pointing that ERα is involved in this effect. Furthermore, 0.05 μM CPF enhanced ERα phosphorylation at the tyrosine 537 (P-Tyr537-ERα) [30].

The phosphorylation at this site was also found increased by the action of other EDs as hexachlorobenzene [36]. P-Tyr537-ERα has been related to the activation of this receptor and proliferation induction by other authors. Our results indicate that the increment of cell proliferation observed when estrogen dependent MCF-7 cells are grown in presence of 0.05 μM CPF is mediated by ERα activation (Figure 2).

We have also described that CPF modulates cell cycle progression and cyclin E and D1 expression. It has been described that many environmental toxic compounds are capable of modifying the cell cycle progression. This deregulation is a hallmark of tumor cells and is crucial in human cancer progression [38].

Taking into account that CPF modulated the proliferation of both cell lines as described above, we wondered if this pesticide could affect cycling in breast cancer cells. 50 μM CPF induced cell accumulation in S phase in MCF-7 cell line whereas, in MDA-MB-231 cells, conduced to a significant increment of cells in G2/M phase at the same concentration. However, CPF at 0.05 μM did not affect the cell cycle distribution in either MCF-7 or MDA-MB-231 cells.

When the expression of cell cycle regulation proteins cyclins D1 and E were analyzed in MCF-7 cell line, we found that 0.05 μM CPF induced an increment in cyclin D1 levels whilst 50 μM CPF produced a significant augment in cyclin E expression. These cyclins were not found altered in MDA-MB-231. Again, the observed effects of CPF are highly dependent of its concentration.

It has been described that cyclin D binds to and activates cyclin dependent kinase (cdk) 4 or cdk6. Active complex cyclin D-cdk4/6 phosphorylates the G1 checkpoint protein Rb, resulting in the release of E2F, a transcription factor sequestered and inhibited by Rb [39]. E2F induces transcription of cyclin E and other genes that are required for S-phase. Cyclin E binds to and activates cdk2 and the complex has to be imported into the nucleus to phosphorylate essential proteins to initiate DNA replication in S-phase.

There, cyclin E also has roles in histone biosynthesis, centrosome duplication, S-phase transcription, pre-mRNA splicing, oncogenic transformation and re-entry into the cell cycle from quiescence [40]. After that, the decrease in cyclin E levels is required for the transition from S to G2/M. If an error in DNA structure is detected, cells are maintained in S phase until DNA is repaired. This mechanism explain our results in MCF-7 cells exposed to 50 μM CPF. Certainly, we have not observed an incorporation of the analog of timidine, bromide-deoxy-uridine (BrdU) in MCF-7 cells exposed to 50 μM CPF that indicates that DNA synthesis is not occurring at this dose while those exposed to 0.05 μM CPF showed a significant increment of BrdU incorporation [30].

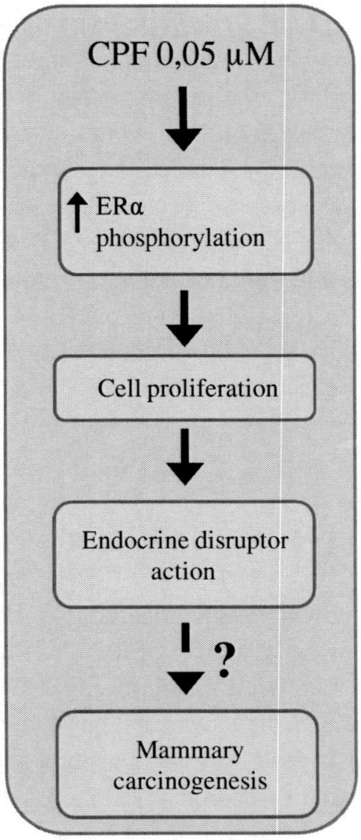

Figure 2.

In order to ascertain the effects of CPF as an ED, we evaluated the action of this pesticide on the mammary gland development. To this aim, 40-day-old virgin female Sprague–Dawley rats received CPF daily for three months. Doses were chosen for studying the effects of CPF at concentrations that do not affect cholinesterase action. For these investigations, we used two relevant concentrations for rat treatment, one of them coincident with the *Acceptable Daily Intake* or ADI (0.01 mg/Kg/day) and the other corresponding to *Non-observed-adverse effect level* or NOAEL (1.00 mg/Kg/day) [41]. At the end of the experiments, the animals were euthanized and mammary gland, liver and blood were analyzed. We could not detect changes in body weight of the animals. Liver weight, histology and biochemical parameters of hepatic function were not modified by CPF treatment at the lower dose. Some necrotic and hemorrhagic foci were found in the hepatic tissue of the animals intoxicated with the higher dose of the pesticide. However, when mammary tissue was studied, many alterations were found in the histological analysis. We detected an increase in the number of mammary hyperplasic foci in the gland from animals intoxicated with CPF at both doses. An increment in sclerosing adenosis benign breast lesions with increased numbers of distorted lobules accompanied by stromal fibrosis were found in animals exposed to the pesticide. Many reports agree that this common proliferative lesion of breast conduces to an approximate doubling of breast cancer risk [42]. Our results obtained in female rats were in concordance with our in vitro experiments where an estrogenic action of CPF has been found in MCF-7 estrogen dependent cells as we above described.

As we previously mentioned, EDs are able to act interfering the endocrine system at different levels. In this regard, we have found that virgin rats exposed to CPF presented a proliferative mammary tissue similar to the pregnant state, lower E2 circulating levels and a prolonged estrous cycle. These results point to CPF as an ED because stimulate mammary gland proliferation and alter the estrogenic balance at systemic levels [43].

Finally, oxidative stress found in breast cancer cells exposed to CPF was also observed in mammary gland from CPF-intoxicated rats. Redox metabolism was evaluated measuring CAT activity, thiol groups' levels and lipid peroxidation in mammary tissues of exposed rats. Our experiments revealed that 1 mg/Kg/day CPF increases enzymatic and non-enzymatic antioxidant defenses and 0.01 mg/Kg/day CPF augment only antioxidant enzyme activity. We have interpreted these results as a response of the redox system to the injury provoked by the pesticide intoxication [43].

Conclusion and Perspectives

Breast cancer is the most common malignant pathology among women around the world. Environmental factors seem to be related with the etiology of this disease but the mechanisms implicated are not completely known until now [44]. Nowadays, a great interest in understanding the role of those factors on cancer risk has awoken among researchers. Despite the increasing evidences in the last years, various questions about the relation between breast cancer risk and pesticide uses are maintained unsolved [45].

Many of these compounds have shown to be EDs acting on steroid receptors; serotonin, dopamine or norepinephrine neurotransmitters receptors, orphan receptors as aryl-hydrocarbon receptors, affecting the biosynthesis or metabolism of estrogens and other numerous mechanisms that converge in endocrine system or reproduction. Molecules known as EDs conform a highly heterogeneous group and includes synthetic chemicals used as industrial solvents or lubricants and their byproducts as polychlorinated biphenyls or dioxins; plastics as bisphenol A; pesticides as methoxychlor, chlorpyrifos or dichlorodiphenyl-trichloroethane; fungicides as vinclozolin; pharmaceutical agents as diethylstilbestrol and heavy metals a chromium or cadmium.

In this way, it is fundamental to know the consequences of the exposure to those substances, particularly in relation to breast cancer risk. We have evaluated the action of CPF as an ED using both *in vivo* and *in vitro* experimental models. We have found that CPF induces cell proliferation via estrogen receptor phosphorylation and oxidative stress in breast cancer cells. Both mechanisms have been implicated in mammary carcinogenesis. We have also found that CPF exposure produces proliferating changes in mammary gland tissue organization. In view of our results, we postulate that CPF alters the endocrine system with consequences in mammary gland development. So, the use of this pesticide has to be revised since it could be act as environmental risk factor for breast cancer (Figure 3). New investigations have to be performed to evaluate CPF effects using relevant environmental concentrations to understand the mechanisms of action of this pesticide that could affect human population or workers health.

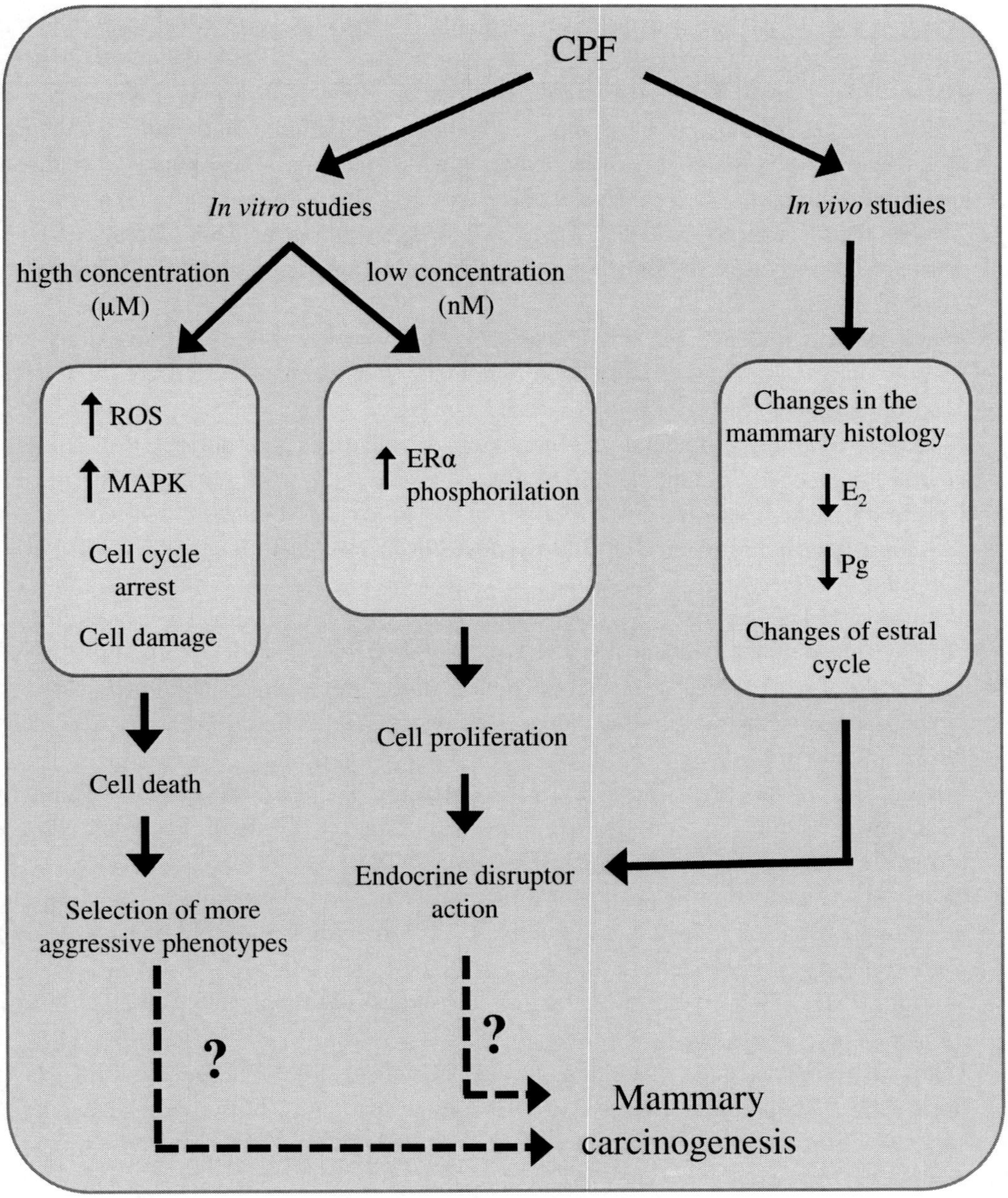

Figure 3.

REFERENCES

[1] Casida, J.E., Quistad, G.B. Organophosphate toxicology: safety aspects of nonacetylcholinesterase secondary targets. *Chem. Res. Toxicol.* 17(8):983-998 (2004).

[2] Tan, D.H., Peng, S.O., Wu, Y.L., Wang, Y.M., Lu, C.F., Ding, W., Wang, Q.X., Yan, C.H. Chlorpyrifos induces delayed cytotoxicity after withdrawal in primary hippocampal neurons through extracellular signal-regulated kinase inhibition. *Biol. Pharm. Bull.* 32:1649–1655 (2009).

[3] Rush, T., Liu, X.Q., Hjelmhaug, J., Lobner, D. Mechanisms of chlorpyrifos and diazinon induced neurotoxicity in cortical culture. *Neuroscience* 166:899–906 (2010).

[4] Eaton, D.L., Daroff, R.B., Autrup, H., Bridges, J., Buffler, P., Costa, L.G., Coyle, J., McKhann, G., Mobley, W.C., Nadel, L., Neubert, D., Schulte-Hermann, R., Spencer, P.S. Review of the toxicology of chlorpyrifos with an emphasis on human exposure and neurodevelopment. *Crit. Rev. Toxicol.* 38 Suppl 2:1-125 (2008).

[5] Deacon, M.M., Murray, J.S., Pilny, M.K., Rao, K.S., Dittenber, D.A., Hanley, T.R. Jr, John J.A. Embryotoxicity and fetotoxicity of orally administered chlorpyrifos in mice. *Toxicol. Appl. Pharmacol.* 54(1):31-40 (1980).

[6] Muto, M.A., Lobelle, F. Jr., Bidanset, J.H., Wurpel, J.N. Embryotoxicity and neurotoxicity in rats associated with prenatal exposure to DURSBAN. *Vet. Hum. Toxicol.* 34(6):498-501(1992).

[7] Slotkin, T.A. Developmental cholinotoxicants: nicotine and chlorpyrifos. *Environ Health Perspect.* 107 Suppl 1:71-80 (1999).

[8] Crumpton, T.L., Seidler, F.J., Slotkin, T.A. Is oxidative stress involved in the developmental neurotoxicity of chlorpyrifos? *Brain Res. Dev. Brain Res.* 121(2):189-195 (2000).

[9] Whitney, K.D., Seidler, F.J., Slotkin, T.A.. Developmental neurotoxicity of chlorpyrifos: cellular mechanisms. *Toxicol. Appl. Pharmacol.* 134(1):53-62 (1995).

[10] Li, W., Casida, J.E. Organophosphorus neuropathy target esterase inhibitors selectively block outgrowth of neurite-like and cell processes in cultured cells. *Toxicol. Lett.* 98(3):139-146 (1998).

[11] Loewy, M., Kirs, V., Carvajal, G., Venturino, A., Pechen de D'Angelo, M. Groundwater contamination by azinphos methyl in the northern Patagonic Region (Argentina). *Sci. Total Environ.* 225:211–218 (1999).

[12] Gore, A.C. Organochlorine pesticides directly regulate gonadotropin-releasing hormone gene expression and biosynthesis in the GT1-7 hypothalamic cell line. *Mol. Cell. Endocrinol.* 192(1-2):157-70 (2002).

[13] Waring, R.H., Harris, R.M. Endocrine disrupters—a threat to women's health? *Maturitas* 68: 111–115 (2011).

[14] Diamanti-Kandarakis, E., Bourguignon, J.P., Giudice, L.C., Hauser, R., Prins, G.S., Soto, A.M., Zoeller, R.T., Gore, A.C. Endocrine-disrupting chemicals: an endocrine society scientific statement. *Endocr Rev.* 30:293–342 (2009).

[15] ANMAT. *Disposition number 2659/2008.* 4–7 (2008).

[16] Sanghi, R., Pillai, M.K.K., Jayalekshmi, T.R., Nair, A. Organochlorine and organophosphorus pesticide residues in breast milk from Bhopal, Madhya Pradesh, India. *Hum. Exp. Toxicol.* 22:73–76 (2003).

[17] Qiao, D., Seidler, F.J., Slotkin, T.A. Developmental neurotoxicity of chlorpyrifos modeled in vitro: comparative effects of metabolites and other cholinesterase inhibitors on DNA synthesis in PC12 and C6 cells. *Environ. Health Perspect.* 109:909–913 (2001).

[18] Howard, M.D., Pope, C.N. In vitro effects of chlorpyrifos, parathion, methyl parathion and their oxons on cardiac muscarinic receptor binding in neonatal and adult rats. *Toxicol.* 170:1–10 (2002).

[19] Song, X., Seidler, F.J., Saleh, J.L., Zhang, J., Padilla, S., Slotkin, T.A. Cellular mechanisms for developmental toxicity of chlorpyrifos: targeting the adenylyl cyclase signaling cascade. *Toxicol. Appl. Pharmacol.* 145:158–174 (1997).

[20] Gupta, S.C., Mishra, M., Sharma, A., Deepack, Balaji, T.G., Kumar, R., Mishra, R.K., Chowdhuri, D.K. Chlorpyrifos induces apoptosis and DNA damage in Drosophila through generation of reactive oxygen species. *Ecotoxicol. Environ. Saf.* 73:1415–1423 (2010).

[21] Clarke, R.B., Sims, A.H., Howell, A. The Origin of Estrogen Receptor a-Positive and a-Negative Breast Cancer. *Horm. Carcinog. V* 79–86 (2008).

[22] Lee, W.J., Colt, J.S., Heineman, E.F., McComb, R., Weisenburger, D.D., Lijinsky, W., Hard, M.H. Agricultural pesticide use and risk of glioma in Nebraska, United States. *Occup. Environ. Med.* 62: 786–792 (2005).

[23] Lee, W.J., Sandler, D.P., Blair, A., Samanic, C., Cross, A.J., Alavanja, M.C. Pesticide use and colorectal cancer risk in the Agricultural Health Study. *Int. J. Cancer* 121:339–346 (2007).

[24] Lee, W.J., Blair, A., Hoppin, J.A., Lubin, J.H., Rusiecki, J.A., Sandler, D.P., Dosemeci, M., Alavanja, M.C. Cancer incidence among pesticide applicators exposed to chlorpyrifos in the Agricultural Health Study. *J. Natl. Cancer Inst.* 96:1781–1789 (2004).

[25] Del Prado Lu, J.L. Multipesticide residue assessment of agricultural soil and water in major farming areas in Benguet, Philippines. *Arch. Environ. Contam. Toxicol.* 59:175–181 (2010).

[26] Jaipieam, S., Visuthismajarn, P., Siriwong, W., Borjan, M., Robson, M. G. Inhalation exposure of organophosphate pesticides by vegetable growers in the Bang-Rieng subdistrict in Thailand. *J. Environ. Public Health* 2009, 452373 (2009).

[27] Jergentz, S., Mugni, H., Bonetto, C., Schulz, R. Assessment of insecticide contamination in runoff and stream water of small agricultural streams in the main soybean area of Argentina. *Chemosphere* 61:817–826 (2005).

[28] Marino, D., Ronco, A. Cypermethrin and chlorpyrifos concentration levels in surface water bodies of the Pampa Ondulada, Argentina. *Bull. Environ. Contam. Toxicol.* 75:820–826 (2005).

[29] Querejeta, G.A., Ramos, L.M., Flores, A.P., Hugher, E.A., Montserrat, J.M.. Environmental pesticide distribution in horticultural and floricultural periurban production units. *Chemosphere* 87:566–572 (2012).

[30] Ventura, C., Núñez, M., Miret, N., Martinel Lamas, D., Randi, A., Venturino, A., Rivera, E., Cocca, C. Differential mechanisms of action are involved in chlorpyrifos effects in estrogen-dependent or -independent breast cancer cells exposed to low or high concentrations of the pesticide. *Toxicol. Lett.* 213:184–193 (2012).

[31] Reuter, S., Gupta, S.C., Chaturvedi, M.M., Aggarwal, B.B., Oxidative stress, inflammation, and cancer: how are they linked? *Free Radic. Biol. Med.* 49:1603–1616 (2010).

[32] Ventura, C., Venturino, A., Miret, N., Randi, A., Rivera, E., Núñez, M., Cocca, C. Chlorpyrifos inhibits cell proliferation through ERK1/2 phosphorylation in breast cancer cell lines. *Chemosphere* 120:343–350 (2015).

[33] Torres, M. Mitogen-activated protein kinase pathways in redox signaling. *Front. Biosci.* 8:369–391 (2003).

[34] Wu, W.S. The signaling mechanism of ROS in tumor progression. *Cancer Metastasis Rev.* 25:695–705 (2006).

[35] Yoon, S.O., Yun, C.H., Chung, A.S. Dose effect of oxidative stress on signal transduction in aging. *Mech. Ageing Dev.* 123:1597–1604 (2002).

[36] García, M.A., Peña, D., Alvarez, L., Cocca, C., Pontillo, C., Bergoc, R., de Pisarev, D.K., Randi, A. Hexachlorobenzene induces cell proliferation and IGF-I signaling pathway in an estrogen receptor alpha-dependent manner in MCF-7 breast cancer cell line. *Toxicol. Lett.* 192(2):195-205 (2010).

[37] Varricchio, L., Migliaccio, A., Castoria, G., Yamaguchi, H., de Falco, A., Di Domenico, M., Giovannelli, P., Farrar, W., Appella, E., Auricchio, F. Inhibition of estradiol receptor/Src association and cell growth by an estradiol receptor alpha tyrosine-phosphorylated peptide. *Mol. Cancer Res.* 5(11):1213-1221 (2007).

[38] Marc, J., Mulner-Lorillon, O., Bellé, R. Glyphosate-based pesticides affect cell cycle regulation. *Biol. Cell* 96(3):245–249 (2004).

[39] Musgrove, E.A., Caldon, C.E., Barraclough, J., Stone, A., Sutherland, R.L. Cyclin D as a therapeutic target in cancer. *Nat. Rev. Cancer.* 11(8):558-572 (2011).

[40] Caldon, C.E., Musgrove, E.A. Distinct and redundant functions of cyclin E1 and cyclin E2 in development and cancer. *Cell Div.* 5:2 (2010).

[41] World Health Organization. WHO specifications and evaluations for public health pesticides. Chlorpyrifos. pp 47 (2009) at <http://www.who.int/whopes/quality/ Chlorpyrifos_WHO_specs_eval_Mar_2009.pdf>

[42] Cichon, M.A., Degnim, A.C., Visscher, D.W., Radisky, D.C. Microenvironmental influences that drive progression from benign breast disease to invasive breast cancer. *J. Mammary Gland. Biol. Neoplasia.* 15(4):389-397 (2010).

[43] Ventura, C., Martinel Lamas, D., Bourguignon, N., Lux-Lantos, V, V., Rivera, E., Cocca, C., Níñez, M. El Clorpirifos afecta el desarrollo de la glándula mamaria de rata e induce el desvalance redox en este tejido. *Rev. Med.* 73:161–162 (2013).

[44] Blair, A., Kazerouni, N. Reactive chemicals and cancer. *Cancer Causes Control* 8:473–490 (1997).

[45] Reynolds, P., Hurley, S.E., Gunier, R.B., Yerabati, S., Quach, T., Hertz, A. Residential Proximity to Agricultural Pesticide Use and Incidence of Breast Cancer in California, 1988–1997. *Environ. Health Perspect.* 113:993–1000 (2005).

In: Chlorpyrifos
Editor: Caitlin Mayes

ISBN: 978-1-63482-111-7

Chapter 3

ACUTE EXPOSURE OF THE FRESHWATER SNAIL *PLANORBARIUS CORNEUS* TO AN ENVIRONMENTAL CONCENTRATION OF CHLORPYRIFOS: RECOVERY STUDIES OF ENZYME ACTIVITIES IN GONADS AND EFFECTS ON REPRODUCTION

***Pamela R. Rivadeneira*[1], *Macarena Agrelo*[1] *and Gisela Kristoff*[1, 2,*]**

[1]Departamento de Química Biológica,
Facultad de Ciencias Exactas y Naturales,
Universidad de Buenos Aires,
Ciudad Universitaria,
Buenos Aires, Argentina
[2]IQUIBICEN- UBA/CONICET,
Ciudad Universitaria, Buenos Aires, Argentina

ABSTRACT

Chlorpyrifos is an organophosphate insecticide widely used in a variety of food crops in Argentina and other countries. In a previous work, we have reported that subchronic exposure (14 days) to an environmental concentration of chlorpyrifos (5 μg L^{-1}) caused severe effects on reproduction of the freshwater hermaphrodite gastropod *Planorbarius corneus*. The aim of the present work was to study the effects of chlorpyrifos on the reproduction of this adult snail after an acute (2 days) exposure. Also B-esterase activities (cholinesterases and carboxylesterases) were measured in the gonads, to study the possible link of the biochemical responses to the reproduction effects. Snails were exposed during 2 days to 5 μg L^{-1} of chlorpyrifos. Water control and

[*] Corresponding author: Gisela Kristoff; Facultad de Ciencias Exactas y Naturales. Universidad de Buenos Aires; Departamento de Química Biológica, IQUIBICEN-UBA/CONICET; Ciudad Universitaria, Pab. II. 4to piso, 1428 Buenos Aires, Argentina. TE: 0054-11-4576-3342; Fax number: 0054-11-4576-3342; E-mail: gkristoff@qb.fcen.uba.ar; giselakristoff@hotmail.com.

solvent control (acetone) group were included in the bioassay. After 2 days of exposure, some snails per group were used to measured enzyme activities and other snails were transferred to water. Then, the bioassay was continued for 14 days. Snails were fed once a week. During the 14 days in water, the number of egg masses was registered and, for each egg mass, the number of egg masses without eggs, the number of eggs per mass, the time for hatching, the hatchings and the survival of the offspring were counted. After 2 days of exposure and after 2 and 14 days in water, snails were placed on ice, the shells were removed and gonad homogenates were realized. Supernatant fraction (6 for each time and for each group) was used to measure cholinesterase and carboxylesterase activities and protein content. After an acute exposure, 5 $\mu g\ L^{-1}$ of chlorpyrifos caused a decrease on cholinesterase activity using acetylthiocholine as substrate (45 %) and carboxylesterase activities (26 % using p-nitrophenyl acetate as substrate and 40 % using p-nitrophenyl butyrate) in gonads. After 14 days in water, no recovery of cholinesterase activity was observed. However, carboxylesterases measured with both substrates returned to control values. Regarding the reproduction effects, only an increase in the number of egg masses without eggs and an increase in the time for hatching were observed. Recovery of carboxylesterases in gonads could be related to the lower effects observed on reproduction in comparison with the subchronic exposure. After 14 days of exposure, 5 $\mu g\ L^{-1}$ of chlorpyrifos had produced a strongly decrease in B-esterase activities and severe effects on reproduction, such as the decreases in the % of hatching and survival of the offspring, the increases in egg masses without eggs and in the time of hatching. Exposures to environmental concentrations of chlorpyrifos caused different toxic effects on the reproduction of *Planorbarius corneus* depending on the time of exposure.

Keywords: Cholinesterases; carboxylesterases; chlorpyrifos; gastropods; reproduction

ABBREVIATIONS

AsCh	acetylthiocholine iodide
CES	carboxylesterases
ChE	cholinesterase
IC_{50}	concentration that causes 50 % of inhibition
DTNB	5,5´-dithio-2-bis-nitrobenzoate
OP	organophosphate insecticide
p-NPA	p-nitrophenyl acetate
p-NPB	p-nitrophenyl butyrate

1. INTRODUCTION

Organophosphorous (OP) and carbamate insecticides are commonly used in agricultural practices since they have less persistence in terrestrial and aquatic environments than organochlorine pesticides (Barron and Woddburn, 1995; Varó et al., 2002). However, they are toxic to non-target species at low concentrations (Varó et al., 2002) and can cause severe, long-term population effects on aquatic non-target species, particularly invertebrates (Barata et al., 2004, Fulton and Key, 2001). The toxicity of these insecticides is mainly due to the inhibition of acetylcholinesterase activity (AChE), the enzyme which hydrolyzes acetylcholine. Chemical inhibition of AChE results in overwhelming of post-synaptic acetylcholine receptors and hyper-stimulation, which leads to physiologic aberrations ranging from behavioral impairment to death (Titon et al., 2011). Ops and carbamates, also inhibit carboxylesterases (CES), which play an important role in the detoxification and tolerance of pesticides (Jokanovic et al., 1996; Sanchez Hernandez, 2007). Cholinesterases (ChE) and CES belong to the group of hydrolases which Aldrige (1953) classified as B-esterases. These esterases are inhibited by Ops, instead of A-esterases which are able to hydrolyze Ops and carbamates and C esterases which do not interact with them (Lagerre et al., 2009).

Chlorpyrifos is a non-systemic OP insecticide widely used in a variety of food crops in Argentina and other countries. The concentrations of chlorpyrifos reported in surface water range between 0.026 and 700 μg L^{-1} (EPA USEPA, 2006; Moore et al., 2002; Wood and Starke, 2002) and its half-life in water ranging from 29 to 74 days (Racke, 1993). The direct toxicity of chlorpyrifos is assumed to be the result of the initial metabolic activation forming chlorpyrifos oxon, with the subsequent irreversible inactivation of ChE at neural junctions (Palma et al., 2009).

It was probed that chlorpyrifos can inhibit ChE activity of several aquatic invertebrates (Rivadeneira et al., 2013). Also, the toxic effect on reproduction has been reported mainly in crustacean species. In *Daphnia magna* it produced a decrease in the number of offspring produced per male (Palma et al., 2009) while in *Daphnia carinata* chlorpyrifos produced a decrease on survival and fecundity in the first generation and an increase in the time for hatching in the second one (Zaliznick and Nugegoda, 2006).

Chronic and subchronic exposures are relative terms especially in relation to the species. Generally, a chronic exposure implies an exposure of at least 10 % of the lifetime. An acute exposure corresponds to until 96 h while the subchronic exposure involves a time between the acute and the chronic exposure (Barile, 2008). In a previous work, we have reported that the subchronic exposure (14 days) to 5 μg L^{-1} of chlorpyrifos caused severe effects on reproduction of the freshwater hermaphrodite gastropod *Planorbarius corneus,* such as the decreases in the % of hatching and in the % of survival of the offspring and the increases in the number of egg masses without eggs and in the time of hatching (Rivadeneira et al., 2013). In laboratory conditions, organism of *P. corneus* can live until 2 years (Collins Baker, 1945). The aim of the present work was to study the effects of chlorpyrifos on the reproduction of adult organisms of *P. corneus* after an acute (2 days) exposure. Also B-esterase activities (cholinesterases and carboxylesterases) were measured in the gonads to study the possible link of the biochemical responses to the reproduction effects.

2. Materials and Methods

2.1. Chemicals

Acetylthiocholine iodide (AcSCh), p-nitrophenyl acetate (p-NPA), p-nitrophenyl butyrate (p-NPB), 5,5´-dithio-2-bis-nitrobenzoate (DTNB), and chlorpyrifos were purchased from Sigma-Aldrich of Argentina S.A. All chemicals used were of analytical reagent grade.

2.2. Organisms

Adult P. corneus snails were purchased from Discus Morón S.R.L., Buenos Aires, Argentina, and then reared in our laboratory under standard conditions in an aerated glass aquaria (17-20 L), at 22 ± 2 ºC, and under a 14:10 (L:D) h photoperiod regime. Adult snails of similar size (12 ± 2 mm and 0.337 ± 0.068 g) were used for all the experiments.

2.3. Bioassays

To perform the bioassays, we used 1 L glass vessels, containing 800 ml for each solution (dechlorinated water; 0.001 % of acetone in dechlorinated water, and 5 μg L^{-1} of chlorpyrifos in dechlorinated water). No mortality was observed either in the control group or in the treated group. All the bioassays were performed at 22 ± 2 ºC under a photoperiod of 14:10 (L:D) h. The following physico-chemical parameters were recorded: total hardness = 67 ± 3 mg $CaCO_3$ L^{-1}; alkalinity = 29 ± 2 mg $CaCO_3$ L^{-1}; pH 7.0 ± 0.2 and conductivity = 250 ± 17 μS. The stock solution of the insecticide was prepared in acetone. Chlorpyrifos concentration was tested by HPLC with UV detector (Cacciatore et al., 2013). Each concentration of chlorpyrifos was obtained by diluting the stock solution with dechlorinated water. To avoid pesticide degradation during the bioassays, the chlorpyrifos solution was changed every day. The concentration of acetone in the bioassays was 0.001 %. The concentration used (5 μg L^{-1}) produced 50 % of inhibition of ChE activity after 48 hours of exposure in the whole organism soft tissue (Cacciatore et al., 2013). All snails were previously (-4 days) placed in vessels containing dechlorinated water (Fig. 1). Then, the snails were exposed 2 days to the insecticide (treatment group) or acetone (solvent control) or continued in dechlorinated water (water control). Bioassays were performed using 18 containers with 6 snails for each group and for each time of sampling. After 48 hours, 6 containers for each group were used to B-esterase determinations. In the other 12 vessels for group, the insecticide solution and the acetone were removed and changed by dechlorinated water. Bioassays continued for 14 days more (Fig. 1). Snails were fed with lettuce every 7 days. ChE and CES were measured at time 0, after 2 days of exposure and after 2 and 14 days in the recovery period (in water) (Fig. 1). At the times of sampling, 6 snails for the same vessel were used to obtain one gonad homogenate. 6 gonad homogenates were obtained for each group and each time of sampling.

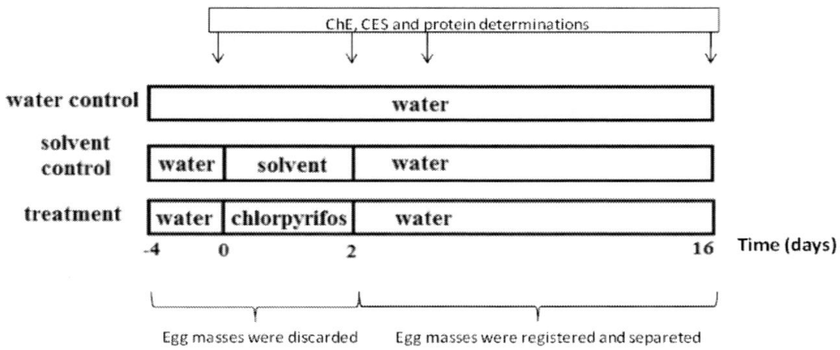

Figure 1. Scheme of the bioassay.

2.4. Homogenates

Animals were placed on ice for 6-8 min for anesthetic effect. The shells were carefully removed and the gonads were separated at 0 °C. The gonads were washed in distilled water, placed on filter paper to drain extra fluids, and weighed. Tissues were homogenized in 20 mM Tris/HCl buffer, pH 7.5, plus 0.5 mM EDTA and centrifuged at 11,000 x g for 20 min at 4 °C. The supernatant fraction was used to determine ChE, CES and the protein content.

2.5. Cholinesterase (ChE) Activity

ChE activity was measured in the supernatant fraction, in 100 mM phosphate buffer, pH 8, 0.2 mM DTNB, and acetylthiocholine iodide (AsCh) 1.5 mM, according to the method of Ellman et al. (1961). AsCh was used in the determination because it is the substrate with which the highest ChE activity is obtained (Cacciatore et al., 2012). Also, the substrate concentration chosen allowed us to work at the V_{max} of the enzyme. Activity was recorded continuously at 412 nm. Rates were corrected for spontaneous hydrolysis of the substrate and non-specific reduction of the chromogen by tissue extracts, and specific activity was expressed as μmol min^{-1} mg $protein^{-1}$.

2.6. Carboxylesterases (CES) Activity

Hydrolysis of p-NPA and p-NPB by CES was determined as previously described in Kristoff et al. (2010). Reactions were performed in 2.5 mL 100 mM phosphate buffer pH 8.0 containing 5% acetone and 1 mM p-NPA or p-NPB. The concentrations of the substrates chosen allowed us to work at the V_{max} of the enzyme. Activity was continuously recorded at 400 nm. Specific activity was calculated using the molar extinction coefficient for p-nitrophenol (18.6 mM^{-1} cm^{-1}).

2.7. Protein Content

Protein content was determined according to the method of Lowry et al. (1951), using bovine serum albumin as standard.

2.8. Reproduction Parameters

The reproduction parameters were obtained as previously described by Rivadeneira et al. (2013). Egg masses laid until the 48 hours of exposure were removed and discarded. Then, every day, during 14 days in water, each egg mass of each group was registered and transferred to an individual glass container with water (Fig. 1). The total number of egg masses of each vessel of each group was counted. In each egg mass, the number of eggs per egg mass, the number of non-embryonated eggs per egg mass, the time for hatching and the hatchings per egg mass were recorded by observing with a magnifying glass. After hatching, the snails were fed with lettuce and after the first month, the survival percentage of the offspring was obtained. Fig. 2 shows a normal egg mass with 1 non-embryonated egg and 7 embryonated eggs.

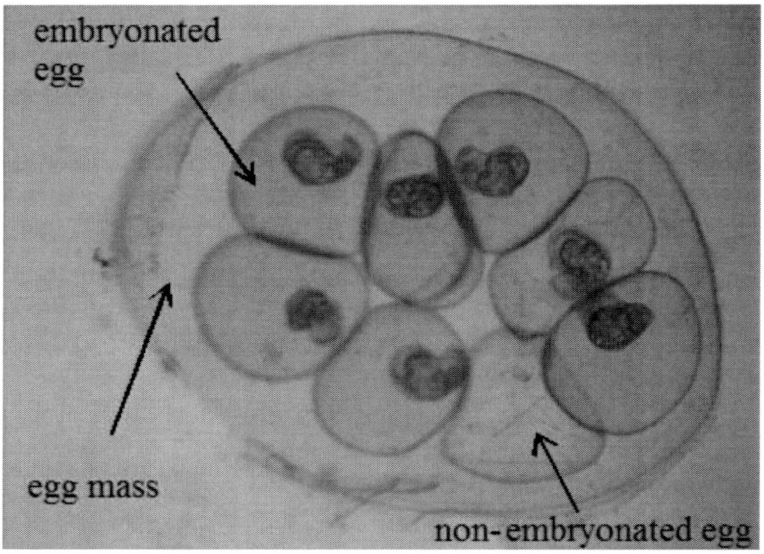

Figure 2. Egg mass of *Planorbarius corneus*. The optical magnification utilized was 30 x. The egg masses measures correspond to 5.5 ± 0.8 mm long and 4.6 ± 0.8 mm width (mean ± DS of 20 egg masses).

2.9. Data Analysis

Results were expressed as mean ± S.D. Data was analyzed by ANOVA. When overall F statistic was significant, pair wise comparisons among treatment groups were performed by Tukey HSD post-test by using VassarStats (http://vassarstats.net). In all cases, the level of significance used was 0.05.

3. Results

3.1. Cholinesterase Activity (ChE)

Fig. 3 shows ChE activity in the gonads of *P. corneus* exposed to chlorpyrifos 48 hours and after 2 and 14 days in water (recovery period). 5 μg L^{-1} of chlorpyriphos produced 45 % of inhibition in gonads. ChE activity remained inhibited after 14 days in water. ChE activity in the control group was not modified during all the bioassay. Control ChE activity in the gonads of *P. corneus* corresponds to 0,421 ± 0,064 μmol/min/mg de protein. No statistically differences were observed between ChE activity in water control and solvent control ($p >$ 0.05).

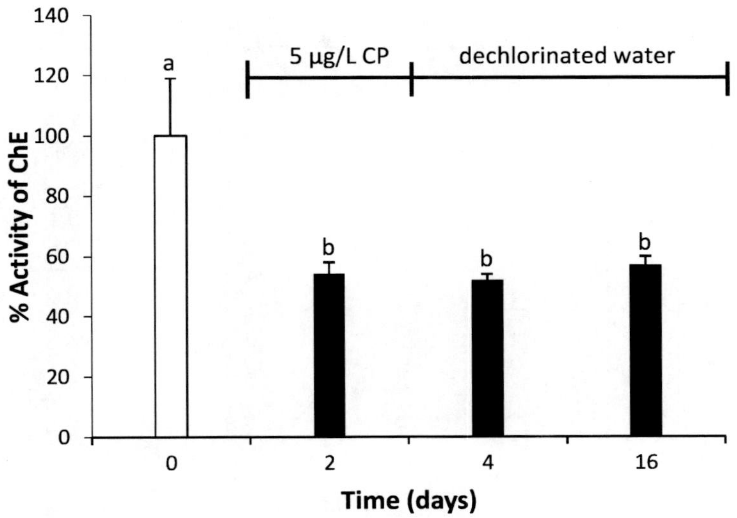

Figure 3. Cholinesterase activity in *Planorbarius corneus* exposed 2 days to chlorpyrifos and in the recovery period in water. Activity was expressed as % of the control. Data correspond to the main ± SD. Data with different letters are statistically different ($p < 0.05$).

3.2. Carboxylesterases Activity

Fig. 4 and 5 show CES activity in the gonads of P. corneus during the bioassay. Acute exposure produced an inhibition of 30 % using p-NPA (Fig. 4) and 40 % using p-NPB (Fig. 5). After 14 days in water CES activity returned to control values. CES activity in the control group was not modified during all the bioassay. Control CES activity in the gonads of P. corneus corresponds to 0,150 ± 0,027 µmol/min/mg protein with p-NPA as substrate and 0,182 ± 0,022 µmol/min/mg protein with p-NPB. No statistically differences were observed between CES activities in water control and solvent control ($p > 0.05$).

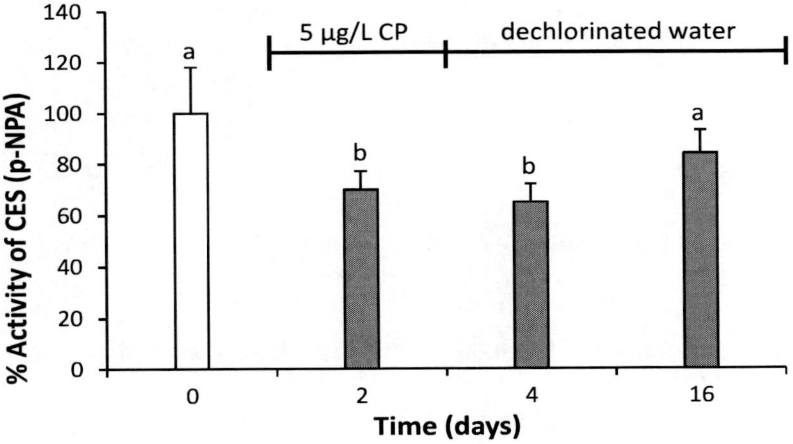

Figure 4. Carboxylesterases activity (CES) measured with p-NPA as substrate in *Planorbarius corneus* exposed 2 days to chlorpyrifos and in the recovery period in water. Activity was expressed as % of the control. Data correspond to the main ± SD. Data with different letters are statistically different ($p < 0.05$).

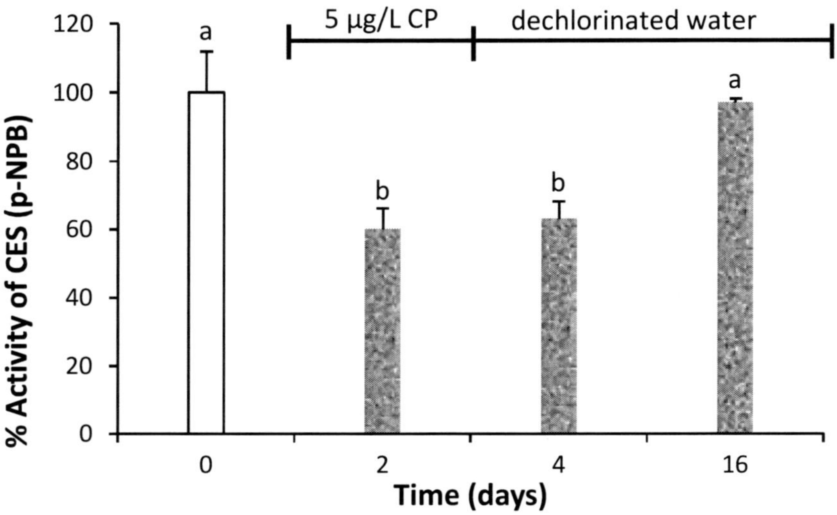

Figure 5. Carboxylesterases activity (CES) measured with p-NPB as substrate in *Planorbarius corneus* exposed 2 days to chlorpyrifos and in the recovery period in water. Activity was expressed as % of the control. Data correspond to the main ± SD. Data with different letters are statistically different ($p < 0.05$).

3.3. Parameters of Reproduction

The exposure of 2 days to 5 µg L^{-1} of chlorpyrifos produced an increase in the number of egg masses without eggs. No differences between control and exposed organisms were found in the number of egg masses and in the number of eggs and non-embryonated eggs per mass (Table 1).

Table 1. Parameters of reproduction in *Planorbarius corneus*

Group	No of egg masses (in 14 days)	No. of egg per mass	No. of non-embryonated eggs per mass	No. of egg masses without eggs
Solvent Control	20 ± 3	12 ± 4	4 ± 3	1 ± 1
Exposed to chlorpyrifos	21 ± 2	13 ± 5	4 ± 3	4 ± 1*

Data correspond to the mean ± SD. Data with * is statistically different to the control ($p < 0.05$).

Regarding hatching and survival of the offspring, only an increase in the time for hatching was observed (Table 2).

Table 2. Hatching and survival of the offspring of *Planorbarius corneus*

Group	Time for hatching (days)	Percentage of hatching	Percentage of survival (1 month)
Solvent control	9 ±2	72 ± 16	76 ± 19
Exposed to chlorpyrifos	16 ± 6*	70 ± 19	75 ± 18

Data correspond to the mean ± SD. Data with * is statistically different to the control ($p < 0.05$).

No statistically differences were observed between water control and solvent control group ($p > 0.05$).

4. Discussion

An acute exposure to 5 μg L^{-1} of chlorpyrifos caused a similar decrease on ChE (45 %) and CES (26 % with p-NPA and 40 % with p-NPB) in gonads. These results were similar to the data reported previously by Cacciatore et al. (2013) in the total soft tissue of *P. corneus*. After 14 days of exposure, Rivadeneira et al. (2013) have reported an inhibition of 90 % in ChE and CES measured with p-NPB and 40 % in CES with p-NPA as substrate.

During the recovery period in P. *corneus* exposed 2 days to chlorpyrifos, no recovery of ChE activity was observed. However, CES measured with both substrates, returned to control values. An inhibition of ChE that remained inhibited for weeks has been previously reported in other organisms. Kristoff et al. (2006) have described in the snail *Biomphalaria glabrata* exposed to 5 mg L^{-1} of the OP azinphos-methyl an inhibition of 50 % of ChE that could not increase after 21 days in water. Ferrari et al. (2004) have reported that 35 days in water were necessary to recover ChE activity in goldfish exposed to azinphos-methyl after an inhibition of 90 %. Similarly, in the oligochaete *Lumbriculus variegatus* exposed to 0.1 mg L^{-1} (90 % inhibition of ChE), the activity continued inhibited until the studied time (21 days). However, if the oligochaetes were exposed to the CI_{50} (0.006 mg L^{-1}) ChE activity returned to control values after 14 days in water (Kristoff et al., 2006). Recovery of ChE depends on the specie, the OP and the initial percentage of inhibition.

Regarding recovery of CES, fewer studies have been reported. Kristoff et al. (2012) described in *B. glabrata* exposed to 2.5 mg L^{-1} of azinphos-methyl, which produced approximately 70 % of inhibition of CES, recovery of the activity in the total soft tissue after 14 days in water.

Concerning the reproduction effects, only an increase in the number of egg masses without eggs and an increase in the time for hatching were observed in the present work. However, after subchronic exposure (14 days), 5 μg L^{-1} of chlorpyrifos produced a greater increase in these parameters and also, a strongly decrease in the % of hatching and survival of the offspring (Rivadeneira et al., 2013).

Tripathi and Singh (2004) and Kristoff et al. (2011) have compared ChE inhibition with reproductive alterations in the freshwater snails *Lymnaea acuminata* and in *B. glabrata* exposed to carbaryl and azinphos-methyl respectively. In both organisms, ChE inhibition greater than 50 % was related with no hatched eggs, while, lower inhibitions of ChE were related with less severe signs on reproduction. In *P. corneus* exposed 2 days to 5 μg L^{-1} of chlorpyrifos, the 45 % of inhibition of ChE could be related with no severe signs on reproduction.

On the other hand, CES, which can influence the sensitivity of the organism to anticholinesterase agents binding or hydrolyzing them (Sanchez-Hernandez, 2007), might be involved in the reproduction process. CES, whose expression increased in the reproductive stage, were found in *Mytilus galloprovincialis* (Mikhailov et al., 1997) and in *Drosophila* (Korochkin 1980; Kubli, 1992). Therefore, the recovery of CES in gonads of *P. corneus* exposed 2 days to chlorpyrifos, could be related with the lower effects observed on reproduction in comparison with the subchronic exposure that courses with a strong inhibition of CES.

In particular, the effects on reproduction have important ecological significance because the survival of the species is determined by the success in giving birth to new individuals. The alteration in the reproductive potential resulting from the presence of contaminants in the environment can lead to malformations in the offspring, a decrease in the number of individuals of a population, or even the extinction of a species (van der Oost et al., 2003). We conclude that the exposure to environmental concentrations of chlorpyrifos caused different toxic effects on the reproduction of *P. corneus* depending on the time of exposure, which is related with ChE and CES inhibitions.

ACKNOWLEDGMENTS

This work was supported by grants from the University of Buenos Aires (20020130200055BA and 20020110100070BA) and CONICET (1122012 0100680CO).

REFERENCES

Aldridge, W. N., 1953. Two types of esterases (A and B) hydrolysing p-nitrophenyl acetate, propionate and butyrate, and a method for their determination. *Biochem., J.* 53, 110-117.

Barata, C., Solayan, A., Porte C., 2004. Role of B-esterases in assessing toxicity of organophosphorus (chlorpyrifos, malathion) and carbamate (carbofuran) pesticides to *Daphnia magna*. *Aquat. Toxicol.,* 66, 125-139.

Barile, F. A. 2008. Principles of Toxicology Testing. Taylor and Francis Group, *CRC Press,* USA.

Barron, M.G., Woodburn, K.B., 1995. Ecotoxicology of chlorpyrifos. *Rev. Environ. Contam. Toxicol.,* 144, 1–93.

Cacciatore, L. C.; Kristoff, G.; Verrengia Guerrero, N., Cochón, A., 2012. Binary mixtures of azinphos-methyl oxon and chlorpyrifos oxon produce *in vitro* synergistic cholinesterase inhibition in *Planorbarius corneus. Chemosphere,* 88, 450-458.

Cacciatore, L. C., Verrengia Guerrero, N. R., Cochón, A., 2013. Cholinesterase and carboxylesterase inhibition in *Planorbarius corneus* exposed to binary mixtures of azinphos-methyl and chlorpyrifos. *Aquat. Toxicol.,* 128-129, 124-134.

Collins Baker, F, 1945. The Molluscan Family Planorbidae. The University of Illinois Press, Urbana, United States of America.

Ellman, G. L., Courtney K. D., Andres, V.Jr., Featherstone, R. M., 1961. A new and rapid colorimetric determination of acetylcholinesterase activity. *Biochem. Pharmacol.,* 7, 88-95.

EPA USEPA, United States Environmental Protection Agency, 2002. Chlorpyrifos Facts, EPA 738- F-01-006. http://www.epa.gov/oppsrrd1 /REDs/factsheets/chlorpyrifos_fs.htm.

Fulton, M. H., Key, P. B., 2001. Acetylcholinesterase inhibition in estuarine fish and invertebrates as an indicator of organophosphorus insecticide exposure and effects. *Environ. Toxicol. Chem.,* 20, 37-45.

Jokanovic, M., 2001. Biotransformation of organophosphorus compounds. *Toxicology,* 166, 139-160.

Korochkin, L. I., 1980. Genetic regulation of isozyme patterns in *Drosophila* during development. In: Market, C. L. (Ed). Isozymes: Current tropics in medical and biological research. New York: Alan R. Liss. 304, 159-202.

Kristoff, G., Cacciatore, L. C., Verrengia Guerrero, N. R., Cochón, A. C., 2011. Effects of the organophosphate insecticide azinphos-methyl on the reproduction activity of *Biomphalaria glabrata. Chemosphere,* 84, 585-591.

Kristoff, G., Chiny Barrionuevo, D., Cacciatore, L. C., Verrengia Guerrero, N. R., Cochón, A., 2012. *In vivo* studies on inhibition and recuperation of B-esterase activity in *Biomphalaria glabrata* exposed to azinphos-methyl: analysis of enzyme, substrate and tissue dependence. *Aquat. Toxicol.,* 112-113, 19-26, 2012.

Kristoff, G., Verrengia Guerrero, N. R., Cochón, A. C., 2010. Inhibition of cholinesterases and carboxylesterases of two invertebrate species, *Biomphalaria glabrata* and *Lumbriculus variegatus*, by the carbamate pesticide carbaryl. *Aquat. Toxicol.* 96, 115-123.

Kristoff, G., Verrengia Guerrero, N. R., Pechén de D'Angelo, A. M., Cochón, A. C., 2006. Inhibition of cholinesterase activity by azinphos-methyl in two freshwater invertebrates: *Biomphalaria glabrata* and *Lumbriculus variegatus*. *Toxicology,* 222, 185-194.

Kubli, E., 1992. The sex peptide. *BioEssays,* 14, 779-784.

Laguerre, C., Sanchez-Hernandez, J., Kohler, H.R., Triebskorn, R., Capowiez, Y., Rault, M., Mazzia, C., 2009. B-type esterases in the snail *Xeropicta derbentina*: an enzymological analysis to evaluate their use as biomarkers of pesticide exposure. *Environ. Pollut.,* 157, 199–207.

Lowry, O.H., Rosebrough, N. J., Farr, A. L., Randall, R. J., 1951. Protein measurement with the folin phenol reagent. *J. Biol. Chem.,* 193, 265-275.

Mikhailov, A. I., Torrado, M., Korochkin, L. I., Kopantzeva, M. A., Méndez, J., 1997. Male predominant carboxylesterase expresion in the reproductive system of molluscs and insects : Inmunochemical and biochemical similarity between *Mytilus* male associated polypeptide (MAP) and Drosophila sex-specific esterase S. *Comp. Biochem. Physiol.,* 118, 197-208.

Moore, M.T., Schulz, R., Cooper, C.M., Smith Jr., S., Rodgers Jr., J.H., 2002. Mitigation of chlorpyrifos run off using constructed wetlands. *Chemosphere,* 46, 827–835.

Palma, P., Palma, V. L., Fernández, R. M., Bohn, A., Soares, A. M. V. M., Barbosa, I. R., 2009. Embryo-toxic effects of environmental concentrations of chlorpyrifos in the crustacean *Daphnia magna. Ecotoxicol. Environ. Saf.,* 72, 1714-1718.

Racke, 1993. Environmental fate of chlorpyrifos. *Rev. Environ. Contam. Toxicol.,* 131, 1-150.

Sanchez-Hernandez, J. C., 2007. Ecotoxicological perspectives of B-esterases in the assessment of pesticide contamination. In: Plattenberg, R. H. (Ed.), Environmental Pollution: New Research. *Nova Science Publishers,* Inc. N.Y., USA, pp. 1-45.

Rivadeneira, P.R., Agrelo, M., Otero, S., Kristoff, G., 2013. Different effects of subchronic exposure to low concentrations of the organophosphate insecticide chlorpyrifos in a freshwater gastropod. *Ecotox. Environ. Saf.,* 90, 82–88.

Tripathi, P. K., Singh, A., 2004. Carbaryl induced alterations in the reproduction and metabolism of the freshwater snail *Lymnaea acuminata. Pest. Biochem. Physiol.,* 79, 1-9.

Triton, F. A., Bammler, T. K., Gallagher, E. P., 2011. Swimming impairment and acetylcholinesterase inhibition in zebrafish exposed to cooper and chlorpyrifos separately, or as mixtures. *Comp. Biochem. Physiol.,* 153, 9-16.

US EPA, United States Environmental Protection Agency, 2001. Interim registration eligibility decision for azinphos-methyl. Case No. 0235.

Van der Oost, R., Beyer, J., Vermeulen, N. P. E., 2003. Fish bioaccumulation and biomarkers in environmental risk assessment: a review. *Environ. Toxicol. Pharmacol.* 13, 57-149.

Varó, I., Navarro, J. C., Amat, F., Guilhermino, L., 2002. Characterization of cholinesterases and evaluation of the inhibitory potential of chlorpyrifos and dichlorvos to *Artemia salina* and *Artemia parthenogenetica*. *Chemosphere,* 48, 563-569.

Varó, I., Amat, F., Navarro, J. C., Barreda, M., Pitarch, E., Serrano, R., 2006. Assessment of the efficacy of *Artemia sp.* (Crustacea) cysts chorion as barrier to chlorpyrifos (organophosphorus pesticide) exposure. Effect on hatching and survival. *Sci. Total Environ.* 366, 148-153.

Zaliznick, L., Nugegoda, D., 2006. Effect of sublethal concentrations of chlorpyrifos on three sucesive generations of *Daphnia carinata*. *Ecotoxicol. Environ. Saf.* 64, 207-214.

In: Chlorpyrifos
Editor: Caitlin Mayes

ISBN: 978-1-63482-111-7

Chapter 4

EFFECT OF CHLORPYRIFOS ON PROLIFERATIVE ACTIVITY OF VARIOUS CELL CULTURES

***Natália Kovalkovičová*[1], *Juraj Pistl*[2,*], *Jana Polláková*[2], *Tomáš Csank*[2] *and Jaroslav Legáth*[1]**

[1] Department of Toxicology, University of Veterinary Medicine and Pharmacy, Košice, Slovak Republic

[2] Department of Microbiology and Immunology, University of Veterinary Medicine and Pharmacy, Košice, Slovak Republic

ABSTRACT

Organophosphates are used frequently in Europe for pest control due to their low price and broad spectrum of activity. In 2003, they accounted for over 59% (4645 tonnes) of insecticide sales in the EU, with chlorpyrifos the top selling insecticide (15.6%, 1226 tonnes).

Chlorpyrifos was also one of the most widely used organophosphates in the US for pest control, but the US Environmental Protection Agency (EPA) imposed a ban on the sale of chlorpyrifos for residential use in December 2001. The primary route of exposure to the general population of children and adults is dietary. Many authors deal with the toxicity of chlorpyrifos today.

In our experimental study different cell cultures, derived from kidney (RK13), liver (WBF344 and Hepa 1c1c7), and insect origin (Sf21) were used to examine the effect of the chlorpyrifos. The most sensitive cells to chlorpyrifos were Sf21 cells with significant suppression of their proliferative activity ranging from 10^{-1} – 10^{-5} M. However, the significant suppression of proliferative activity was also recorded in mammalian cell cultures: liver cells Hepa 1c1c7 (10^{-1} – 10^{-3} M), WBF344 (10^{-1} – 10^{-2} M), and kidney cells RK13 (10^{-1} M). Cytopathic effect and the LDH leakage into the medium were observed in RK 13 (10^{-1} – 10^{-3} M) > WBF344, and Hepa 1c1c7 cells (10^{-1} – 10^{-2} M) > the

* Corresponding author: Juraj Pistl, prof., DVM, PhD., Department of Microbiology and Immunology, University of Veterinary Medicine and Pharmacy, Komenského 73, 041 81 Košice, Slovak Republic. Tel: +421 915 984 588, e-mail: juraj.pistl@uvlf.sk.

insect cell line Sf21 (10^{-1} M), compared to solvent control. The results indicate that chlorpyrifos exposure caused cell-type dependent decrease in cell proliferation and the cell membrane damage.

Keywords: Insecticide, organophosphate, chlorpyrifos, cell cultures, cell proliferation, LDH

INTRODUCTION

Pests, diseases and weeds of major food crops are currently controlled mainly with the use of plant protection products (PPPs - products, which are commonly referred to as "pesticides") which offer, in many cases, the only satisfactory method of limiting yield losses, maintaining high product quality (Oerke and Dehne, 2004; Cooper and Dobson, 2007), and improving the nutritional value of food and sometimes its safety (Boxall, 2001; Narayanasamy, 2006). Pesticides are also applied in forestry, public health, homes, and gardens (Aktar et al., 2009). They help to control hundreds of weed species, more than one million species of harmful insects and roughly 1,500 plant diseases (NACA, 1993; Ware and Whitacre, 2004).

The annual application of synthetic pesticides to food crops in the EU exceeds 140,000 tonnes (Eurostat statistical books, 2007), an amount that corresponds to 280 grams per EU citizen per year. Despite European policies to reduce pesticide use, EU statistics data show that the annual pesticide consumption has not decreased (Eurostat statistical books, 2007). The US Environmental Protection Agency (EPA) information on US pesticide usage, from 2007, reports that over 1 billion tons of pesticides are used in the US every year. This is 22% of the estimated 5.2 billion pounds of pesticides used worldwide.

Agricultural use accounted for 80% of pesticide use in the US. There are over 350,000 current and historic pesticide products registered in the United States, and the pesticide business is a 12.5 billion dollar industry in the US alone (http://www.sustainabletable.org /263/pesticides).

Although pesticides are at the top of the list of dangerous pollutants, modern pesticides are fast-acting, some of them may be unique in their action against a specific pest, they can control large infestations, they are generally easy to apply, and lead to increased crop yield by reducing crop losses. Thus, from this point of view, pesticides can be considered as an economic, labor-saving, and efficient tool of pest management with great popularity in most sectors of the agricultural production (Cooper and Dobson, 2007; Damalas, 2009). On the other hand, there is now overwhelming evidence that some of these chemicals do pose a potential risk to humans and other life forms and unwanted side effects to the environment (Burger et al., 2008; Mariyono, 2008). No segment of the population is completely protected against exposure to pesticides and potentially serious health effects may occur in humans and animals (Aktar et al., 2009; Damalas and Eleftherohorinos, 2011).

Harmful environmental impacts of pesticide use include: loss of biodiversity and elimination of key species (e.g., bees), water pollution, soil contamination, and pest resistance, resulting in the need for increased application of pesticides, or formulation of alternate pesticides. Many of these effects depend on the toxicity of the pesticide, the measures taken during its application, the dosage applied, the adsorption on soil colloids, the weather conditions prevailing after application, and how long the pesticide persists in the environment (Damalas and Eleftherohorinos, 2011).

The risk assessment of the impact of pesticides either on human health or on the environment is not an easy or particularly accurate process because of differences in the periods and levels of exposure, the types of pesticides used (regarding toxicity and persistence), and the environmental characteristics of the areas where pesticides are usually applied (Damalas and Eleftherohorinos, 2011).

Pesticide Exposure in Humans

Although pesticides are developed through very strict regulation processes to function with reasonable certainty and minimal impact on human health and the environment, serious concerns have been raised about health risks resulting from occupational exposure and from residues in food and drinking water.

It is estimated that man and animals are exposed to more than 1,000 pesticides every day because of their widespread use and persistence in different environmental compartments - in the air, water, soil, and plants.

Occupational exposure to pesticides often occurs in the case of agricultural workers in open fields and greenhouses, workers in the pesticide industry, and exterminators of house pests (Wilson and Tisdell, 2001; Maroni et al., 2006; Atreya, 2007; Martinez-Valenzuela et al., 2009; Soares and Porto, 2009). Evidently, workers who mix, load, transport, and apply formulated pesticides are normally considered to be the group that will receive the greatest exposure because of the nature of their work and are therefore at the highest risk for possible acute intoxications (Fenske and Day, 2005).

In developing countries, farmers face great risks of exposure due to the use of toxic chemicals that are banned or restricted in other countries, incorrect application techniques, poorly maintained or totally inappropriate spraying equipment, inadequate storage practices, and often the reuse of old pesticide containers for food and water storage (Ecobichon, 2001; Asogwa and Dongo, 2009).

Exposure of the general population to pesticides occurs primarily through eating food and drinking water contaminated with pesticide residues, whereas substantial exposure can also occur in or around the home, when living close to a workplace that uses pesticides or even when workers bring home contaminated articles (Wilson and Tisdell, 2001; Jaga and Dharmani, 2003; Maroni et al., 2006; Soares and Porto, 2009).

As a result of pesticide use in or around the home, individuals can be exposed during the preparation and application of pesticides or even after the applications are completed, whereas delayed exposure can occur through inhalation of residual air concentrations or exposure to residues found on surfaces, clothing, bedding, food, dust, discarded pesticide containers, or application equipment. Pesticides have been measured in residential environments, most notably in indoor dust (Rudel et al., 2001; Rudel et al., 2003; Colt et al., 2004; Colt et al., 2005; Bradman et al., 2007; Harnly et al., 2009; Roberts et al., 2009; Ward et al., 2009; Gunier et al., 2011). Several studies indicate that pesticide residues persist indoors due to the lack of sunlight, rain, temperature extremes, microbial action, and other factors that facilitate degradation (Roberts et al., 2009).

Carpet dust is a good environmental medium for assessing long-term exposure in the home (Lioy et al., 2002; Gunier et al., 2011; Quiros-Alcala et al., 2011). Previous studies have reported that concentrations of agricultural pesticides in carpet dust are higher in residences closer to treated fields, and in farm homes (Lu et al., 2000; Curwin et al., 2005; Ward et al., 2006; Harnly et al., 2009).

Also, accidental poisoning with pesticides in the home is a possibility from pesticide use around the house or garden. Exposure is likely to occur from pesticide spills, improper use, poor storage, or because of incorrect usage caused by not reading the pesticide label.

Pesticide mishandling, such as transferring the products from their original packages into household containers, and also the lack of compliance with instructions on the label can also be sources of exposure (Jaga and Dharmani, 2003).

Non-occupational exposure originating from pesticide residues in food, air, and drinking water generally involves low doses and is chronic (or semi-chronic). Approximately 300 different pesticides have been reported as contaminants of food products of European origin (Commission staff working document, 2012). Up to 50% of fruits, vegetables, and cereals grown in the European Union are known to contain pesticide residues (Commission staff working document, 2012). Nonetheless, one out of twenty food items is known to exceed a current EU legal limit for an individual pesticide.

Further, over 25% of fruits, vegetables, and cereals contain detectable residues of at least two pesticides (Commission staff working document, 2012). Processed food and baby food are also commonly contaminated. In addition, other sources, such as contaminated drinking water, dusts, and spray drift contribute to human exposures.

The total level of population exposures to pesticides is unknown, but data from US population studies show that the majority of the population has detectable concentrations of pesticide metabolites in the urine (Mage et al., 2004; Bjorling-Poulsen et al., 2008).

Health Effects Associated with Pesticide Exposure

Worldwide it is estimated that approximately 1.8 billion people engage in agriculture and most use pesticides to protect the food and commercial products that they produce.

Others use pesticides occupationally for public health programs, and in commercial applications, while many others use pesticides for lawn and garden applications, and in and around the home (Alavanja, 2009).

Obviously, exposure to pesticides poses a continuous health hazard, especially in the agricultural working environment. By their very nature, most pesticides show a high degree of toxicity because they are designed to kill certain organisms and thus create some risk of harm.

Within this context, pesticide use has raised serious concerns not only of potential effects on human health, but also about impacts on wildlife and sensitive ecosystems (Stoate et al., 2001; Pistl et al., 2003; Pistl et al., 2004; Berny, 2007; Power, 2010).

The World Health Organization has reported that roughly three million pesticide poisonings occur annually, resulting in 220,000 deaths worldwide (WHO, 1992). Agricultural pesticide poisoning is a major public health problem in the developing world, killing at least 250,000 – 370,000 people each year (Gunnell and Eddleston, 2003; Gunnell et al., 2007; Dawson et al., 2010). World-wide deaths and chronic diseases due to pesticide poisoning are about 1 million per year (Environews Forum, 1999).

It is well accepted that *acute poisonings* cause health effects, such as *seizures, rashes*, and *gastrointestinal illness* (Solomon et al., 2007). Excluding acute poisonings, *contact dermatitis* is thought to be the most common health effect of pesticides, either through irritant or allergic mechanisms. Along with *eye injuries*, it is the health effect most likely to be seen and might be the only indicator of exposure.

Chronic effects, such as *cancer* and adverse *reproductive outcomes,* have also been studied extensively, and the results have been interpreted in various ways, as evidence that pesticides are safe or are a cause for concern because they can be detrimental to human health. Genotoxic potential is a primary risk factor for long-term health effects such as cancer and reproductive health outcomes (Bolognesi, 2003).

The review by Alavanja et al. (2004) summarized studies examining the link between pesticide exposure and *cancer*. *Non-Hodgkin lymphoma* has been one of the most extensively studied cancers. Associations between Non-Hodgkin lymphoma and exposures to organophosphates have been reported. *Leukemia* has also been studied extensively. Similar associations have been shown with *prostate cancer, multiple myeloma*, and *soft tissues sarcomas*. There is less supportive literature of an association between pesticides and other types of cancer, although there is some literature of an association between chlorinated compounds and *breast* and *testicular cancer* and *Hodgkin disease* (McCauley et al., 2006).

Occupational exposures to pesticides, and adverse *reproductive effects,* have been reviewed by Hanke and Jurewicz (2004). Employment in agriculture appears to be associated with the risk of infertility or congenital malformations in offspring, particularly orofacial cleft, as well as musculoskeletal and nervous system defects.

Many pesticides have been identified as *endocrine disruptors*, of these 46% are insecticides, 21% herbicides, and 31% fungicides; some of them were withdrawn from general use many years ago but are still found in the environment (e.g., DDT and atrazine in several countries) (Mnif et al., 2011).

Exposure to pesticides has been associated with *altered immune response* (Pistl et al., 2003; Pistl et al., 2004), which causes an increased susceptibility of exposed organism to viral, bacterial, parasitic infections, and to tumors. It is known that farmers have high risk of Hodgkin´s disease, melanomas, multiple myeloma, and leukemia (Lee et al., 2007; Dennis et al., 2010; Kokouva et al., 2011). An excessive incidence of non-Hodgkin's lymphoma (a cancer of the immune system) has been reported among farmers and other occupational groups working with pesticides (Pistl et al., 2004; Chiu and Blair, 2009; Hohenadel et al., 2011).

In some cases, it has been suggested that *neurological disorders* may be connected to pesticide exposure. Long-term effects of pesticides on the nervous system include cognitive and psychomotor dysfunction, and neurodegenerative and neurodevelopmental effects (Jett, 2011; Hargreaves, 2012). Pesticide poisonings result in well-described acute and chronic *neurotoxic syndromes* (Sanborn et al., 2007). There is also extensive literature supporting the association of *Parkinson's disease* and other neurologic diseases and pesticide exposure (Brown et al., 2006; Blanc-Lapierre et al., 2012; Freire and Koifman, 2012; van der Mark et al., 2012).

Young children are particularly vulnerable to adverse health effects that may result from pesticide exposures. Epidemiologic evidence of the neurodevelopmental toxicity of pesticide exposure during pregnancy is growing (Rauh et al., 2006; Engel et al., 2007; Eskenazi et al., 2007; Bjorling-Poulsen et al., 2008; Harari et al., 2010; Munoz-Quezada et al., 2013). *In utero* and/or postnatal chronic exposures to organophosphorus pesticides have been associated with poorer neurodevelopment in children (Rauh et al., 2006; Engel et al., 2007; Eskenazi et al., 2007; Marks et al., 2010; Ross et al., 2013), altered fetal growth, low birth weight (Whyatt et al., 2004; Sathyanarayana et al., 2010; Wohlfahrt-Veje et al., 2011), and shortened gestational duration (Eskenazi et al., 2004). Epidemiologic studies have observed an association between residential proximity to use of agricultural pesticides and fetal death, neural tube defects, autism, Parkinson's disease, and childhood cancer (Searles Nielsen et al., 2010).

In the review by Kamel and Hoppin (2004) of the health effects of pesticide exposure, the authors report that chronic pesticide exposure is associated with a broad range of nonspecific symptoms, including *headache, dizziness, fatigue, weakness, nausea, chest tightness, difficulty in breathing, insomnia, confusion*, and *difficulty concentrating*. Many of the studies indicate that pesticide exposure is associated with deficits in *cognitive function*. Of particular interest are the studies performed in areas where organophosphates are widely used (Bjorling-Poulsen et al., 2008).

The results on toxicity characterization of the 276 legally marketed active substances in Europe indicate that 32 out of the 78 fungicides, 25 out of the 87 herbicides, and 24 out of the 66 insecticides are related to at least one health effect (e.g., carcinogenic, endocrine disruptor, reproductive and developmental toxicity, acute toxicity) (Karabelas et al., 2009).

Pesticide Exposure in Animals

The circumstances involving lethal exposures in animals can be classified as due to accidents with the approved use, misuse, and deliberate abuse of pesticides. In the first case, animal poisonings occur when pesticides are applied to the approved target and at the proper application rate (Augspurger et al., 1996; Pain et al., 2004; Ogada, 2014).

In the second case, pesticides are not applied according to the specified conditions for use (Greig-Smith et al., 1994; Guitart et al., 1996). In the last case, pesticides are used in a deliberate or illegal attempt to poison animals, and secondary poisonings in scavengers can occur because of the high doses of the pesticide usually employed (Allen et al., 1996; Berny et al., 1997).

Nowadays, there are fewer cases of acute poisoning caused by high doses of toxic substances but there are more and more cases of chronic poisoning because of environmental pollution. Diseases caused by chronic exposure to toxic substances are very often non-specific, and manifested as a consequence of the immune system weakening or by decreases in weight gain in economically important animal species (Pistl et al., 2003; Pistl et al., 2004).

Moreover, the accumulation of exogenous substances in tissues of economically important animal species also affects the safety and quality of food of animal origin.

The pesticides most frequently involved in animal poisonings are *insecticides* and *rodenticides* (de Snoo et al., 1999; Guitart et al., 1999; Motas-Guzman et al., 2003; Berny, 2007; Martinez-Haro et al., 2008).

INSECTICIDES

Some 10,000 species of the more than 1 million species of insects are crop-eating, and of these, approximately 700 species worldwide cause most of the insect damage to man's crops, in the field and in storage (Ware and Whitacre, 2004).

Therefore, the most commonly applied pesticides are insecticides.

Insecticides are agents of chemical or biological origin that control insects. Control may result from killing the insect or otherwise preventing it from engaging in behaviors deemed destructive. Insecticides may be *natural* or *manmade* and are applied to target pests in a myriad of formulations and delivery systems (sprays, baits, slow-release diffusion, etc.). The science of biotechnology has, in recent years, even incorporated bacterial gene coding for insecticidal proteins into various crop plants that cause death to unsuspecting pests that feed on them (Ware and Whitacre, 2004).

Insecticides are commonly used in agricultural, public health, and industrial applications, as well as household and commercial uses (e.g., control of roaches and termites). In the United States, insecticides accounted for 12% of total pesticides applied to the surveyed crops and the most commonly used insecticides are *the organophosphates, pyrethroids* and *carbamates* (Figure 1; http://www.epa.gov/caddis/ssr_ins_int.html).

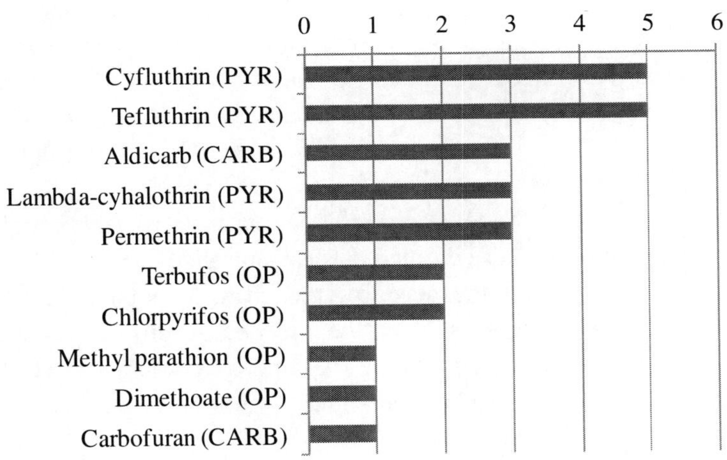

http://www.epa.gov/caddis/ssr_ins_int.html.
PYR: pyrethroid; CARB: carbamate; OP: organophosphate.

Figure 1. US insecticide use in 2001 in millions of acres.

A Brief History and Present Time of Insecticides

Pesticides are not a modern invention. The first recorded use of insecticides was in 2500 B.C. by the Sumarians, who used *sulphur compounds* to control insects and mites. Pliny the Elder (A.D. 23-79) recorded most of the earlier insecticide uses in his *Natural History*. Included among these were the use of *gall from a green lizard* to protect apples from worms and rot. Later, a variety of materials were used with questionable results: *extracts of pepper and tobacco, soapy water, whitewash, vinegar, turpentine, fish oil, brine, lye* among many others. Other methods arose as time went on, such as controlling body lice in China with *mercury* and *arsenical compounds* in 1200 B.C., burning sulphur to kill insects and controlling ants with *honey* and *arsenic* in 1600. *Elemental sulphur*, still in use today as an insecticide, together with other inorganic chemicals, such as *arsenic, mercury* and *lead* are frequently referred as the *"first generation"* pesticides, which were applied to crops for many centuries, until the beginning of the 20th century (Committee on the Future Role of Pesticides in US Agriculture, 2000). The use of the first generation of pesticides was largely abandoned because they were either too ineffective or too toxic, also showing a great tendency to accumulate in soil to the point of inhibiting plant growth. Moreover, with the passage of time, pests became resistant and tolerant to these pesticides.

Writings from ancient Greece and Rome show that religion, folk magic, and the use of what may be termed chemical methods were tried for the control of plant diseases, weeds, insects, and animal pests. As there was no chemical industry, any products used had to be of either plant or animal derivation or, if of a mineral nature, easily obtainable or available. Thus, for example, *smokes* are recorded as being used against insects, as were various plant extracts such as *bitter lupin* or *wild cucumber*. *Tar* was also used on tree trunks to trap crawling insects.

The modern history of chemical insecticides in the United States dates from 1867, when *Paris green* proved effective against the Colorado potato beetle. Within a decade, *Paris green* and *kerosene oil emulsion* were being employed against a variety of chewing and sucking insects. In the early part of the 20th century, *fluorine compounds* and *plant-derived insecticides* were developed. Except for plant derivatives, such as *nicotine, pyrethrin*, and *rotenone*, early insecticides were almost all inorganic chemicals. *Pyrethrum*, which is derived from the dried flowers of *Chrysanthemum cinerariaefolium* "Pyrethrum daisies", has been used as an insecticide for over 2000 years.

Persians used the powder to protect stored grain and later, crusaders brought information back to Europe, which claimed that dried round daisies controlled head lice.

Expanding agriculture to meet the needs of the growing population necessitated the application of more effective pesticides; thus, the *"second-generation"* pesticides, as they came to be known, were developed. These pesticides, comprising *synthetic organic chemicals*, made their debut in the 1930s (Committee on the Future Role of Pesticides in US Agriculture, 2000).

Up until the 1940s, *inorganic substances* were still widely used in pest control, then early organics such as *nitrophenols, chlorophenols, creosote, naphthalene,* and *petroleum oils* were used for insect pests. The drawback for many of these products was their high rates of application, lack of selectivity and phytotoxicity.

World War II opened the *Modern Era of Chemical* control, with the introduction of a new concept of insect control, synthetic organic insecticides. The modern era of pesticides essentially began with the introduction, in 1939, of the insecticide *DDT (dichlorodiphenyl trichloroethane)*. The astonishing efficacy of DDT led to the development of a variety of chlorinated hydrocarbons, such as *gamma-lindane* and *toxaphene,* which were quickly adopted in the agricultural practice (Daly et al., 1998). The growth in synthetic pesticides accelerated in the 1940s with the discovery of the effects of *DDT, aldrin, dieldrin, endrin, chlordane, parathion,* and *captan*. These products were effective and inexpensive with DDT being the most popular, because of its broad-spectrum activity. DDT was widely used, appeared to have low toxicity to mammals, and reduced insect-born diseases, like malaria, yellow fever and typhus; consequently, in 1949, Dr. Paul Muller won the Nobel Prize in Medicine for discovering its insecticidal properties. However, in 1946, resistance to DDT by houseflies was reported and, because of its widespread use, there were reports of harm to non-targeted plants and animals, and problems with residues. Serious concerns about the human safety and biological effects of *DDT* led 86 countries to ban its use in 1972.

In the 1940s, manufacturers also began producing large amounts of other synthetic pesticides, like *the organophosphates*, which, despite their greater toxicity to mammals and other non-target species, enjoyed considerable popularity because of their broad spectrum efficacy and rather low cost (Casida and Quistad, 1998; Casida and Durkin, 2013a). *Organophosphates* attack insect nervous systems, much like the chlorinated hydrocarbons, but are much quicker to break down into non-toxic substances.

A large and versatile group, the organophosphates include *parathion*, with one of the highest mammalian toxicities, and *malathion,* with one of the lowest.

Pesticide use to date has increased 50-fold since 1950. Currently there are thousands of synthetic pesticide products made up of more than 1,000 different chemicals and combinations thereof (Miller, 2002). On the other hand, potential health risks of pesticides identified in toxicological and epidemiological studies include cancer, genetic malformations, neuro-developmental disorders, and damage of the immune system (Pistl et al., 2004; Sanborn et al., 2007; McKinlay et al., 2008).

Throughout most of the 1950s, consumers and most policy makers were not overly concerned about the potential health risks in using pesticides. Food was cheaper because of the new chemical formulations and with the new pesticides; there were no documented cases of people dying or being seriously hurt by their "normal" use. There were some cases of harm from misuse of the chemicals. But the new pesticides seemed rather safe, especially compared to the forms of *arsenic* that had killed people in the 1920s and 1930s.

The urgent ban to major “second generation” pesticides for valid reasons, apparently created the need for a re-evaluation of pesticide applicability, and for the development of a legal framework regulating their production and use.

By the end of the 1960s, this need culminated in the creation of the *Environmental Protection Agency* (EPA) in the U.S., which was the first governmental agency taking action against pesticide usage at that time (Walker et al., 2003).

Rachel Carson’s book - *Silent Spring*, published in 1962 (Carson, 1962), first drew attention to the hazard of the widespread extensive use of pesticides for the environment (namely birds), and for human health. The book resulted in big modifications to the policy on pesticides, leading to a ban on DDT and certain other pesticides.

In the 1960s, with increasing worldwide interest in environmental protection, chemical insecticides became objects of scientific and popular protest. Critics charged that chemical insecticides were dangerous and self-defeating, provoking the development of resistance by target pests, sabotaging ecological systems, and poisoning people and other organisms, as well as the environment. In response, governments have restricted or proscribed many of the most dangerous insecticides, including many chlorinated hydrocarbon standbys: *DDT, benzene hexachloride, lindane, aldrin, dieldrin, chlordane, heptachlor, endrin,* and *toxaphene* - all powerful, broad-spectrum contact and stomach poisons.

Research into pesticides continued in the 1970s and 1980s. For insecticides, there was the synthesis of a *3*rd *generation of pyrethroids*, the introduction of *avermectins, benzoylureas,* and B_t *(Bacillus thuringiensis)* as a spray treatment. As many of the agrochemicals introduced at this time had a single mode of action, thus making them more selective, problems with resistance occurred. Management strategies were introduced to combat this negative effect.

In the 1990s, research activities concentrated on finding new members of existing families, which have greater selectivity and better environmental and toxicological profiles. In addition, new families of agrochemicals have been introduced to the market, such as *chloronicotinyl, spinosyn, fiprole* and *diacylhydrazine* insectides. Many of the new agrochemicals can be used at grams rather than the kilograms per hectare (Committee on the Future Role of Pesticides in U.S. Agriculture, 2000).

New insecticide chemistry has allowed better resistance management and improved selectivity. This period also saw the refinement of mature products in terms of use patterns with the introduction of newer, more user-friendly, and environmentally safe formulations. Integrated pest management systems, which use all available pest control techniques in order to discourage the development of pest populations and reduce the use of pesticides and other interventions to levels that are economically justified, have also contributed to reducing pesticide use.

Today the pest management toolbox has expanded to include the use of genetically engineered crops designed to produce their own insecticides or exhibit resistance to broad-spectrum herbicide products or pests. These include varieties of corn and cotton, resistant to corn borer and bollworm, respectively.

In addition, the use of Integrated Pest Management (IPM) systems, which discourage the development of pest populations and reduce the use of agrochemicals, have also become more widespread. These changes have altered the nature of pest control and have the potential to reduce and/or change the nature of agrochemicals used (Committee on the Future Role of Pesticides in US Agriculture, 2000).

Very drastic changes have occurred in the list of legally marketed pesticides, over the last years, because of the legislation on marketed pesticides. For example, in the EU, approximately 704 active substances were banned, of which 26% were insecticides, 23% herbicides, and 17% fungicides (Karabelas et al., 2009).

Today several hundred chemical insecticidal agents are registered and licensed in more than 10,000 formulations.

Promptly effective, easy to use, and readily available, chemicals have become the modern weapons of choice against insects, contributing to stable food and fiber productivity, to human and animal health, and to the comfort and quality of human life.

Classification

Insecticides are classified based on their structure and mode of action. Many insecticides act upon the nervous system of the insect (e.g., cholinesterase inhibition) (Casida and Durkin, 2013b) while others act as growth regulators or endotoxins (http://www.epa.gov/caddis /ssr_ins_int. html). Table 1 lists the major classes of insecticides and their modes of action (Radcliffe et al., 2009).

Table 1. Insecticide types and their modes of action

Insecticide type	Mode of Action
Organochlorine	Most act on neurons by causing a sodium/potassium imbalance preventing normal transmission of nerve impulses, while some act on the GABA (γ-aminobutyric acid) receptor preventing chloride ions from entering the neurons causing a hyper-excitable state characterized by tremors and convulsions; usually broad-spectrum insecticides that have been taken out of use.
Organophosphate	Cause acetylcholinesterase (AChE) inhibition and accumulation of acetylcholine at neuromuscular junctions causing rapid twitching of voluntary muscles and eventually paralysis; broad-range insecticide, generally the most toxic of all pesticides to vertebrates.
Organosulfur	Exhibit ovicidal activity (i.e., they kill the egg stage); used only against mites with very low toxicity to other organisms.
Carbamates	Cause acetylcholinesterase (AChE) inhibition causing central nervous system effects (i.e. rapid twitching of voluntary muscles and eventually paralysis); very broad-spectrum toxicity and highly toxic to fish.
Formamidines	Inhibit the enzyme monoamine oxidase that degrades neurotransmitters causing an accumulation of these compounds; affected insects become quiescent and die; used in the control of organophosphate- and carbamate-resistant pests.
Dinitrophenols	Act by uncoupling or inhibiting oxidative phosphorylation preventing the formation of adenosine triphosphate (ATP); All types have been withdrawn from use.
Organotins	Inhibit phosphorylation at the site of dinitrophenol uncoupling, preventing the formation of ATP; used extensively against mites on fruit trees, formerly used as an antifouling agent and molluscicide; very toxic to aquatic life.
Pyrethroids	Acts by keeping open the sodium channels in neuronal membranes affecting both the peripheral and central nervous systems causing a hyper-excitable state causing such symptoms as tremors, incoordination, hyperactivity and paralysis; effective against most agricultural insect pests; extremely toxic to fish.
Nicotinoids	Act on the central nervous system causing irreversible blockage of the postsynaptic nicotinergic acetylcholine receptors; used in the control of sucking insects, soil insects, whiteflies, termites, turf insects, and the Colorado potato beetle; have generally low toxicity to mammals, birds, and fish.
Spinosyns	Acts by disrupting binding of acetylcholine in nicotinic acetylcholine receptors at the postsynaptic cell; effective against caterpillars, lepidopteran larvae, leaf miners, thrips, and termites; regarded for its high level of specificity.
Pyrazoles	Inhibits mitochondrial electron transport at the NADH-CoQ reductase site leading to disruption of ATP formation; effective against psylla, aphids, whitefly and thrips; results of testing on one type (fipronil) indicate no effects on the clams, oysters, or fish, with marginal effects on shrimp.
Pyridazinones	Interrupt mitochondrial electron transport at Site 1; mainly used as a miticide; display toxicity to aquatic arthropods and fish.
Quinazolines	Acts on the larval stages of most insect by inhibiting or blocking the synthesis of chitin in the exoskeleton; developing larvae exhibits rupture of the malformed cuticle or death by starvation; not registered in US.

Table 1. (Continued)

Insecticide type	Mode of Action
Botanicals	Depending upon the type can have various effects: Pyrethrum – affects both the central and peripheral nervous systems, stimulating nerve cells to produce repetitive discharges and eventually leading to paralysis; commonly used to control lice. Nicotine – mimics acetylcholine (Ach) in the central nervous system ganglia, causing twitching, convulsions and death; used most to control aphids and caterpillars. Rotenone – acts as a respiratory enzyme inhibitor; used as a piscicide that kills all fish at doses non-toxic to fish food organisms. Limonene – affects the sensory nerves of the peripheral nervous system; used to control fleas, lice, mites, and ticks, while remaining virtually non-toxic to warm-blooded animals and only slightly toxic to fish. Neem – reduces feeding and disrupts molting by inhibiting biosynthesis or metabolism of ecdysone, the juvenile molting hormone; commonly used against moth and butterfly larvae.
Synergists/ Activators	Inhibit cytochrome P-450 dependent polysubstrate monooxygenases (PSMOs) preventing the degradation of toxicants, enhancing the activity of insecticides when used in concert; synergists and activators are not in themselves considered toxic or insecticidal.
Antibiotics	Act by blocking the neurotransmitter GABA at the neuromuscular junction; feeding and egg laying stop shortly after exposure while death may take several days; most promising use of these materials is the control of spider mites, leafminers, and other difficult to control greenhouse pests.
Fumigants	Act as narcotics that lodge in lipid-containing tissues inducing narcosis, sleep, or unconsciousness; pest affected depends on particular compound.
Inorganics	Mode of action is dependent upon type of inorganic: may uncouple oxidative phosphorylation (arsenicals), inhibit enzymes involved in energy production, or act as desiccants; pest group depends on compound (e.g., sulfur for mites, boric acid for cockroaches).
Biorational	Grouped as biochemicals (hormones, enzymes, pheromones natural agents such as growth regulators) or microbials (viruses, bacteria, fungi, protozoa and nematodes), acting as either attractants, growth regulators or endotoxins; known for very low toxicity to non-target species.
Benzoylureas	Act as insect growth regulators by interfering with chitin synthesis; greatest value is in the control of caterpillars and beetle larvae but is also registered for gypsy moth and mushroom fly; some types are known for their impacts on invertebrates (reduced emergent species) and early life stages of sunfish (reduced weight).

Radcliffe et al., 2009.

Most insecticides are much more toxic to insects than to mammals. In many cases, the mechanism of this selective toxicity lies in differential actions on the target receptors/channels. However, there are some exceptions, for example, the selective toxicity of organophosphate insecticide *malathion* is due primarily to the difference in detoxifying enzymes between mammals and insects. With the exception of organophosphates and carbamate, insecticides that inhibit cholinesterases (Casida and Durkin, 2013b), most other insecticides act directly on neuronal receptors and/or ion channels, thereby causing symptoms of poisoning (Narahashi et al., 2007).

ORGANOPHOSPHATES

Organophosphates (OPs) are commonly used as pesticides in agriculture, industry, and around the home/garden throughout the world. They also have residential and indoor applications for pest control, especially for cockroaches and termites (Gurunathan et al., 1998; Aprea et al., 2000; Whyatt and Barr, 2001; Morgan et al., 2005; Van den Hazel and Zuurbier, 2005; Becker et al., 2006; Saunders et al., 2012). In addition, these chemicals are used as parasiticides in veterinary medicine (Gupta, 2012).

Most organophosphate pesticides registered for use in the United States are used as insecticides. In 1999, 60 million pounds of organophosphate pesticides were used in agriculture and about 17 million pounds were used in nonagricultural applications (US EPA, 2003a,b).

History of Organophosphates

The first organophosphate compound, *tetraethyl pyrophosphate*, was synthesized in 1854 by Philipe de Clermont. In 1932, Lange and Kruger described the synthesis of *dimethyl* and *diethyl phosphorofluoridate*. Based on the chemistry of these compounds, Gerhard Schrader (a chemist at the I.G. Farbenindustrie) led the exploration of the organophosphate class of compounds that could be used as insecticides.

One of the earliest organophosphate insecticides synthesized by Schrader was *parathion*, which is still used worldwide.

Before World War II, the German Ministry of Defense developed highly toxic organophosphate compounds of G series *(tabun, sarin,* and *soman)* and *diisopropyl phosphorofluoridate*.

In the 1950s, organophosphate compounds with super toxicity, such as *VX* and *VR*, were synthesized in the United Kingdom and the Soviet Union. After World War II, thousands of organophosphates have been synthesized in the search for compounds with species selectivity, i.e. more toxic to insects and less toxic to mammals (e.g. *malathion*). Today, more than 100 organophosphates are in use for a variety of purposes, such as protection of crops, grains, gardens, homes, and public health (Gupta, 2012).

Nowadays, the common use of organophosphate insecticides results in their accumulation, environmental pollution, and acute and chronic poisoning events (Rahimi et al., 2006). For this reason, the use of organophosphate insecticides has to be strictly controlled and restricted. Accordingly, the majority of countries have strong regulations on the application of pesticides. Also, the applied insecticides and their by-products in the environment, water, and food are monitored, applying different bioanalytical techniques.

Basic Structure of Organophosphorus Compounds

Organophosphates are esters or thiols derived from phosphoric, phosphonic, phosphinic or phosphoramidic acid. There are at least 13 types of organophosphates (Gupta, 2012). The basic structure of organophosphorus compounds is shown in Figure 2 (Greene and Pohanish, 2005).

Despite differences in chemical structures, all organophosphates share one thing: they all have a pentavalent phosphorus atom and a characteristic phosphoryl bond (P = O) or thiophosphoryl bond (P = S). Essentially, organophosphates are esters of phosphoric acid with varying combinations of oxygen, carbon, sulfur, and/or nitrogen attached. Of course, the chemistry of these compounds is much more complex.

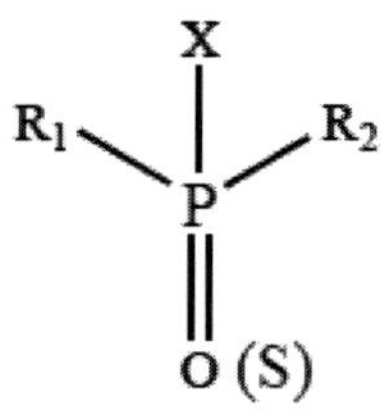

Greene and Pohanish, 2005.

Figure 2. General structural formula of organophosphates.

The organophosphates that are derivatives of phosphoric or phosphonic acid possess anticholinesterase activity, unlike those that are derivatives of phosphinic acid. Usually, organophosphate compounds have two alkyl substituents and an additional substituents group (leaving group), which is more labile to hydrolysis than the alkyl group (Marrs, 1993).

Some organophosphates (such as *dichlorvos, monocrotophos,* and *trichlorfon*) are direct acethylchlinesterase (AChE) inhibitors, while those of phosphorothioates type (such as *bromophos, diazinon, fenthion*, and *parathion*) possess minimal or no anticholinesterase (anti-AChE) activity and require desulfuration to the analogous oxon before acquiring anti-AChE activity. Also, organophosphates, which are used as defoliants (*S,S,S-tributyl phosphorotrithioate* and *S,S,S-tributyl phosphorotrithioite*), and herbicides (*glyphosate* and *gluphosinate*), are of very low mammalian toxicity (Gupta, 2012).

Oxo forms of organophosphate insecticides are highly and almost equally toxic to warm-blooded as well as cold-blooded organisms. On the other hand, thio forms are converted into the oxo forms by mixed function oxidases. The activation proceeds in cold-blooded organisms but this is not common in warm-blooded organisms where dealkylation into non-toxic compounds takes place (Pohanka, 2011).

Thus, numerous derivatives of highly toxic insecticides have been prepared to reduce the toxicity towards warm-blooded organisms and retain toxicity to insects, thereby enhancing their specificity.

Examples of effective and commonly used organophosphate insecticides, which are relatively safe for warm-blooded organisms, are: *malathion, chlorpyrifos, fenitrothion, pirimiphos-methyl, dimethoate, phosalone* (Pohanka, 2011).

Use of Organophosphates

Organophosphorus Nerve Agents/Gases

Organophosphorus nerve agents include *tabun (GA), sarin (GB), soman (GD), cyclosarin (GF), VX,* and *VR*. Sarin, soman, and cyclosarin are phosphonofluoridates, and VX is a phosphonothioate. Soman has four, while sarin and VX have two isoforms, which significantly differ in toxicity and irreversible AChE inactivation rate. Based on the acute toxicity, VX is the most toxic compound among all the nerve agents (Gupta, 2005).

These compounds are highly toxic and pose continuous threats for the lives of humans as well as animals, because they can be used as chemical weapons of mass destruction. These agents have been used by dictators and terrorists. These compounds produce toxicity by directly inhibiting AChE, and are much more potent than organophosphate pesticides, as they cause lethality to animals in the submilligram range (Watson et al., 2006; Yanagisawa, 2014).

Acute poisoning by a nerve agent leads to *contraction of pupils, profuse salivation, convulsions, involuntary urination and defecation,* and *eventual death by asphyxiation,* as control is lost over respiratory muscles. Moreover, the effects of nerve agents are very long lasting and cumulative (increased by successive exposures), and survivors of nerve agent poisoning usually suffer *chronic neurological damage* that can lead to continuing psychiatric effects.

Irreversible Acetylcholinesterase Inhibitors As Therapeutic Agents

Organophosphates, except for their use as toxic compounds, have been applied in ophthalmology as therapeutic agents in the treatment of *chronic glaucoma,* e.g. *diisopropyl fluorophosphate* and *echothiophate*.

Organophosphate compounds may be used in the therapy of *neurological damages* such as Alzheimer´s disease and Parkinson's disease. An example is *trichlorfon (metrifonate),* which used to be applied as a pesticide, and has medicine implementation analogous to carbamate rivastigmine (Colovic et al., 2013).

Organophosphorus Insecticides

The majority of organophosphate compounds are used as pesticides. The organophosphates have two distinctive features: they are generally more toxic to vertebrates than other classes of insecticides, and most are chemically unstable or nonpersistent *(Ware and Whitacre, 2004)*.

Organophosphate insecticides are used in agricultural and domestic pest control, accounting for 50% of global insecticidal use (Casida and Quistad, 2004).

Commonly used organophosphate insecticides have included *ethyl parathion, malathion, methyl parathion, chlorpyrifos, diazinon, dichlorvos, phosmet, fenitrothion, tetrachlorvinphos, azinphos methyl, pirimiphos-methyl, dimethoate, phosalon*e (Greene and Pohanish, 2005). Their use is, however, accompanied by widespread toxicity in non-target organisms, including man.

Nowadays, organophosphates are one of the most common causes of poisoning worldwide occurring as a result of agricultural use, suicide, or accidental exposure.

Toxicokinetics of Organophosphates

Organophosphate pesticides can be absorbed by all routes, including *inhalation, ingestion*, and *dermal absorption*. Ingestion encounters with contaminated feed/food with pesticides residue, while dermal exposure is more relevant when these insecticides are used as ectoparasiticides in the form of dust, dip, or oily solution. Inhalation of airborne insecticides occurs during or soon after aerial spray, particularly due to chemical drift.

Once the insecticide reaches a portal of entry, it is available for absorption. It is established that following absorption, these insecticides are well distributed in tissues throughout the body. Organophosphate insecticides may follow either activation or detoxification, or both. Activation implies that the metabolite is more toxic than the parent compound, e.g. the conversion of malathion to malaoxon. This process is often referred to as "lethal synthesis". On the other hand, detoxification implies that the metabolite is less toxic than the parent compound (Gupta, 2012).

Mechanism of Action

Organophosphates produce their toxicity by virtue of irreversible inhibition of acetylcholinesterase (AChE) enzyme, which terminates the action of the neurotransmitter acetylcholine (ACh) at the synapses in nervous tissue and at the neuromuscular junctions. These chemicals are referred to as "anticholinesterases" (Casida and Durkin, 2013b).

Acetylcholinesterase inactivation, induced by various inhibitors, leads to acetylcholine accumulation, hyperstimulation of nicotinic and muscarinic receptors, and disrupted neurotransmission. Hence, acetylcholinesterase inhibitors, interacting with the enzyme as their primary target, are applied as relevant drugs and toxins (Colovic et al., 2013).

According to the mode of action, AChE inhibitors can be divided into two groups: *irreversible* and *reversible*. Reversible inhibitors, competitive or noncompetitive, mostly have therapeutic applications, while toxic effects are associated with irreversible AChE activity modulators (Colovic et al., 2013).

Actually, organophosphates are substrate analogues to ACh, and like natural substrates, enter the active site covalently binding to serine – OH group. As in acetylation, organophosphate is split and the enzyme is phosphorylated. While the acyl enzyme is quickly hydrolyzed to regenerate the free enzyme, dephosphorylation is very slow (on the order of days), and the phosphorylated enzyme cannot hydrolyze the neurotransmitter. The inhibition of the enzyme leads to accumulation of ACh in the synaptic cleft resulting in over-stimulation of nicotinic and muscarinic ACh receptors and impeded neurotransmission. Some of the organophosphates with strong AChE inhibiting potential are also used as nerve agents or nerve gases (Gupta, 2012; Casida and Durkin, 2013b).

Additionally, many studies suggest that both acute and chronic intoxication with organophosphates *disturb the redox processes* changing the activities of antioxidative enzymes and causing enhancement of lipid peroxidation in many organs, and there is little correlation between organ damage and the degree of organophosphate induced AChE inhibition. Indeed, in acute, and rather subchronic or chronic organophosphate exposition, *induction of oxidative stress* has been reported as the main mechanism of its toxicity (Dandapani et al., 2003; Ranjbar et al., 2005).

There is some evidence that organophosphates may affect *liver, kidney, muscles, immune,* and *hematological system*, causing many human body disorders (Abdollahi et al., 2004).

Although organophosphate insecticides are being replaced by what are now better and more effective chemotypes, they play the major role in providing food, fiber and health protection for an ever-expanding human population (Casida and Durkin, 2013a).

However, many organophosphates are extremely toxic and lack species selectivity, and therefore their inadvertent/accidental use continues to pose a threat to the environment, human and animal health, wildlife, and aquatic systems (Gupta, 2012).

Clinical Signs in Humans

The typical symptoms of acute poisoning are agitation, muscle weakness, muscle fasciculations, miosis, hypersalivation, and sweating. Severe poisonings may cause respiratory failure, unconsciousness, confusion, convulsions, and/or death (Costa, 2006).

The toxicity of organophosphates is not limited to the acute phase, but chronic effects have long been noted. Actually, repeated or prolonged exposure to organophosphates may result in the same effects as acute exposure, including the delayed symptoms. The effects, reported in workers repeatedly exposed, include impaired memory and concentration, disorientation, severe depressions, irritability, confusion, headache, speech difficulties, delayed reaction times, nightmares, sleepwalking, and drowsiness or insomnia. Influenza-like conditions with headache, nausea, weakness, loss of appetite, and malaise have also been reported. Neurotransmitters such as ACh are profoundly important in the brain's development, and many organophosphates have neurotoxic effects on developing organisms, even from low levels of exposure, causing various diseases of the nervous and immune systems (Ray and Richards, 2001; Casida and Quistad, 2004; Corsini et al., 2013).

Organophosphates can produce delayed neurotoxic effects in humans and chickens, called organophosphate-induced delayed polyneuropathy (OPIDP). It is associated with phosphorylation, and further dealkylation (aging) of a protein in neurons called neuropathy target esterase, subsequently leading to this syndrome. The symptoms of this neuropathy are paralysis and ataxia, and appear between 14 and 24 days after the poisoning (Sogorb and Vilanova, 2002; Gupta, 2005). OPIDP is characterized by distal degeneration of long- and large diameter motor and sensory axons of both peripheral nerves and spinal cord (Moretto and Lotti, 2005). Among all animal species, the hen appears to be the most sensitive and is therefore used as an animal model.

Organophosphate insecticide-induced intermediate syndrome (IMS) is usually observed in individuals who have ingested a massive dose of an organophosphate insecticide, either accidentally or in a suicide attempt. IMS is clearly a separate clinical entity from acute toxicity and delayed neuropathy. Clinically, IMS is characterized by acute paralysis and weakness in the areas of several cranial motor nerves, neck flexors, facial, extraocular, palatal, nuchal, proximal limb, and respiratory muscles 24 – 96 h after poisoning. Generalized weakness, depressed deep tendon reflexes, ptosis, and diplopia are also evident. These symptoms may last for several days or weeks depending on the organophosphate involved. To date, organophosphates that are known to cause IMS include bromophos, chlorpyrifos, diazinon, dicrotophos, dimethoate, disulfoton, fenthion, malathion, merphos, methamidophos, methyl parathion, monocrotophos, omethoate, parathion, phosmet, and trichlorfon (Gupta, 2012).

Clinical Signs in Animals

Most animal poisoning cases in the field are acute in nature. Onset of clinical signs usually occurs within 15 min to 1 h, which is soon followed by the signs of maximal severity, although these timings tend to vary depending upon the organophosphate compound and its dose, and species.

For example, onset of clinical signs is delayed with chlorpyrifos (Dursban) and dimethoate (Rogor).

Clinical signs observed in poisoned animals can be divided into *local* and *systemic* effects. The *local effects* involve the eyes and the lungs, owing to their exposure to vapors or droplets of the insecticides.

These effects, however, are of significance in the case of animals only when exposure is via spraying. The *systemic effects* are primarily on the brain, skeletal muscles, lungs, heart, and other organs.

The clinical signs can also be classified as *muscarinic, nicotinic*, and *central.*

Muscarinic receptor-associated effects are manifested by *vomiting, abdominal and chest pain, salivation, lacrimation, urination, diarrhea* (SLUD), *miosis* (pinpoint pupils), *tracheobronchial secretion, lung edema,* and *cyanosis.*

The *nicotinic* receptor-associated effects are produced on autonomic ganglia and skeletal muscles, and the affected animals show *twitching of muscles, tremors, followed by convulsions, and seizures*. This condition may lead to *paralysis.*

The *central effects* include *apprehension, stimulation, followed by depression.* The affected animals may also show *restlessness, ataxia, stiffness of the neck, and coma.* Death occurs due to respiratory failure and cardiac arrest (Gupta, 2012).

Chlorpyrifos – Uses, Physical-Chemical Properties, Toxicity, Exposure, Health and Environmental Risks

Chlorpyrifos - the Most Widely Used Organophosphorus Insecticide in the World

Organophosphates are used frequently in many countries for pest control due to their low price and broad spectrum of activity (Saulsbury et al., 2009; Saunders et al., 2012). In 2003, they accounted for over 59% (4645 tonnes) of insecticide sales in the EU, with chlorpyrifos the top selling insecticide (15.6%, 1226 tonnes) (Eurostat statistics).

Chlorpyrifos was also one of the most widely used organophosphates in the U.S. for pest control (Gurunathan et al., 1998), but the U.S. Environmental Protection Agency (EPA) imposed a ban on the sale of chlorpyrifos for residential use (to control cockroaches, fleas, and termites) and animal use, including flea and tick control, in the United States in December 2001 (US EPA 2002, 2008; CPF cancellation order, 2006). Most nonagricultural uses were also phased out in the European Union in 2005 (Official Journal of the European Union Legislation, 2007). Therefore, the primary route of exposure to the general population of children and adults today is dietary. Children and adults, including pregnant females, who are farm workers or living on or near farms may be exposed through additional pathways, including dermal or inhalation routes (Li et al., 2012; Mink et al., 2012).

Chlorpyrifos [O, O-diethyl-*O*-(3,5,6-trichloro-2-pyridyl) phosphorothionate was introduced into the market in 1965. Chlorpyrifos is an *organophosphorus insecticide, acaricide, and miticide* currently used to control insect pests on a variety of food and feed crops, most commonly corn. Today, it is registered in about 100 countries worldwide and more than 50 crops are protected from insect infestation with its products. It is effective in controlling cutworms, corn rootworms, cockroaches, grubs, flea beetles, flies, termites, fire ants, and lice in grain, cotton field, fruit, nut and vegetable crops, as well as on lawns and ornamental plants. It is also registered for direct use on sheep and turkeys, for horse site treatment, dog kennels, domestic dwellings, farm buildings, storage bins, and commercial establishments (BCPC, 2006; Solomon et al., 2014).

Physical-Chemical Properties

The compound is a colorless to white crystalline solid, has a mild mercaptan-like odor, and is only slightly soluble in water but soluble in most organic solvents. Chlorpyrifos is present as the active ingredient in a variety of formulations and delivery systems, including basic formulations such as emulsifiable concentrates, granular and wettable powders, water-dispersible granulars, micro-encapsulated suspensions, and gel-based products (Eaton et al., 2008). Further physicochemical and other substance-related data are given in Table 2 (Eaton et al., 2008).

Table 2. Physical-chemical properties of chlorpyrifos

CAS no.	2921-88-2
Chemical formula	$C_9H_{11}Cl_3NO_3PS$
Molecular weight	350.57
Melting point	41–42 °C
Boiling point	Decomposes at about 160 °C
Density at 43.5°C	1.398 g/cm^3
Water solubility at 20°C	0.7 mg/L
Water solubility at 25°C	2.0 mg/L
Organic solvent solubility	79% w/w in isooctane 43% w/w in methanol readily in others
Partition coefficients	Log *K*ow 4.82 Log *K*oc 3.73
Vapor pressure at 20°C and 25°C	1.87×10^{-5} mm Hg
Henry's law constant at 25°C	1.23×10^{-5} atm-m^3/mol
Conversion factors at 25°C	1ppm = 14.3 mg/ m^3 1 mg/ m^3 = 0.07 ppm

Eaton et al., 2008.

Environmental Degradation of Chlorpyrifos

Following application to crops, chlorpyrifos quickly binds to soil and plants. Though it typically degrades rapidly in the environment, residual levels of chlorpyrifos can last for long periods of time (Eaton et al., 2008).

Chlorpyrifos undergoes a number of degradation pathways while in the environment (ATSDR, 1997). Chlorpyrifos and its metabolites are susceptible to photodegradation, with a half-life of approximately 3 days; in the presence of hydroxyl radicals in the atmosphere the half-life is lowered to about 6 hours. Upon entering surface water, chlorpyrifos degradation is associated with abiotic hydrolysis or photosensitized oxidation. In soil, photodegradation plays a role in hydrolysis, dechlorination, and oxidation of chlorpyrifos. However, in indoor environments, chlorpyrifos can persist for several months because of the relative lack of sunlight, water, and/or soil microorganisms that contribute to its rapid degradation in the outdoor environment (ATSDR, 1997).

The use of chlorpyrifos for agricultural and gardening purposes lead to accumulation in soil, water, and on food such as vegetables and fruit (Gurunathan et al., 1998; Aprea et al., 2000; Morgan et al., 2005; Van den Hazel and Zuurbier, 2005; Becker et al., 2006; Saunders et al., 2012), in indoor air, including house dust, and on surfaces including toys (Gurunathan et al., 1998; Morgan et al., 2005).

Ecotoxicological effects of chlorpyrifos were described by Moore et al. (2014); Giesy et al. (2014); Giddings et al. (2014) and Williams et al. (2014).

Past and Current Exposures

Over the past 40 years, numerous studies have assessed potential human exposures to chlorpyrifos. Exposure pathways include ingestion, inhalation, and dermal exposure. Dietary exposures to trace levels of chlorpyrifos on food products appear to be the main source of non-occupational exposures to chlorpyrifos, both past and present. In the past, inhalation exposure to chlorpyrifos in indoor air likely contributed significantly to total exposures following indoor application of chlorpyrifos for residential pest control. However, this pathway of exposure is no longer relevant in the United States and the EU, as residential use of chlorpyrifos has been discontinued. Chlorpyrifos can be detected in outdoor air following agricultural applications, but the concentrations are such that community exposures via this pathway are likely to be very low, relative to diet. Inhalation exposure in the occupational environment may contribute substantially to total exposure for applicators, and potentially for farm workers and those family members with whom they live. Dermal exposure, and secondary ingestion of house dust in children, may have contributed a small amount to total exposures before 2001, but is not likely to be a significant pathway for exposure today (Eaton et al., 2008).

Population behavior influences interaction between the environment/ exposure setting and the extent of exposure. For chlorpyrifos, there are three key exposure settings: *occupational* (manufacturing, farming and greenhouse workers), *ambient, and indoor* (Saunders et al., 2012).

The general public, especially children, are mainly at risk from ambient and indoor residential exposure (Gurunathan et al., 1998; Aprea et al., 2000; Morgan et al., 2005; Becker et al., 2006).

Oral exposure can arise, particularly from fruit and vegetables consumed as part of the normal diet, but also from water, milk and derived products (Gurunathan et al., 1998; Morgan et al., 2005). Non-dietary oral exposure (contact with soil and household objects) is an important exposure route for younger children due to their behavior patterns with respect to play at floor level and on/with other surfaces and toys (Fenske et al., 1990). Flea control products persist on a pet's fur and could be transferred to children (Ames et al., 1989). Inhalation of indoor air is another route with house dust a critical component. Dermal exposure is also possible through this route (Lu and Fenske, 1999).

Exposure during pregnancy is an area of concern given the high percentage of women using pest control during pregnancy and the vulnerability of the fetus during development. Fetal exposure occurs through transplacental transfer with the placenta failing to act as a barrier to lipophilic chlorpyrifos (Whyatt and Barr, 2001).

Hence, exposures of pregnant women, infants, and children in the general population to chlorpyrifos occur primarily through the diet and are estimated to be less than 1×10^{-2} μg/kg/d in children and $4–6 \times 10^{-3}$ μg/kg/d in adults (Eaton et al., 2008). Therefore, it is unlikely that inhalation, dermal, or secondary oral (nondietary) chlorpyrifos exposures contribute significantly in the general population today (Eaton et al., 2008).

Toxicokinetics

The metabolism and toxicokinetics of chlorpyrifos were described in detail elsewhere (Timchalk et al., 2002; Timchalk et al., 2006; Eaton et al., 2008; Crane et al., 2012; Marty et al., 2012). Chlorpyrifos is absorbed by all routes of exposure. Absorption, distribution, metabolism, and excretion (ADME) studies demonstrate that over 70% of orally administered chlorpyrifos is absorbed (Timchalk et al., 2002; Eaton et al., 2008). Following oral exposure, rats excreted 90% of ingested chlorpyrifos through the urine and 10% in the feces (Blodgett, 2006).

Dermally administered chlorpyrifos was well absorbed in mice (>60 %) but poorly absorbed in humans (< 3 %) (US EPA, 2011).

In ADME studies, using radiolabeled test material; concentrations of radioactivity in all tissues were low, indicating that chlorpyrifos and its metabolites do not selectively accumulate in any tissue. Although some chlorpyrifos may be stored in fat tissue, bioaccumulation is not expected to be significant due to an elimination half-life in humans of less than three days. The disposition of chlorpyrifos was unaffected by prior exposure and no significant dose or sex-related differences have been observed in the metabolism of chlorpyrifos (Timchalk et al., 2002; US EPA, 2011).

Once absorbed, chlorpyrifos is rapidly metabolized and rapidly excreted (27 h half-life in humans) primarily in the urine, as *3,5,6-trichloropyridinol* (TCP) and conjugates that hydrolyse to *3,5,6-TCP* (Timchalk et al., 2002; US EPA, 2011). Chlorpyrifos is metabolized to a number of metabolites including *chlorpyrifos-oxon (CPO)*, the primary toxic active metabolite, *TCPy*, and *diethylphosphate (DEP)* and *diethylthiophosphate (DETP)*. These metabolites or their glucuronic or sulfate conjugates are excreted in the urine (Barr and Angerer, 2006).

Paraoxonase 1/arylesterase (PON1) is a key organophosphate detoxifying enzyme (Costa et al., 2013).

The half-lives of chlorpyrifos in blood (Nolan et al., 1984), adipose tissue (Kamrin, 1997), and its metabolites in urine are 24 h, 62 h, and 15 – 30 h (Nolan et al., 1984; Kamrin, 1997; Griffin et al., 1999), respectively.

Absorption and elimination characteristics for chlorpyrifos and urinary biomarkers of pesticide exposure are summarized in Table 3 (Egeghy et al., 2011).

Table 3. Absorption and elimination characteristics for chlorpyrifos and urinary biomarkers of pesticide exposure

Compound	Absorption of parent compound			Elimination of metabolites		
	Oral	Dermal	Inhalation	Oral	Dermal	Inhalation
Chlorpyrifos	Volunteer studies: 70% of oral dose excreted in urine as TCPy (Nolan et al., 1984), 93% of oral dose excreted in urine as dialkyl-phosphates (Griffin et al., 1999).	Volunteer studies: 1.3% of dermal dose excreted in urine as TCPy (Nolan et al., 1984). 1% of dermal dose excreted as dialkyl-phosphates (Griffin et al., 1999), 1.2–4.3% of dermal dose excreted as TCPy (Meuling et al., 2005).	No Information.	Volunteer study, 27 h Oral (Nolan et al., 1984). Volunteer study, approx 15.5 h oral (Griffin et al., 1999).	Volunteer study, 27 h dermal (Nolan et al., 1984). Volunteer study, approx 30 h dermal (Griffin et al., 1999). Volunteer study, approx 41 h dermal (Meuling et al., 2005).	No Information.

TCPy or 3,5,6-trichloro-2-pyridinol: metabolite of chlorpyrifos.
Egeghy et al., 2011.

Mechanism of Action

Chlorpyrifos and other organophosphates are currently regulated based on red blood cell (RBC) or brain acetylcholinesterase (AChE) inhibition; the latter is considered to be a primary mode of action for toxicity, especially related to the acute clinical signs of neurotoxicity following short-term high-dose exposures (Li et al., 2012).

In order to become an effective AChE inhibitor, chlorpyrifos is converted to *chlorpyrifos oxon (CPO)* through oxidative metabolism, which then binds to and inhibits AChE. This binding produces inhibition of AChE leading to increased acetylcholine (ACh) levels within synaptic clefts, hyperstimulation of the cholinergic system, and adaptive decreases in muscarinic acetylcholine receptor (mAChR) binding (Eaton et al., 2008).

In addition to AChE inhibition in the brain, spinal cord, and peripheral nervous system, CPO binds to and inhibits red blood cell (RBC) AChE, and plasma butyrylcholinesterase

(BuChE). Plasma butyrylcholinesterase (BuChE) inhibition may occur at exposure levels below those that provide brain or RBC AChE inhibition and is used as a marker of exposure in occupational settings. Although the physiological role of BuChE and RBC AChE is not clear, levels of inhibition of the activity of these enzymes are currently used as critical toxicity endpoints for human health risk assessment, with RBC AChE inhibition considered more relevant than BuChE inhibition.

Age-related differences between adults and offspring in magnitude, time of onset, and/or recovery of brain and RBC AChE inhibition were reported following acute dermal, oral, or subcutaneous administration (Pope and Chakraborti, 1992; Moser et al., 1998; Moser and Padilla, 1998; Abu-Qare et al., 2001). While previous studies have shown levels of chlorpyrifos that are safe in adult animals, recent evidence indicates young animals and humans may be more sensitive to chlorpyrifos toxicity. These differences may be attributable, in part, to age differences in metabolic enzymatic activity, especially at higher doses (Timchalk et al., 2006; Eaton et al., 2008; Saunders et al., 2012). Although neonatal pups were more sensitive than adults to higher acute chlorpyrifos doses (> 10 mg/kg), there was less or an absence of differential sensitivity to chlorpyrifos induced inhibition of AChE at lower (< 10 mg/kg/d) repeated exposures (Liu et al., 1999; Zheng et al., 2000). Following repeated gestational exposures, maternal brain and RBC AChE inhibition was greater in dams than fetus (Lassiter et al., 1998; Lassiter et al., 1999; Mattsson et al., 2000).

Additive inhibitory action of chlorpyrifos and polycyclic aromatic hydrocarbons on acetylcholinesterase activity *in vitro* was observed (Jett et al., 1999).

Other modes of action for chlorpyrifos toxicity have also been proposed, including alterations of presynaptic cholinergic functions or noncholinergic neurochemical processes (Liu and Pope, 1998; Pope, 1999; Rauh et al., 2006; Slotkin and Seidler, 2007; Udarbe Zamora et al., 2008).

Toxicity

Toxicity of chlorpyrifos in different species is present in Table 4.

Table 4. Toxicity of chlorpyrifos in different species

Chemical	Species	Oral LD_{50} (mg/kg)	Dermal LD_{50} (mg/kg)	Inhalation LC_{50}	References
Chlorpyrifos	Dow-Wistar rats males females	163 135			McCollister, 1974 Gupta, 2012
	Sherman rats males females	118 155			Gaines, 1969
	Mice	60			El-Sebae, 1978
	Guinea Pigs, male	500 – 504			McCollister, 1974

Table 4. (Continued)

Chemical	Species	Oral LD_{50} (mg/kg)	Dermal LD_{50} (mg/kg)	Inhalation LC_{50}	References
	Rabbit	1000			Kidd and James, 1991
	Chicken	32			McCollister, 1974
	Sheep	800			Kidd and James, 1991
	Rabbit		1000 – 2000		Gupta, 2012
	Rat		> 2000	> 0.2 mg/l (4 h)	Kidd and James, 1991

Chlorpyrifos is considered moderately toxic, and is an EPA class II toxicant, i.e. oral dose LD_{50} is 50 – 500 mg/kg (Barr and Angerer, 2006). Chlorpyrifos is highly toxic to amphibians (Yin et al., 2009).

Clinical Signs

Symptoms of acute poisoning include brain activation, epileptiformic convulsions, muscular tremors, which lead ultimately to flaccid paralysis, increased sweating, salivation, profound bronchial secretion, bronchoconstriction, increased activity of the intestine and diarrhea, miosis, hypertension, lowered body temperature, and hyperglycemia (Savolainen, 2001).

Although the mechanism of acute chlorpyrifos toxicity involves acetylcholinesterase (AChE) inhibition, other mechanisms unrelated to AChE inhibition, including the induction of oxidative stress, have been implicated (Gultekin et al., 2001; Ambali et al., 2007; Ambali et al., 2010a; Ambali et al., 2010b; Ambali et al., 2011a; Ambali et al., 2011b). As a lipophilic molecule, chlorpyrifos easily passes through the cells into the cytoplasm (Uzun et al., 2010). Once inside the cell, chlorpyrifos induces damage to the cellular molecules (Ncibi et al., 2008). Oxidative damage primarily occurs through production of reactive oxygen species (ROS), which causes damage to macromolecules such as lipids, proteins, and DNA. Under normal circumstances, the body copes with oxidative assault through the repair of the damage or the invocation of the indigenous antioxidant enzymatic and nonenzymatic systems to reduce the pro-oxidation states. However, in situations of increased and accelerated oxidative challenge by chlorpyrifos, as previously reported (Gultekin et al., 2001; Ambali et al., 2010a; Ambali et al., 2010b; Ambali et al., 2011a; Ambali et al., 2011b), the natural antioxidant mechanisms are overwhelmed thereby resulting in damage.

Chlorpyrifos is a neurotoxin and suspected endocrine disruptor (Song et al., 1998; Slotkin, 1999; Roy et al., 2005; Slotkin et al., 2007; Viswanath et al., 2010; Cardona et al., 2013; Slotkin et al., 2013). However, chlorpyrifos demonstrates no potential to interact with the estrogen, androgen, or thyroid pathways at doses below the dose levels that inhibit cholinesterase. Therefore, regulatory exposure limits for chlorpyrifos, which are based on cholinesterase inhibition, are sufficient to protect against potential endocrine alterations (Juberg et al., 2013).

Chlorpyrifos has been associated with asthma, acute, reproductive and developmental toxicity (Eskenazi et al., 2007; Saulsbury et al., 2009; Li et al., 2012; Mink et al., 2012).

Slotkin (2011) found that early-life chlorpyrifos exposure leads to prediabetes.

Anemia and alteration in other hematological parameters have been recorded following repeated chlorpyrifos exposure (Goel et al., 2006; Ambali et al., 2010a; Ambali et al., 2011a).

Chlorpyrifos has been shown not to be mutagenic (Gollapudi et al., 1995), carcinogenic (Young and Grandjean, 1988), or teratogenic (Deacon et al., 1980; Breslin et al., 1996), nor does it adversely affect reproduction (Breslin et al., 1996). Recently, the genotoxic potential of chlorpyrifos was studied by Sandhu et al. (2013); Muller et al. (2014); Wang et al. (2014); Ismail et al. (2014).

Chlorpyrifos exposure has been shown to induce immunotoxicity via the induction of apoptosis partly mediated through the activation of caspase 3 (Nakadai et al., 2006). Chronic chlorpyrifos exposure has been associated with abnormality of the immune system including depression of T-lymphocytes (Thrasher et al., 1993). Free radical-induced oxidative damage, which has been widely implicated in the molecular mechanism of chlorpyrifos cytotoxicity, is an initiator of apoptosis (Corcoran et al., 1994; McConkey et al., 1994). Influence of chlorpyrifos on the profile of subpopulations of immunoactive cells and their phagocytic activity, in an experimental *in vivo* rat model, was studied by Noworyta-Głowacka et al. (2012). A study by Ojha and Srivastava (2014) revealed that chlorpyrifos induced oxidative DNA damage in rat lymphocytes.

Chronic effects of human chlorpyrifos exposures include many of the symptoms of acute toxicity, including plasma and red blood cell cholinesterase depression (Kamrin, 1997). In most instances, symptoms subsided after eliminating the exposure. However, newer evidence suggests some persistent neurologic effects after moderate to low-level non-acute exposures (Kamel and Hoppin, 2004; Kamel et al., 2005; Young et al., 2005).

Exposure to chlorpyrifos in utero has been associated with decreased birth weight and length (Perera et al., 2003; Whyatt et al., 2004). Exposure to chlorpyrifos coupled with decreased paraoxonase (PON) activity has been linked to decreased head circumference (Berkowitz et al., 2004).

There are also some literature data about tissue specific toxicities such as hepatotoxicity and nephrotoxicity of chlorpyrifos after *in vivo* exposure (Khan, 2006; Tripathi and Srivastav, 2010; Mansour and Mossa, 2011).

In Vitro Effects on Proliferative Activity of Various Cell Cultures (Experimental Study)

The aim of this experimental study was to screen the direct effect of chlorpyrifos in mammalian kidney and liver cell cultures in comparison with insect cell cultures under *in vitro* conditions.

Material and Methods

The Insecticide Tested

Chemical data on tested insecticide chlorpyrifos are depicted in Figure 3 (BCPC, 2006).

Chlorpyrifos was dissolved in dimethylsulfoxide (DMSO, Lachema, Brno, Czech Republic), of which the final concentration in the maintenance medium was 1%. DMSO is a commonly used solvent of test materials designated for the treatment of cells in culture. The basic molar concentrations of insecticide, freshly prepared before each experiment, were 10^{-1} – 10^{-5} M. These insecticide concentrations were added to cell cultures at the rate of 1% of total cell volume; i.e. the actual dose was 100x lower than the basic one.

After chlorpyrifos exposure cell viability, proliferation and lactate dehydrogenase release were evaluated in various cell cultures.

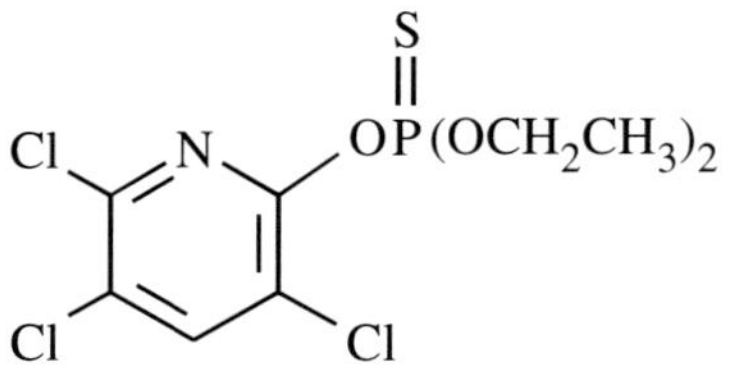

CAS No. 2921-88-2.
Molecular formula: $C_9H_{11}Cl_3NO_3PS$.
Mol. wt.: 350.6.
Chemical family: organophosphates.
Use: insecticide.
Producer, Purity: Supelco, Bellefonte, US; 99.5%.

Figure 3. Chemical data on chlorpyrifos (BCPC, 2006).

Cell Cultures

Cell lines RK13 (rabbit kidney), WBF344 (rat liver), Hepa 1c1c7 (murine hepatoma cell line), and the IPLBSF-21 (the pupal ovarian cells of the fall army worm, *Spodoptera frugiperda* – Sf21) were used in the study. RK13 were cultured in minimal essential medium (MEM) supplemented with 10% (v/v) fetal calf serum (FCS) and 1% (v/v) antibiotic solution of Penicillin G Sodium Salt and Streptomycin Sulphate (Gibco, Invitrogen, Corp., US) at 37 °C. WBF344 and Hepa 1c1c7 cells were cultured in Dulbecco´s modified Eagle´s medium (D-MEM) with 10% (v/v) FCS, 1% (v/v) antibiotic solution of Penicillin G Sodium Salt, and Streptomycin Sulphate (Gibco, Invitrogen, Corp., US) in humidified 5% CO_2 at 37°C. Sf-900 II SFM medium was used for insect cell line (Gibco, Invitrogen, Corp., US) supplemented with 1% (v/v) antibiotic solution of Penicillin G Sodium Salt and Streptomycin Sulphate (Gibco, Invitrogen, Corp., US). Sf21 cells were cultured at 27°C.

Cell Density and Cytopathic Effect

Cell density and cytopathic effect were determined by the standard counting technique using an inverted microscope (Carl Zeiss, Germany). The cytopathic effect was determined on the base of the cell disintegration.

Proliferation Test (PT)

Cells were seeded in 100 μl of cell culture medium in a 96-multiwell culture plate (Corning, Inc., US) at a density of $2x10^5$/ml. After 24 h incubation different concentrations of tested insecticide diluted in DMSO were then added in amount of 1% of culture volume, and treated cells were incubated for 48 h. There were five replicates of each treatment.

After the exposure period, a colorimetric immunoassay was used to quantify cell proliferation (Cell Proliferation ELISA Kit, BrdU-colorimetric, Roche Diagnostics, GmbH, Germany).

The mean optical densities were converted into percentages of residual cell viability, expressed as a percentage of proliferative activity (% PA):

$$\% \text{ PA} = \frac{OD^{insecticide}}{OD^{DMSO}} \text{ x } 100;$$

where $OD^{insecticide}$ is the mean value of OD of cells treated with insecticide and OD^{DMSO} is the mean value of OD of cells treated with the solvent control (DMSO), measured at 450 nm.

Cytotoxicity Assay

Cells were seeded in 100 μl of complete medium in a 96-multiwell culture plate (Corning, Inc., US) at a density of 2 $x10^5$/ml and incubated for 24 h. Growth medium was changed to maintenance medium with 1% (v/v) FCS, and different concentrations of tested insecticide diluted in DMSO were then added in amount of 1% of culture volume. Cells were incubated for additional 24 h. There were five replicates of each treatment. After the exposure period, a non-radioactive colorimetric assay was used to quantify cytotoxicity/cytolysis by measuring LDH activity released from damaged cells (Cytotoxicity Detection Kit^{PLUS}, Roche Diagnostics, GmbH, Germany).

To calculate percent cytotoxicity in each plate, low control (LC) and high control (HC) were set up, and the percentage of cytotoxicity was calculated according to the formula:

$$\text{Cytotoxicity (\%)} = \frac{OD^{insecticide} - OD^{LC}}{OD^{HC} - OD^{LC}} \text{ x } 100$$

where $OD^{insecticide}$ is the mean value of OD of cells treated with insecticide, OD^{LC} (low control) is the mean value of OD cells treated with the solvent control (DMSO) and OD^{HC} (high control) is the mean value of OD cells treated with the lysis buffer (the maximum releasable LDH activity in the cells) measured at 450 nm.

Statistical Analysis

Data were analyzed by ANOVA, followed by Dunnett´s test. Results are presented as mean ± SD, $p < 0.05$ was considered to be statistically significant.

Results and Discussion

Traditionally, animal models have been used to examine the toxic potential of pesticides. With the discovery of a greater number of new pesticides, the feasibility of using *in vivo* models to screen for toxicity becomes impractical with respect to time, ethical, and cost considerations. *In vitro* cell culture systems have proven to be useful for toxicity prediction of target organs on chemicals or drug exposure (Davila et al., 1998).

Cell-based systems have become popular to study interactions of chemicals with intact cells because they offer high-level integration. Previously different *in vitro* cell systems were developed for studies of toxicity (Davila et al., 1998; Li et al., 2003). The liver is known to be the main site of xenobiotic biotransformation due to the ability of this organ to express a plethora of enzymes, both quantitatively and qualitatively.

Therefore, cell lines of liver origin are widely used in biomedical research involving xenobiotic metabolism (Gomez-Lechon et al., 2003; Mersch-Sundermann et al., 2004; Castell et al., 2006). For our study we used WBF344 established from a single cloned non-parenchymal epithelial cell isolated from a normal male adult rat liver and Hepa 1c1c7, murine hepatoma cell line. Kidney is the main organ for elimination of xenobiotics from the body, which is why the RK13 cell line was used in the present study.

Cell lines used in this study were also used as model systems for testing of various xenobiotics, e.g. WBF344 for food and perfume additives (Tang et al., 2007), RK13 for mycotoxins (Rumora et al., 2002), Hepa 1c1c7 for many environmental contaminants (Thornley et al., 2011), industrial chemicals (Park et al., 2002), plant extracts and pigments (Lee et al., 2009), new cosmetic ingredients (Natsch and Emter, 2008), and also pesticides (Hsu et al., 1998; Han et al., 2007; Chen et al., 2008). Continuous cell line Sf21, developed from ovarian tissue of the *Spodoptera frugiperda*, a moth species that is an agricultural pest on corn and other grass species, was chosen for evaluation of the direct effect of insecticide chlorpyrifos on insect cells. Recently, this cell line has been used for testing of insect cell extracts (Staljanssens et al., 2011) or novel insecticides (Powell et al., 2011).

Cytotoxic effect of insecticide chlorpyrifos on cell cultures determined by proliferation activity is shown in Table 5. Proliferative activity of mammalian cell cultures was significantly suppressed as follows: liver cells Hepa 1c1c7 (10^{-1} – 10^{-3} M), WBF344 (10^{-1} – 10^{-2} M), and kidney cells RK13 (10^{-1} M; Table 5). Sf21 cells were the most sensitive to this insecticide with significant suppression of their proliferative activity ranging from 10^{-1} – 10^{-5} M ($p < 0.01$), with PA = 0.2±0.7 – 87.5±6 % (Table 5).

There are a lot of literature data about cytotoxic effects of chlorpyrifos on various cells, e.g. oocytes (Nandi et al., 2011); embryos (Palma et al., 2009); eggs (Scheil et al., 2010); nervous system cells (Song et al., 1998; Das and Barone, 1999; Qiao et al., 2001; Qiao et al., 2003; Qiao et al., 2005; Jameson et al., 2006; Slotkin et al., 2007; Lassiter et al., 2009; Saulsbury et al., 2009; Slotkin et al., 2012, 2013; Raszewski et al., 2014; Campanha et al., 2014); thyroid (De Angelis et al., 2009); peripheral blood lymphocytes (Jamil et al., 2004; Oostingh et al., 2009); bronchial epithelial cells (Oostingh et al., 2009); Jurkat human T cells (Li et al., 2009); mesenchymal stem cells (Hoogduijn et al., 2006), and human breast cell lines (Rich et al., 2012). Literary data about the effects of chlorpyrifos on liver and kidney cells under *in vitro* conditions are rare, however, several *in vivo* studies showed that chlorpyrifos might damage liver and kidneys after long-term exposure (Dahamna et al., 2004; Hoogduijn et al., 2006; Khan, 2006; Tripathi and Srivastav, 2010; Mansour and Mossa,

2011). Das et al. (2008) and Nawaz et al. (2014) found that chlorpyrifos is hepatotoxic only at 100 μM. Chen et al. (2013) studied potential synergistic toxicity of cadmium and chlorpyrifos in a representative human hepatocyte cell line Hep G2 by assessing the cell viability with the MTT assay. After 24 h, no toxicity was observed to Hep G2 cells treated with up to 32 mM Cd^{2+}, and up to 1,280 mM chlorpyrifos, respectively.

Table 5. Cytotoxic effect of insecticide chlorpyrifos on cell cultures tested determined by proliferation test

				Concentration of chlorpyrifos in M (mean values)			
Cell culture	Test parameters	10^{-1}	10^{-2}	10^{-3}	10^{-4}	10^{-5}	Solvent control (DMSO)
RK13	OD_{450}±SD	0.03±0.006	0.25±0.04	0.34±0.03	0.28±0.03	0.27±0.01	0.33±0.02
	%PA±SD	9.1±6.1**	75.8±36.4	103±33.3	84.9±24.3	81.8±3	-
WBF 344	OD_{450}±SD	0.04±0.02	0.06±0.01	0.29±0.02	0.27±0.02	0.28±0.02	0.27±0.03
	%PA±SD	14.8±22.2**	22.2±3.7**	107.4±25.9	92.6±25.9	111.1±3	-
Hepa 1c1c7	OD_{450}±SD	0.08±0.01	0.14±0.01	0.23±0.01	0.25±0.02	0.27±0.01	0.3±0.02
	%PA±SD	26.7±10**	46.7±13.3**	76.6±6.7**	83.3±16.7	90±10	-
Sf21	OD_{450}±SD	0.008±0.001	0.0006±0.001	0.28±0.007	0.25±0.003	0.27±0.04	0.32±0.01
	%PA±SD	2.5±0.6**	0.2±0.7**	87.5±6**	78.1±3**	71.9±3**	-

OD_{450} – optical density; SD – standard deviation; % PA – percentage of proliferative activity; ** – statistical significance < 0.01; * – statistical significance < 0.05; RK13 – Rabbit Kidney cell line; WBF344 – epithelial rat liver cell line; Hepa 1c1c7 – murine hepatoma cell line; SF21 – *Spodoptera frugiperda* pupal ovarian tissue; M – molar concentration; DMSO – dimethylsulfoxide.

The concomitant exposure of the Cd^{2+} and chlorpyrifos mix at 10 mM exerted great impairment to Hep G2 cells, causing approximately 50% reduction in cell viability, compared to the control or the individual treatment by Cd^{2+} or chlorpyrifos ($p < 0.001$). Moreover, remarkable morphological alternations representing cell death were observed for the cells upon the combined treatment, as these cells became rounder and smaller than the cells without treatment or treated with only one chemical. Nawaz et al. (2014) found that the mixture of the herbicide atrazine and chlorpyrifos may result in greater than additive effects on vertebrate liver cells.

Our results showed that chlorpyrifos significantly influenced *in vitro* proliferation of the mammalian liver Hepa 1c1c7, WBF344, and kidney RK13 cells. Sf21 cells were the most sensitive to this insecticide with significant suppression of their proliferative activity ranging from $10^{-1} – 10^{-5}$ M.

Cytopathic effect, determined on the basis of cell disintegration, and the LDH leakage into the medium, was observed in RK 13 in concentrations of $10^{-1} – 10^{-3}$ M, in Hepa 1c1c7 and WBF344 cells in the concentrations of $10^{-1} – 10^{-2}$ M compared to the solvent control. On the other hand, in the insect cell line Sf21 these effects were only observed in the concentration of 10^{-1} M (Table 6). After exposure to lower concentrations of the pesticide, non-significant differences in percentages of cytotoxicity, in studied cell cultures, in comparison with the control were determined.

Table 6. Cytotoxic effect of insecticide chlorpyrifos determined by evaluation of LDH activity released from cells

Cell culture	Test parameters	Concentration of chlorpyrifos in M (mean values)				Solvent control (DMSO)
		10^{-1}	10^{-2}	10^{-3}	10^{-4}	
RK13	OD_{450}±SD	2.38±0.18	2.71±0.06	1.67±0.07	1.46±0.16	1.24±0.16
	% Cytotoxicity±SD	99.1±20.8**	121.4±7.6**	53.3±6.6*	37.1±4.1	22.27±8.3
WBF344	OD_{450}±SD	1.44±0.03	1.33±0.04	0.9±0.01	0.9±0.01	0.94±0.06
	% Cytotoxicity±SD	54.5±2.6**	46.9±3.9**	17.2±0.3	17.2±1	23.43±4
Hepa 1c1c7	OD_{450}±SD	2.6±0.02	2.54±0.02	1.85±0.11	1.70±0.04	1.83±0.17
	% Cytotoxicity±SD	84.8±1.6**	81.4±1.9**	42.4±7.3	33.9±2.8	41.25±10.6
Sf21	OD_{450}±SD	0.87±0.04	0.71±0.04	0.66±0.04	0.67±0.03	0.62±0.08
	% Cytotoxicity±SD	22±3.5*	11.8±2	8.6±1.4	9.2±1	6±3

OD_{450} – optical density; SD – standard deviation; ** – statistical significance < 0.01; * – statistical significance < 0.05; RK13 – Rabbit Kidney cell line; WBF344 – epithelial rat liver cell line; Hepa 1c1c7 – murine hepatoma cell line; SF21 – *Spodoptera frugiperda* pupal ovarian tissue; M – molar concentration; DMSO – dimethylsulfoxide.

The quantitative relationship between LDH release and the loss of animal cell viability or proliferation was described by Wen et al. (2006); Rasouli and Asgari (2008); Wang et al. (2009); Aly and Dominica (2009); El-Shenawy (2010); Cuello et al. (2010). LDH is an important glycolytic enzyme, which is present in the cells of almost all body tissues, and changes in the enzyme activity may provide direct and indirect evidence of the cellular damage and can indicate the toxic mechanism (Sutiakova et al., 2009). The LDH release is a parameter, which reflects the membrane integrity. Chen et al. (2013) assessed the level of lactate dehydrogenase in human hepatocyte cell line Hep G2. Chlorpyrifos alone, at the concentration of 10 mM, could not bring harm to these cells, but the LDH release was significantly increased in cells upon the binary exposure of Cd^{2+} and chlorpyrifos ($p < 0.01$), compared to the control or the individual treatment by Cd^{2+} or chlorpyrifos. Wu et al. (1992) observed the correlation of LDH activity, with loss of cell viability in Sf-9 insect cell cultures. Our study showed a significant proliferative suppression of Sf21 cells, but without correlation with LDH leakage into the maintenance medium and cytopathic effect. Cell proliferation at these cells was significantly suppressed but without cell destruction (except exposure to 10^{-1} M of chlorpyrifos). On the other hand, a cytopathic effect was observed in RK 13 > Hepa 1c1c7 and WBF344 cells. DMSO, which is a commonly used solvent for test materials designated for the treatment of cells in culture, did not cause substantial alterations in the studied cultures.

Conclusion

Pesticides are chemical compounds used extensively in modern day agriculture and households for the control of pests. Pesticides also play a role in preventing the spread of infectious diseases.

Though there can be benefits to using pesticides, inappropriate use can counter productively, increase pest resistance, and kill the natural enemies of the pests. Some pesticides are found to be highly persistent in nature, thereby causing contamination of soil, ground, and surface water. Their toxic effects are manifested in different ways such as bioaccumulation, bio-magnification, chronic toxicity, acute immune response, allergic reaction, and mutagenic, teratogenic and carcinogenic effects, organ toxicity, etc.

Several indicators have been used to assess the potential risk of pesticides to human and animal health and the environment. However, their use indicated reduced certainty, suggesting the need for development of alternative models that should increase the accuracy and reliability of pesticide risk assessment, and thus contribute to reduction of the possible adverse effects of pesticides on human and animal health and the environment.

In vitro cell culture systems have proven useful for toxicity prediction of target organs on chemicals or drug exposure. Therefore, the experimental study was designed to assess the toxic potential of the organophosphate insecticide chlorpyrifos on various model cell cultures. The results showed that chlorpyrifos, under *in vitro* conditions, suppressed cell proliferation of mammalian cells at the concentration range 10^{-1} (kidney cell line) – 10^{-3}M (liver cell lines), and of insect cells at the concentrations $10^{-1} – 10^{-5}$ M. Chlorpyrifos exposure caused cell-type dependent decrease in cell proliferation and the cell membrane damage. These results also showed that kidney and liver cell cultures might be useful as a supplementary tool in toxicity screening before testing substances on animals *in vivo*. On the other hand, the insect cell line could be an optimal method for *in vitro* evaluation of insecticide potential.

Acknowledgments

This study was supported by VEGA Grants No. 1/0855/12, 1/0287/11 and the National Reference Laboratory for Pesticides, University of Veterinary Medicine and Pharmacy in Košice, Slovakia.

References

Abdollahi, M., Ranjbar, A., Shadnia, S., Nikfar, S., Rezaie, A. (2004). Pesticides and oxidative stress: A review. *Medical Science Monitor*, 10, RA141-RA147.

Abu-Qare, A. W., Abdel-Rahman, A., Brownie, C., Kishk, A. M., Abou-Donia, M. B. (2001). Inhibition of cholinesterase enzymes following a single dermal dose of chlorpyrifos and methyl parathion, alone and in combination, in pregnant rats. *Journal of Toxicology and Environmental Health-Part A*, 63, 173-189.

Aktar, W., Sengupta, D., Chowdhury, A. (2009): Impact of pesticides use in agriculture: their benefits and hazards, pp. 1: *Interdisciplinary Toxicology*.

Alavanja, M. C. R. (2009). Introduction: Pesticides use and exposure extensive worldwide. *Reviews on Environmental Health*, 24, 303-309.

Alavanja, M. C. R., Hoppin, J. A., Kamel, F. (2004). Health effects of chronic pesticide exposure: Cancer and neurotoxicity. *Annual Review of Public Health,* 25, 155-197.

Allen, G. T., Veatch, J. K., Stroud, R. K., Vendel, C. G., Poppenga, R. H., Thompson, L., Shafer, J. A., Braselton, W. E. (1996). Winter poisoning of coyotes and raptors with Furadan-laced carcass baits. *Journal of Wildlife Diseases,* 32, 385-389.

Aly, H. A. A., Domenech, O. (2009). Cytotoxicity and mitochondrial dysfunction of 2,3,7,8-tetrachlorodibenzo-p-dioxin (TCDD) in isolated rat hepatocytes. *Toxicology Letters*, 191, 79-87.

Ambali, S., Akanbi, D., Igbokwe, N., Shittu, M., Kawu, M., Ayo, J. (2007). Evaluation of subchronic chlorpyrifos poisoning on hematological and serum biochemical changes in mice and protective effect of vitamin C. *Journal of Toxicological Sciences*, 32, 111-120.

Ambali, S. F., Abubakar, A. T., Shittu, M., Yaqub, L. S., Anafi, S. B., Abdullahi, A. (2010a). Chlorpyrifos-induced alteration of hematological parameters in Wistar rats: Ameliorative effect of zinc. *Research Journal of Environmental Toxicology*, 4, 55-66.

Ambali, S. F., Abubakar, A. T., Shittu, M., Yaqub, L. S., Kobo, P. I., Giwa, A. (2010b). Ameliorative Effect of Zinc on Chlorpyrifos-Induced Erythrocyte Fragility in Wistar Rats. *New York Science Journal*, 3, 117-122.

Ambali, S. F., Ayo, J. O., Esievo, K. A., Ojo, S. A. (2011a). Hemotoxicity induced by chronic chlorpyrifos exposure in Wistar rats: mitigating effect of vitamin C. *Vet. Med. Int.*, 2011, 945439.

Ambali, S. F., Ayo, J. O., Ojo, S. A., Esievo, K. A. N. (2011b). Ameliorative effect of vitamin C on chronic chlorpyrifos-induced erythrocyte osmotic fragility in Wistar rats. *Human and Experimental Toxicology*, 30, 19-24.

Ames, R. G., Brown, S. K., Rosenberg, J., Jackson, R. J., Stratton, J. W., Quenon, S. G. (1989). Health Symptoms and Occupational Exposure to Flea Control Products among California Pet Handlers. *American Industrial Hygiene Association Journal*, 50, 466-472.

Aprea, C., Strambi, M., Novelli, M. T., Lunghini, L., Bozzi, N. (2000). Biologic monitoring of exposure to organophosphorus pesticides in 195 Italian children. *Environmental Health Perspectives*, 108, 521-525.

Asogwa, E. U., Dongo, L. N. (2009). Problems associated with pesticide usage and application in Nigerian cocoa production: A review. *African Journal of Agricultural Research,* 4, 675-683.

Atreya, K. (2007). Pesticide use knowledge and practices: A gender differences in Nepal. *Environmental Research*, 104, 305-311.

ATSDR (1997): Toxicological profile for chlorpyrifos, pp. 217: Agency for Toxic Substances and Disease Registry, Atlanta, GA.

Augspurger, T., Smith, M. R., Meteyer, C. U., Converse, K. A. (1996). Mortality of passerines adjacent to a North Carolina cornfield treated with granular carbofuran. *Journal of Wildlife Diseases*, 32, 113-116.

Barr, D. B., Angerer, J. (2006). Potential uses of biomonitoring data: A case study using the organophosphorus pesticides chlorpyrifos and malathion. *Environmental Health Perspectives*, 114, 1763-1769.

BCPC (2006): Chlorpyrifos. In: C. D. C. Tomlin (Ed.), *A world compendium: The pesticide manual* (14th edition, pp. 186-187). Alton, Hempshire, United Kingdom: British Crop Protection Council.

Becker, K., Seiwert, M., Angerer, J., Kolossa-Gehring, M., Hoppe, H. W., Ball, M., Schulz, C., Thumulla, J., Seifert, B. (2006). GerES IV Pilot Study: Assessment of the exposure of German children to organophosphorus and pyrethroid pesticides. *International Journal of Hygiene and Environmental Health,* 209, 221-233.

Berkowitz, G. S., Wetmur, J. G., Birman-Deych, E., Obel, J., Lapinski, R. H., Godbold, J. H., Holzman, I. R., Wolff, M. S. (2004). In utero pesticide exposure, maternal paraoxonase activity, and head circumference. *Environmental Health Perspectives*, 112, 388-391.

Berny, P. (2007). Pesticides and the intoxication of wild animals. *Journal of Veterinary Pharmacology and Therapeutics,* 30, 93-100.

Berny, P. J., Buronfosse, T., Buronfosse, F., Lamarque, F., Lorgue, G. (1997). Field evidence of secondary poisoning of foxes (Vulpes vulpes) and buzzards (Buteo buteo) by bromadiolone, a 4-year survey. *Chemosphere*, 35, 1817-1829.

Bjorling-Poulsen, M., Andersen, H. R., Grandjean, P. (2008). Potential developmental neurotoxicity of pesticides used in Europe. *Environmental Health*, 7.

Blanc-Lapierre, A., Bouvier, G., Garrigou, A., Canal-Raffin, M., Raherison, C., Brochard, P., Baldi, I. (2012). Chronic central nervous system effects of pesticides: State-of-the-art. *Revue D Epidemiologie Et De Sante Publique*, 60, 389-400.

Blodgett, D. J. (2006): Organophosphate and carbamate insecticides. In: Peterson, M. E., Talcott, P. A.: *Small animal toxicology*, 2nd Edition edition, pp. 941-955). St. Louis, Missouri: Saunders.

Bolognesi, C. (2003). Genotoxicity of pesticides: a review of human biomonitoring studies. *Mutation Research-Reviews in Mutation Research*, 543, 251-272.

Boxall, R. A. (2001). Post-harvest losses to insects - a world overview (Reprinted). *International Biodeterioration and Biodegradation*, 48, 137-152.

Bradman, A., Whitaker, D., Quiros, L., Castorina, R., Henn, B. C., Nishioka, M., Morgan, J., Barr, D. B., Harnly, M., Brisbin, J. A., Sheldon, L. S., Mckone, T. E., Eskenazi, B. (2007). Pesticides and their metabolites in the homes and urine of farmworker children living in the Salinas Valley, CA. *Journal of Exposure Science and Environmental Epidemiology*, 17, 331-349.

Breslin, W. J., Liberacki, A. B., Dittenber, D. A., Quast, J. F. (1996). Evaluation of the developmental and reproductive toxicity of chlorpyrifos in the rat. *Fundamental and Applied Toxicology*, 29, 119-130.

Brown, T. P., Rumsby, P. C., Capleton, A. C., Rushton, L., Levy, L. S. (2006). Pesticides and Parkinson's disease - Is there a link? *Environmental Health Perspectives,* 114, 156-164.

Burger, J., de Mol, F., Gerowitt, B. (2008). The "necessary extent" of pesticide use-Thoughts about a key term in German pesticide policy. *Crop Protection,* 27, 343-351.

Campanha, H. M., Carvalho, F., Schlosser, P. M. (2014). Active and peripheral anionic sites of acetylcholinesterase have differential modulation effects on cell proliferation, adhesion, and neuritogenesis in the NG108-15 cell line. *Toxicology Letters*, 230, 2, 122-131.

Cardona, D., Lopez-Granero, C., Canadas, F., Llorens, J., Flores, P., Pancetti, F., Sanchez-Santed, F. (2013). Dose-dependent regional brain acetylcholinesterase and acylpeptide hydrolase inhibition without cell death after chlorpyrifos administration. *Journal of Toxicological Sciences*, 38, 193-203.

Carson, R. (1962). *Silent Spring edition*. Boston, Massachusetts: Houghton Mifflin.

Casida, J. E., Durkin, K. A. (2013a). Anticholinesterase insecticide retrospective. *Chem. Biol. Interact.*, 203, 221-5.

Casida, J. E., Durkin, K. A. (2013b). Neuroactive insecticides: targets, selectivity, resistance, and secondary effects. *Annu. Rev. Entomol.*, 58, 99-117.

Casida, J. E., Quistad, G. B. (1998). Golden age of insecticide research: past, present, or future? *Annu. Rev. Entomol.*, 43, 1-16.

Casida, J. E., Quistad, G. B. (2004). Organophosphate toxicology: safety aspects of nonacetylcholinesterase secondary targets. *Chem. Res. Toxicol.*, 17, 983-98.

Castell, J. V., Jover, R., Martínez-Jiménez, C. P., Gómez-Lechón, M. J. (2006). Hepatocyte cell lines: Their use, scope, and limitations in drug metabolism studies (Review). *Expert Opinion on Drug Metabolism and Toxicology,* 2, 2, 183-212.

Chen, L. Q., Qu, G. B., Sun, X., Zhang, S. P., Wang, L., Sang, N. G., Du, Y. G., Liu, J., Liu, S. J. (2013). Characterization of the Interaction between Cadmium and Chlorpyrifos with Integrative Techniques in Incurring Synergistic Hepatoxicity. *Plos One*, 8.

Chen, P. J., Moore, T., Nesnow, S. (2008). Cytotoxic effects of propiconazole and its metabolites in mouse and human hepatoma cells and primary mouse hepatocytes. *Toxicology in Vitro*, 22, 1476-1483.

Chiu, B. C., Blair, A. (2009). Pesticides, chromosomal aberrations, and non-Hodgkin's lymphoma. *J. Agromedicine*, 14, 250-5.

Colovic, M. B., Krstic, D. Z., Lazarevic-Pasti, T. D., Bondzic, A. M., Vasic, V. M. (2013). Acetylcholinesterase Inhibitors: Pharmacology and Toxicology. *Current Neuropharmacology*, 11, 315-335.

Colt, J. S., Lubin, J., Camann, D., Davis, S., Cerhan, J., Severson, R. K., Cozen, W., Hartge, P. (2004). Comparison of pesticide levels in carpet dust and self-reported pest treatment practices in four US sites. *J. Expo Anal. Environ. Epidemiol.*, 14, 74-83.

Colt, J. S., Severson, R. K., Lubin, J., Rothman, N., Camann, D., Davis, S., Cerhan, J. R., Cozen, W., Hartge, P. (2005). Organochlorines in carpet dust and non-Hodgkin lymphoma. *Epidemiology*, 16, 516-25.

Commision staff working document (2012): Monitoring of Pesticide Residues in Products of Plant Origin in the European Union, Norway, Iceland and Liechtenstein, 2012. Available: http://ec.europa.eu/food/fvo/special reports/pesticide_residues/report_2012_en.pdf.

Committee on the Future Role of Pesticides in US Agriculture (2000): Board on Agriculture and Natural Resources, Board on Environmental Studies and Toxicology, National Research Council. The future role of pesticides in US agriculture. Washington, D.C: National Academy Press. Available: http://www.nap.edu/catalog.php?record_id =9598#toc.

Cooper, J., Dobson, H. (2007). The benefits of pesticides to mankind and the environment. *Crop Protection*, 26, 1337-1348.

Corcoran, G. B., Fix, L., Jones, D. P., Moslen, M. T., Nicotera, P., Oberhammer, F. A., Buttyan, R. (1994). Apoptosis: molecular control point in toxicity. *Toxicol. Appl. Pharmacol.*, 128, 169-81.

Corsini, E., Sokooti, M., Galli, C. L., Moretto, A., Colosio, C. (2013). Pesticide induced immunotoxicity in humans: a comprehensive review of the existing evidence. *Toxicology*, 307, 123-35.

Costa, L. G. (2006). Current issues in organophosphate toxicology. *Clinica Chimica Acta,* 366, 1-13.

Costa, L. G., Giordano, G., Cole, T. B., Marsillach, J., Furlong, C. E. (2013). Paraoxonase 1 (PON1) as a genetic determinant of susceptibility to organophosphate toxicity. *Toxicology*, 307, 115-122.

CPF cancellation order. http://www.webcitation.org/query.php?url=http:// www.epa.gov /EPA-PEST/2000/December/Day-06/p30917.htmandrefdoi= 10.1186/1476-069x-8-12.

Crane, A. L., Klein, K., Zanger, U. M., Olson, J. R. (2012). Effect of CYP2B6*6 and CYP2C19*2 genotype on chlorpyrifos metabolism. *Toxicology*, 293, 115-22.

Cuello, S., Goya, L., Madrid, Y., Campuzano, S., Pedrero, M., Bravo, L., Camara, C., Ramos, S. (2010). Molecular mechanisms of methylmercury-induced cell death in human HepG2 cells. *Food Chem. Toxicol.*, 48, 1405-11.

Curwin, B. D., Hein, M. J., Sanderson, W. T., Nishioka, M. G., Reynolds, S. J., Ward, E. M., Alavanja, M. C. (2005). Pesticide contamination inside farm and nonfarm homes. *Journal of Occupational and Environmental Hygiene,* 2, 357-367.

Dahamna, S., Sekfali, N., Walker, C. H. (2004). Biochemical indicators of the hepatotoxic effects of pesticides. Commun. *Agric. Appl. Biol. Sci.*, 69, 821-8.

Daly, H., Doyen, J. T., Prucel, A. H. (1998). *Introduction to Insect Biology and Diversity* (2nd edition). New York, New York: Oxford University Press.

Damalas, C. A. (2009). Understanding benefits and risks of pesticide use. *Scientific Research and Essays*, 4, 945-949.

Damalas, C. A., Eleftherohorinos, I. G. (2011). Pesticide exposure, safety issues, and risk assessment indicators. *International Journal of Environmental Research and Public Health,* 8, 1402-1419.

Dandapani, M., Zachariah, A., Kavitha, M. R., Jeyaseelan, L., Oommen, A. (2003). Oxidative damage in intermediate syndrome of acute organophosphorus poisoning. *Indian Journal of Medical Research*, 117, 253-259.

Das, K. P., Barone, S. (1999). Neuronal differentiation in PC12 cells is inhibited by chlorpyrifos and its metabolites: Is acetylcholinesterase inhibition the site of action? *Toxicology and Applied Pharmacology*, 160, 217-230.

Das, P. C., Cao, Y., Rose, R. L., Cherrington, N., Hodgson, E. (2008). Enzyme induction and cytotoxicity in human hepatocytes by chlorpyrifos and N, N-diethyl-m-toluamide (DEET). *Drug Metabol. Drug Interact.*, 23, 237-260.

Davila, J. C., Rodriguez, R. J., Melchert, R. B., Acosta, D. (1998). Predictive value of in vitro model systems in toxicology. *Annual Review of Pharmacology and Toxicology*, 38, 63-96.

Dawson, A. H., Eddleston, M., Senarathna, L., Mohamed, F., Gawarammana, I., Bowe, S. J., Manuweera, G., Buckley, N. A. (2010). Acute Human Lethal Toxicity of Agricultural Pesticides: A Prospective Cohort Study. *Plos Medicine*, 7.

De Angelis, S., Tassinari, R., Maranghi, F., Eusepi, A., Di Virgilio, A., Chiarotti, F., Ricceri, L., Venerosi Pesciolini, A., Gilardi, E., Moracci, G., Calamandrei, G., Olivieri, A., Mantovani, A. (2009). Developmental exposure to chlorpyrifos induces alterations in thyroid and thyroid hormone levels without other toxicity signs in CD-1 mice. *Toxicological Sciences,* 108, 311-319.

De Snoo, G. R., Scheidegger, N. M. I., de Jong, F. M. W. (1999). Vertebrate wildlife incidents with pesticides: a European survey. *Pesticide Science*, 55, 47-54.

Deacon, M. M., Murray, J. S., Pilny, M. K., Rao, K. S., Dittenber, D. A., Hanley, T. R., Jr., John, J. A. (1980). Embryotoxicity and fetotoxicity of orally administered chlorpyrifos in mice. *Toxicol. Appl. Pharmacol.*, 54, 31-40.

Dennis, L. K., Lynch, C. F., Sandler, D. P., Alavanja, M. C. R. (2010). Pesticide Use and Cutaneous Melanoma in Pesticide Applicators in the Agricultural Heath Study. *Environmental Health Perspectives*, 118, 812-817.

Eaton, D. L., Daroff, R. B., Autrup, H., Bridges, J., Buffler, P., Costa, L. G., Coyle, J., McKhann, G., Mobley, W. C., Nadel, L., Neubert, D., Schulte-Hermann, R., Spencer, P. S. (2008). Review of the toxicology of chlorpyrifos with an emphasis on human exposure and neurodevelopment. *Critical Reviews in Toxicology*, 38, 1-125.

Ecobichon, D. J. (2001). Pesticide use in developing countries. *Toxicology*, 160, 27-33.

Egeghy, P. P., Cohen Hubal, E. A., Tulve, N. S., Melnyk, L. J., Morgan, M. K., Fortmann, R. C., Sheldon, L. S. (2011). Review of pesticide urinary biomarker measurements from selected US EPA children's observational exposure studies. *Int. J. Environ. Res. Public Health*, 8, 1727-54.

El-Sebae, A. H., Ahmed, N. S., Soliman, S. A. (1978). Effect of pre-exposure on acute toxicity of organophosphorus insecticides to white mice. *J. Environ. Sci. Health B*, 13, 11-24.

El-Shenawy, N. S. (2010). Effects of insecticides fenitrothion, endosulfan, and abamectin on antioxidant parameters of isolated rat hepatocytes. *Toxicol. In Vitro,* 24, 1148-57.

Engel, S. M., Berkowitz, G. S., Barr, D. B., Teitelbaum, S. L., Siskind, J., Meisel, S. J., Wetmur, J. G., Wolff, M. S. (2007). Prenatal organophosphate metabolite and organochlorine levels and performance on the Brazelton Neonatal Behavioral Assessment Scale in a multi-ethnic pregnancy cohort. *Am. J. Epidemiol.*, 165, 1397-404.

Environews Forum (1999). Killer environment: *Environ. Health Perspect.*, A62-3.

Eskenazi, B., Harley, K., Bradman, A., Weltzien, E., Jewell, N. P., Barr, D. B., Furlong, C. E., Holland, N. T. (2004). Association of in utero organophosphate pesticide exposure and fetal growth and length of gestation in an agricultural population. *Environ. Health Perspect.*, 112, 1116-24.

Eskenazi, B., Marks, A. R., Bradman, A., Harley, K., Barr, D. B., Johnson, C., Morga, N., Jewell, N. A. (2007). Organophosphate pesticide exposure and neurodevelopment in young Mexican-American children. *Environmental Health Perspectives,* 115, 792-798.

Eurostat statistical books (2007): The use of plant protection products in the European Union. Data 1992–2003. Available: http://epp.eurostat.ec. europa.eu/cache /ITY_OFFPUB/KS-76-06-669/EN/KS-76-06-669-EN. PDF.

Eurostat statistics. Avaiable: http://epp.eurostat.ec.europa.eu/tgm/table.do?tab=table andinit=1andplugin=0andlanguage=enandpcode=tag00087.

Fenske, R. A., Black, K. G., Elkner, K. P., Lee, C. L., Methner, M. M., Soto, R. (1990). Potential exposure and health risks of infants following indoor residential pesticide applications. *Am. J. Public Health*, 80, 689-93.

Fenske, R. A., Day, E. W., Jr. (2005): Assessment of exposure for pesticide handlers in agricultural, residential and institutional environments. In: C. A. Franklin and J. P. Worgan (Eds.), *Occupational and Residential Exposure Assessment for Pesticides edition*, pp. 11-43). Chichester, United Kingdom: John Wiley and Sons.

Freire, C., Koifman, S. (2012). Pesticide exposure and Parkinson's disease: Epidemiological evidence of association. *Neurotoxicology*, 33, 947-971.

Gaines, T. B. (1969). Acute toxicity of pesticides. *Toxicol. Appl. Pharmacol.*, 14, 515-34.

Giddings, J. M., Williams, W. M., Solomon, K. R., Giesy, J. P. (2014). Risks to aquatic organisms from use of chlorpyrifos in the United States. *Reviews of environmental contamination and toxicology*, 231, 119-162.

Giesy, J. P., Solomon, K. R., Cutler, G. C., Giddings, J. M., Mackay, D., Moore, D. R., Purdy, J., Williams, W. M. (2014). Ecological risk assessment of the uses of the organophosphorus insecticide chlorpyrifos, in the United States. *Reviews of environmental contamination and toxicology,* 231, 1-11.

Goel, A., Dani, V., Dhawan, D. K. (2006). Role of zinc in mitigating the toxic effects of chlorpyrifos on hematological alterations and electron microscopic observations in rat blood. *Biometals*, 19, 483-492.

Gollapudi, B. B., Mendrala, A. L., Linscombe, V. A. (1995). Evaluation of the Genetic Toxicity of the Organophosphate Insecticide Chlorpyrifos. *Mutation Research-Genetic Toxicology*, 342, 25-36.

Gomez-Lechon, M., Donato, M., Castell, J., Jover, R. (2003). Human hepatocytes as a tool for studying toxicity and drug metabolism. *Curr. Drug Metab.,* 4, 292-312.

Greene, S. A., Pohanish, R. P. (2005). *Sittig's Handbook of Pesticides and Agricultural Chemicals edition*. Oxford, United Kingdom: William Andrew Publishing.

Greig-Smith, P. W., Thompson, H. M., Hardy, A. R., Bew, M. H., Findlay, E., Stevenson, J. H. (1994). Incidents of poisoning of honeybees (Apis mellifera) by agricultural pesticides in Great Britain 1981-1991. *Crop Protection,* 13, 567-581.

Griffin, P., Mason, H., Heywood, K., Cocker, J. (1999). Oral and dermal absorption of chlorpyrifos: a human volunteer study. *Occup. Environ. Med.,* 56, 10-3.

Guitart, R., Manosa, S., Guerrero, X., Mateo, R. (1999). Animal poisonings: the 10-year experience of a veterinary analytical toxicology laboratory. *Vet. Hum. Toxicol.*, 41, 331-5.

Guitart, R., Mateo, R., Gutierrez, J. M., To-Figueras, J. (1996). An outbreak of thiram poisoning on Spanish poultry farms. *Vet. Hum. Toxicol.*, 38, 287-8.

Gultekin, F., Delibas, N., Yasar, S., Kilinc, I. (2001). In vivo changes in antioxidant systems and protective role of melatonin and a combination of vitamin C and vitamin E on oxidative damage in erythrocytes induced by chlorpyrifos-ethyl in rats. *Arch. Toxicol.*, 75, 88-96.

Gunier, R. B., Ward, M. H., Airola, M., Bell, E. M., Colt, J., Nishioka, M., Buffler, P. A., Reynolds, P., Rull, R. P., Hertz, A., Metayer, C., Nuckols, J. R. (2011). Determinants of agricultural pesticide concentrations in carpet dust. *Environ. Health Perspect.*, 119, 970-6.

Gunnell, D., Eddleston, M. (2003). Suicide by intentional ingestion of pesticides: a continuing tragedy in developing countries. *Int. J. Epidemiol.,* 32, 902-9.

Gunnell, D., Eddleston, M., Phillips, M. R., Konradsen, F. (2007). The global distribution of fatal pesticide self-poisoning: systematic review. *BMC Public Health*, 7, 357.

Gupta, R. C. (2005). *Toxicology of Organophosphate and Carbamate Compounds* (1st edition). Amsterdam, The Netherlands: Academic Press/ Elsevier.

Gupta, R. C. (2012). *Veterinary Toxicology. Basic and Clinical Principles* (2nd edition). Amsterdam, The Netherlands: Academic Press/Elsevier.

Gurunathan, S., Robson, M., Freeman, N., Buckley, B., Roy, A., Meyer, R., Bukowski, J., Lioy, P. J. (1998). Accumulation of chlorpyrifos on residential surfaces and toys accessible to children. *Environmental Health Perspectives,* 106, 9-16.

Han, E. H., Jeong, T. C., Jeong, H. G. (2007). Methoxychlor suppresses the 2,3,7,8-tetrachlorodibenzo-p-dioxin (TCDD)-inducible CYP1A1 expression in murine Hepa-1c1c7 cells. *Journal of Toxicology and Environmental Health-Part a-Current Issues*, 70, 1304-1309.

Hanke, W., Jurewicz, J. (2004). The risk of adverse reproductive and developmental disorders due to occupational pesticide exposure: An overview of current epidemiological evidence. *International Journal of Occupational Medicine and Environmental Health,* 17, 223-243.

Harari, R., Julvez, J., Murata, K., Barr, D., Bellinger, D. C., Debes, F., Grandjean, P. (2010). Neurobehavioral deficits and increased blood pressure in school-age children prenatally exposed to pesticides. *Environ. Health Perspect.*, 118, 890-6.

Hargreaves, A. J. (2012). Neurodegenerations induced by organophosphorous compounds. *Adv. Exp. Med. Biol.*, 724, 189-204.

Harnly, M. E., Bradman, A., Nishioka, M., McKone, T. E., Smith, D., McLaughlin, R., Kavanagh-Baird, G., Castorina, R., Eskenazi, B. (2009). Pesticides in dust from homes in an agricultural area. *Environ. Sci. Technol.,* 43, 8767-74.

Hohenadel, K., Harris, S. A., McLaughlin, J. R., Spinelli, J. J., Pahwa, P., Dosman, J. A., Demers, P. A., Blair, A. (2011). Exposure to multiple pesticides and risk of non-Hodgkin lymphoma in men from six Canadian provinces. *Int. J. Environ. Res. Public Health*, 8, 2320-30.

Hoogduijn, M. J., Rakonczay, Z., Genever, P. G. (2006). The effects of anticholinergic insecticides on human mesenchymal stem cells. *Toxicol. Sci.*, 94, 342-50.

Hsu, C. H., Quistad, G. B., Casida, J. E. (1998). Phosphine-induced oxidative stress in Hepa 1c1c7 cells. *Toxicol. Sci.*, 46, 204-10.

http://www.epa.gov/caddis/ssr_ins_int.html.

http://www.epa.gov/caddis/ssr_ins_int.html.

http://www.sustainabletable.org/263/pesticides.

Ismail, M., Khan, Q. M., Ali, R., Ali, T., Mobeen, A. (2014). Genotoxicity of chlorpyrifos in freshwater fish Labeo rohita using Alkaline Single-Cell Gel Electrophoresis (Comet) assay. *Drug and Chemical Toxicology*, 37, 466-471.

Jaga, K., Dharmani, C. (2003). Sources of exposure to and public health implications of organophosphate pesticides. *Rev. Panam. Salud Publica*, 14, 171-85.

Jameson, R. R., Seidler, F. J., Qiao, D., Slotkin, T. A. (2006). Chlorpyrifos affects phenotypic outcomes in a model of mammalian neurodevelopment: critical stages targeting differentiation in PC12 cells. *Environ. Health Perspect.*, 114, 667-72.

Jamil, K., Shaik, A. P., Mahboob, M., Krishna, D. (2004). Effect of organophosphorus and organochlorine pesticides (monochrotophos, chlorpyriphos, dimethoate, and endosulfan) on human lymphocytes in-vitro. *Drug Chem. Toxicol.*, 27, 133-44.

Jett, D. A. (2011). Neurotoxic pesticides and neurologic effects. *Neurol. Clin.*, 29, 667-77.

Jett, D. A., Navoa, R. V., Lyons, M. A., Jr. (1999). Additive inhibitory action of chlorpyrifos and polycyclic aromatic hydrocarbons on acetylcholinesterase activity in vitro. *Toxicol. Lett.*, 105, 223-9.

Juberg, D. R., Gehen, S. C., Coady, K. K., LeBaron, M. J., Kramer, V. J., Lu, H., Marty, M. S. (2013). Chlorpyrifos: weight of evidence evaluation of potential interaction with the estrogen, and rogen, or thyroid pathways. *Regul. Toxicol. Pharmacol.*, 66, 249-63.

Kalamarakis, A. E., Markellou, E. (2007). Efficacy evaluation of plant protection products at the EU level: Data requirements and evaluation principles. *Journal of Pesticide Science*, 32, 1-9.

Kamel, F., Engel, L. S., Gladen, B. C., Hoppin, J. A., Alavanja, M. C. R., Sandler, D. P. (2005). Neurologic symptoms in licensed private pesticide applicators in the Agricultural Health Study. *Environmental Health Perspectives,* 113, 877-882.

Kamel, F., Hoppin, J. A. (2004). Association of pesticide exposure with neurologic dysfunction and disease. *Environmental Health Perspectives*, 112, 950-958.

Kamrin, M. A. (1997). *Pesticide Profiles: Toxicity, Environmental Impact, and Fate* (1st edition). Boca Renton, Florida: CRC Press.

Karabelas, A. J., Plakas, K. V., Solomou, E. S., Drossou, V., Sarigiannis, D. A. (2009). Impact of European legislation on marketed pesticides - A view from the standpoint of health impact assessment studies. *Environment International,* 35, 1096-1107.

Khan, S. M. (2006). Protective effect of black tea extract on the levels of lipid peroxidation and antioxidant enzymes in liver of mice with pesticide-induced liver injury. *Cell Biochemistry and Function*, 24, 327-332.

Kidd, H., James, D. R. (1991). *The Agrochemicals Handbook* (3rd edition). Cambridge, United Kingdom: Royal Society of Chemistry Information Services.

Kokouva, M., Bitsolas, N., Hadjigeorgiou, G. M., Rachiotis, G., Papadoulis, N., Hadjichristodoulou, C. (2011). Pesticide exposure and lymphohaematopoietic cancers: a case-control study in an agricultural region (Larissa, Thessaly, Greece). *BMC Public Health*, 11.

Lassiter, T. L., Barone, S., Moser, V. C., Padilla, S. (1999). Gestational exposure to chlorpyrifos: Dose response profiles for cholinesterase and carboxylesterase activity. *Toxicological Sciences*, 52, 92-100.

Lassiter, T. L., MacKillop, E. A., Ryde, I. T., Seidler, F. J., Slotkin, T. A. (2009). Is fipronil safer than chlorpyrifos? Comparative developmental neurotoxicity modeled in PC12 cells. *Brain Research Bulletin*, 78, 313-322.

Lassiter, T. L., Padilla, S., Mortensen, S. R., Chanda, S. M., Moser, V. C., Barone, S. (1998). Gestational exposure to chlorpyrifos: Apparent protection of the fetus? *Toxicology and Applied Pharmacology*, 152, 56-65.

Lee, S. J., Nam, K. W., Mar, W. (2009). Induction of quinone reductase activity by psoralidin isolated from Psoralea corylifolia in mouse hepa 1c1c7 cells. *Archives of Pharmacal Research*, 32, 1061-1065.

Lee, W. J., Sandler, D. P., Blair, A., Samanic, C., Cross, A. J., Alavanja, M. C. R. (2007). Pesticide use and colorectal cancer risk in the Agricultural Health Study. *International Journal of Cancer*, 121, 339-346.

Li, A. A., Lowe, K. A., McIntosh, L. J., Mink, P. J. (2012). Evaluation of Epidemiology and Animal Data for Risk Assessment: Chlorpyrifos Developmental Neurobehavioral Outcomes. *Journal of Toxicology and Environmental Health-Part B-Critical Reviews*, 15, 109-184.

Li, Q., Kobayashi, M., Kawada, T. (2009). Chlorpyrifos induces apoptosis in human T cells. *Toxicology*, 255, 53-57.

Li, W., Choy, D. F., Lam, M. S., Morgan, T., Sullivan, M. E., Post, J. M. (2003). Use of cultured cells of kidney origin to assess specific cytotoxic effects of nephrotoxins. *Toxicology in Vitro*, 17, 107-113.

Lioy, P. J., Freeman, N. C. G., Millette, J. R. (2002). Dust: A metric for use in residential and building exposure assessment and source characterization. *Environmental Health Perspectives*, 110, 969-983.

Liu, J., Olivier, K., Pope, C. N. (1999). Comparative neurochemical effects of repeated methyl parathion or chlorpyrifos exposures in neonatal and adult rats. *Toxicology and Applied Pharmacology*, 158, 186-196.

Liu, J., Pope, C. N. (1998). Comparative presynaptic neurochemical changes in rat striatum following exposure to chlorpyrifos or parathion. *Journal of Toxicology and Environmental Health-Part A*, 53, 531-544.

Lu, C., Fenske, R. A., Simcox, N. J., Kalman, D. (2000). Pesticide exposure of children in an agricultural community: evidence of household proximity to farmland and take home exposure pathways. *Environ. Res.*, 84, 290-302.

Lu, C. S., Fenske, R. A. (1999). Dermal transfer of chlorpyrifos residues from residential surfaces: Comparison of hand press, hand drag, wipe, and polyurethane foam roller measurements after broadcast and aerosol pesticide applications. *Environmental Health Perspectives*, 107, 463-467.

Mage, D. T., Allen, R. H., Gondy, G., Smith, W., Barr, D. B., Needham, L. L. (2004). Estimating pesticide dose from urinary pesticide concentration data by creatinine correction in the Third National Health and Nutrition Examination Survey (NHANES-III). *Journal of Exposure Analysis and Environmental Epidemiology,* 14, 457-465.

Mansour, S. A., Mossa, A. T. H. (2011). Adverse effects of exposure to low doses of chlorpyrifos in lactating rats. *Toxicology and Industrial Health*, 27, 213-224.

Mariyono, J. (2008). Direct and indirect impacts of integrated pest management on pesticide use: a case of rice agriculture in Java, Indonesia. *Pest Management Science*, 64, 1069-1073.

Marks, A. R., Harley, K., Bradman, A., Kogut, K., Barr, D. B., Johnson, C., Calderon, N., Eskenazi, B. (2010). Organophosphate Pesticide Exposure and Attention in Young Mexican-American Children: The CHAMACOS Study. *Environmental Health Perspectives*, 118, 1768-1774.

Maroni, M., Fanetti, A. C., Metruccio, F. (2006). Risk assessment and management of occupational exposure to pesticides in agriculture. *Med. Lav.*, 97, 430-7.

Marrs, T. C. (1993). Organophosphate poisoning. *Pharmacol. Ther.*, 58, 51-66.

Martinez-Haro, M., Mateo, R., Guitart, R., Soler-Rodriguez, F., Perez-Lopez, M., Maria-Mojica, P., Garcia-Fernandez, A. J. (2008). Relationship of the toxicity of pesticide formulations and their commercial restrictions with the frequency of animal poisonings. *Ecotoxicology and Environmental Safety,* 69, 396-402.

Martinez-Valenzuela, C., Gomez-Arroyo, S., Villalobos-Pietrini, R., Waliszewski, S., Calderon-Segura, M. E., Felix-Gastelum, R., Alvarez-Torres, A. (2009). Genotoxic biomonitoring of agricultural workers exposed to pesticides in the north of Sinaloa State, Mexico. *Environment International,* 35, 1155-1159.

Marty, M. S., Andrus, A. K., Bell, M. P., Passage, J. K., Perala, A. W., Brzak, K. A., Bartels, M. J., Beck, M. J., Juberg, D. R. (2012). Cholinesterase inhibition and toxicokinetics in immature and adult rats after acute or repeated exposures to chlorpyrifos or chlorpyrifos-oxon. *Regulatory Toxicology and Pharmacology*, 63, 209-224.

Mattsson, J. L., Maurissen, J. P. J., Nolan, R. J., Brzak, K. A. (2000). Lack of differential sensitivity to cholinesterase inhibition in fetuses and neonates compared to dams treated perinatally with chlorpyrifos. *Toxicological Sciences*, 53, 438-446.

McCauley, L. A., Anger, W. K., Keifer, M., Langley, R., Robson, M. G., Rohlman, D. (2006). Studying health outcomes in farmworker populations exposed to pesticides. *Environmental Health Perspectives*, 114, 953-960.

McCollister, S. B., Kociba, R. J., Humiston, C. G., McCollister, D. D. (1974). Studies on the acute and long-term oral toxicity for chlorpyrifos (0,0-diethyl-0(3,5,6-trichloro-2-pyridyl) phosphorothioate). *Food Cosmet. Toxicol.*, 12, 45-61.

McConkey, D. J., Jondal, M. B., Orreniu, S. G. (1994): Chemical-induced apoptosis in the immune system. In: J. H. Dean, M. I. Luster, A. E. Munson, and I. Kimber (Eds.), Immunotoxicology and Immunopharmacology. *Target Organ Toxicology Series* (2nd edition, pp. 473-485). Philadelphia, Pennsylvania: Lippincott-Raven.

McKinlay, R., Plant, J. A., Bell, J. N. B., Voulvoulis, N. (2008). Endocrine disrupting pesticides: implications for risk assessment. *Environment International*, 34, 168-183.

Mersch-Sundermann, V., Knasmuller, S., Wu, X. J., Darroudi, F., Kassie, F. (2004). Use of a human-derived liver cell line for the detection of cytoprotective, antigenotoxic and cogenotoxic agents. *Toxicology*, 198, 329-340.

Meuling, W. J. A., Ravensberg, L. C., Roza, L., van Hemmen, J. J. (2005). Dermal absorption of chlorpyrifos in human volunteers. *International Archives of Occupational and Environmental Health*, 78, 44-50.

Miller, G. T. (2002). *Living in the Environment* (12th edition). Belmont, California: Wadsworth Thomson Learning.

Mink, P. J., Kimmel, C. A., Li, A. A. (2012). Potential Effects of Chlorpyrifos on Fetal Growth Outcomes: Implications for Risk Assessment. *Journal of Toxicology and Environmental Health-Part B-Critical Reviews,* 15, 281-316.

Mnif, W., Hassine, A. I. H., Bouaziz, A., Bartegi, A., Thomas, O., Roig, B. (2011). Effect of Endocrine Disruptor Pesticides: A Review. *International Journal of Environmental Research and Public Health,* 8, 2265-2303.

Moore, D. R., Teed, R. S., Greer, C. D., Solomon, K. R., Giesy, J. P. (2014). Refined avian risk assessment for chlorpyrifos in the United States. *Rev. Environ. Contam. Toxicol.*, 231, 163-217.

Moretto, A., Lotti, M. (2005): Peripheral Nervous System Effects and Delayed Neuropathy. In: R. C. Gupta (Ed.), *Toxicology of Organophosphate and Carbamate Compounds* (1st edition, pp. 361-370). Amsterdam, The Netherlands: Academic Press/Elsevier.

Morgan, M. K., Sheldon, L. S., Croghan, C. W., Jones, P. A., Robertson, G. L., Chuang, J. C., Wilson, N. K., Lyu, C. W. (2005). Exposures of preschool children to chlorpyrifos and its degradation product 3,5,6-trichloro-2-pyridinol in their everyday environments. *Journal of Exposure Analysis and Environmental Epidemiology*, 15, 297-309.

Moser, V. C., Chanda, S. M., Mortensen, S. R., Padilla, S. (1998). Age- and gender-related differences in sensitivity to chlorpyrifos in the rat reflect developmental profiles of esterase activities. *Toxicological Sciences*, 46, 211-222.

Moser, V. C., Padilla, S. (1998). Age- and gender-related differences in the time course of behavioral and biochemical effects produced by oral chlorpyrifos in rats. *Toxicology and Applied Pharmacology*, 149, 107-119.

Motas-Guzman, M., Maria-Mojica, P., Romero, D., Martinez-Lopez, E., Garcia-Fernandez, A. J. (2003). Intentional poisoning of animals in southeastern Spain - A review of the veterinary toxicology service from Murcia, Spain. *Veterinary and Human Toxicology*, 45, 47-50.

Muller, M., Hess, L., Tardivo, A., Lajmanovich, R., Attademo, A., Poletta, G., Simoniello, M. F., Yodice, A., Lavarello, S., Chialvo, D., Scremin, O. (2014). Neurologic dysfunction and genotoxicity induced by low levels of chlorpyrifos. *Neurotoxicology*, 45, 22-30.

Munoz-Quezada, M. T., Lucero, B. A., Barr, D. B., Steenland, K., Levy, K., Ryan, B., Iglesias, V., Alvarado, S., Concha, C., Rojas, E., Vega, C. (2013). Neurodevelopmental effects in children associated with exposure to organophosphate pesticides: A systematic review. *Neurotoxicology*, 39, 158-168.

National Agricultural Chemicals Association (1993). From lab to label - the research, testing, and registration of agricultural chemicals. Washington, D.C.

Nakadai, A., Li, Q., Kawada, T. (2006). Chlorpyrifos induces apoptosis in human monocyte cell line U937. *Toxicology*, 224, 202-209.

Nandi, S., Gupta, P. S. P., Roy, S. C., Selvaraju, S., Ravindra, J. P. (2011). Chlorpyrifos and endosulfan affect buffalo oocyte maturation, fertilization, and embryo development in vitro directly and through cumulus cells. *Environmental Toxicology*, 26, 57-67.

Narahashi, T., Zhao, X., Ikeda, T., Nagata, K., Yeh, J. Z. (2007). Differential actions of insecticides on target sites: basis for selective toxicity. *Human and Experimental Toxicology*, 26, 361-366.

Narayanasamy, P. (2006). *Postharvest pathogens and disease management edition.* Somerset, New Jersey: John Wiley and Sons.

Natsch, A., Emter, R. (2008). Skin sensitizers induce antioxidant response element dependent genes: application to the in vitro testing of the sensitization potential of chemicals. *Toxicol. Sci.*, 102, 110-9.

Nawaz, A., Razpotnik, A., Rouimi, P., de Sousa, G., Cravedi, J. P., Rahmani, R. (2014) Cellular impact of combinations of endosulfan, atrazine, and chlorpyrifos on human primary hepatocytes and HepaRG cells after short and chronic exposures. *Cell Biol. Toxicol.*, 30, 17-29.

Ncibi, S., Ben Othman, M., Akacha, A., Krifi, M. N., Zourgui, L. (2008). Opuntia ficus indica extract protects against chlorpyrifos-induced damage on mice liver. *Food Chem. Toxicol.*, 46, 797-802.

Nolan, R. J., Rick, D. L., Freshour, N. L., Saunders, J. H. (1984). Chlorpyrifos: pharmacokinetics in human volunteers. *Toxicol. Appl. Pharmacol.*, 73, 8-15.

Noworyta-Glowacka, J., Bankowski, R., Siennicka, J., Wiadrowska, B., Ludwicki, J. K. (2012). Influence of chlorpyrifos on the profile of subpopulations of immunoactive cells and their phagocytic activity in an experimental in vivo model. *Annals of Agricultural and Environmental Medicine*, 19, 483-485.

Oerke, E. C., Dehne, H. W. (2004). Safeguarding production - losses in major crops and the role of crop protection. *Crop Protection*, 23, 275-285.

Official Journal of the European Union Legislation L. 2007. 50(part 216): 17–21.

Ogada, D. L. (2014): *The power of poison: Pesticide poisoning of Africa's wildlife, pp. 1-20: Annals of the New York Academy of Sciences.*

Ojha, A., Srivastava, N. (2014). In vitro studies on organophosphate pesticides induced oxidative DNA damage in rat lymphocytes. *Mutation Research - Genetic Toxicology and Environmental Mutagenesis,* 761, 10-17.

Oostingh, G. J., Wichmann, G., Schmittner, M., Lehmann, I., Duschl, A. (2009). The cytotoxic effects of the organophosphates chlorpyrifos and diazinon differ from their immunomodulating effects. *Journal of Immunotoxicology*, 6, 136-145.

Pain, D. J., Gargi, R., Cunningham, A. A., Jones, A., Prakash, V. (2004). Mortality of globally threatened Sarus cranes Grus antigon from monocrotophos poisoning in India. *Science of the Total Environment*, 326, 55-61.

Palma, P., Palma, V. L., Fernandes, R. M., Bohn, A., Soares, A. M. V. M., Barbosa, I. R. (2009). Embryo-toxic effects of environmental concentrations of chlorpyrifos on the crustacean Daphnia magna. *Ecotoxicology and Environmental Safety*, 72, 1714-1718.

Park, J. C., Park, B. J., Lee, D. H., Suh, H., Kim, D. G., Kwon, O. H. (2002). Evaluation of the cytotoxicity of polyetherurethane (PU) film containing zinc diethyldithiocarbamate (ZDEC) on various cell lines. *Yonsei Medical Journal*, 43, 518-526.

Perera, F. P., Rauh, V., Tsai, W. Y., Kinney, P., Camann, D., Barr, D., Bernert, T., Garfinkel, R., Tu, Y. H., Diaz, D., Dietrich, J., Whyatt, R. M. (2003). Effects of transplacental exposure to environmental pollutants on birth outcomes in a multiethnic population. *Environmental Health Perspectives,* 111, 201-205.

Pistl, J., Kovalkovicova, N., Holovska, V., Legath, J., Mikula, I. (2003). Determination of the immunotoxic potential of pesticides on functional activity of sheep leukocytes in vitro. *Toxicology*, 188, 73-81.

Pistl, J., Kovalkovičová, N., Legáth, J., Mikula, I., Holovská, V. (2004). Imunotoxikológia vo veterinárnej medicíne. *Základné princípy a metódy stanovenia imunotoxickosti xenobiotík edition*. Košice, Slovakia: Univerzita veterinárskeho lekárstva v Košiciach.

Pohanka, M. (2011). Cholinesterases, a Target of Pharmacology and Toxicology. *Biomedical Papers-Olomouc*, 155, 219-229.

Pope, C. N. (1999). Organophosphorus pesticides: do they all have the same mechanism of toxicity? *J. Toxicol. Environ. Health B Crit. Rev*., 2, 161-81.

Pope, C. N., Chakraborti, T. K. (1992). Dose-Related Inhibition of Brain and Plasma Cholinesterase in Neonatal and Adult-Rats Following Sublethal Organophosphate Exposures. *Toxicology*, 73, 35-43.

Powell, G. F., Ward, D. A., Prescott, M. C., Spiller, D. G., White, M. R., Turner, P. C., Earley, F. G., Phillips, J., Rees, H. H. (2011). The molecular action of the novel insecticide, Pyridalyl. *Insect. Biochem. Mol. Biol*., 41, 459-69.

Power, A. G. (2010). Ecosystem services and agriculture: tradeoffs and synergies. *Philosophical Transactions of the Royal Society B-Biological Sciences,* 365, 2959-2971.

Qiao, D., Seidler, F. J., Slotkin, T. A. (2001). Developmental neurotoxicity of chlorpyrifos modeled in vitro: comparative effects of metabolites and other cholinesterase inhibitors on DNA synthesis in PC12 and C6 cells. *Environ. Health Perspect*., 109, 909-13.

Qiao, D., Seidler, F. J., Slotkin, T. A. (2005). Oxidative mechanisms contributing to the developmental neurotoxicity of nicotine and chlorpyrifos. *Toxicol. Appl. Pharmacol*., 206, 17-26.

Qiao, D., Seidler, F. J., Violin, J. D., Slotkin, T. A. (2003). Nicotine is a developmental neurotoxicant and neuroprotectant: stage-selective inhibition of DNA synthesis coincident with shielding from effects of chlorpyrifos. *Developmental Brain Research*, 147, 183-190.

Quiros-Alcala, L., Bradman, A., Nishioka, M., Harnly, M. E., Hubbard, A., McKone, T. E., Ferber, J., Eskenazi, B. (2011). Pesticides in house dust from urban and farmworker households in California: an observational measurement study. *Environmental Health*, 10.

Radcliffe, E. D., Hutchinson, W. D., Cancelado, R. E. (2009). *Integrated Pest Management. Concepts, Tactics, Strategies and Case Studies edition*. Cambridge, United Kingdom: Cambridge University Press.

Rahimi, R., Nikfar, S., Abdollahi, M. (2006). Increased morbidity and mortality in acute human organophosphate-poisoned patients treated by oximes: a meta-analysis of clinical trials. *Human and Experimental Toxicology*, 25, 157-162.

Ranjbar, A., Solhi, H., Mashayekhi, F. J., Susanabdi, A., Rezaie, A., Abdollahi, M. (2005). Oxidative stress in acute human poisoning with organophosphorus insecticides; a case control study. *Environmental Toxicology and Pharmacology*, 20, 88-91.

Raszewski, G., Lemieszek, M. K., Łukawski, K., Juszczak, M., Rzeski, W. (2014). Chlorpyrifos and cypermethrin induce apoptosis in human neuroblastoma cell line SH-SY5Y. *Basic and Clinical Pharmacology and Toxicology (in press)*.

Rasouli, M., Asgari, H. R. (2008). Rat liver perfusion and hepatocytes isolation. *Journal of Mazandaran University of Medical Sciences*, 18, 71-80.

Rauh, V. A., Garfinkel, R., Perera, F. P., Andrews, H. F., Hoepner, L., Barr, D. B., Whitehead, R., Tang, D., Whyatt, R. W. (2006). Impact of prenatal chlorpyrifos exposure on neurodevelopment in the first 3 years of life among inner-city children. *Pediatrics*, 118, E1845-E1859.

Ray, D. E., Richards, P. G. (2001). The potential for toxic effects of chronic, low-dose exposure to organophosphates. *Toxicology Letters*, 120, 343-351.

Rich, J. D., Gabriel, S. M., Schultz-Norton, J. R. (2012). In vitro effects of herbicides and insecticides on human breast cells. *ISRN Toxicol.*, 2012, 232461.

Roberts, J. W., Wallace, L. A., Camann, D. E., Dickey, P., Gilbert, S. G., Lewis, R. G., Takaro, T. K. (2009). Monitoring and reducing exposure of infants to pollutants in house dust. *Rev. Environ. Contam. Toxicol.*, 201, 1-39.

Ross, S. M., McManus, I. C., Harrison, V., Mason, O. (2013). Neurobehavioral problems following low-level exposure to organophosphate pesticides: a systematic and meta-analytic review. *Critical Reviews in Toxicology*, 43, 21-44.

Roy, T. S., Sharma, V., Seidler, F. J., Slotkin, T. A. (2005). Quantitative morphological assessment reveals neuronal and glial deficits in hippocampus after a brief subtoxic exposure to chlorpyrifos in neonatal rats. *Brain Res. Dev. Brain Res.*, 155, 71-80.

Rudel, R. A., Brody, J. G., Spengler, J. D., Vallarino, J., Geno, P. W., Sun, G., Yau, A. (2001). Identification of selected hormonally active agents and animal mammary carcinogens in commercial and residential air and dust samples. *J. Air Waste Manag. Assoc.*, 51, 499-513.

Rudel, R. A., Camann, D. E., Spengler, J. D., Korn, L. R., Brody, J. G. (2003). Phthalates, alkylphenols, pesticides, polybrominated diphenyl ethers, and other endocrine-disrupting compounds in indoor air and dust. *Environmental Science and Technology*, 37, 4543-4553.

Rumora, L., Kovacic, S., Rozgaj, R., Cepelak, I., Pepeljnjak, S., Zanic Grubisic, T. (2002). Cytotoxic and genotoxic effects of fumonisin B1 on rabbit kidney RK13 cell line. *Arch. Toxicol.*, 76, 55-61.

Sanborn, M., Kerr, K. J., Sanin, L. H., Cole, D. C., Bassil, K. L., Vakil, C. (2007). Non-cancer health effects of pesticides - Systematic review and implications for family doctors. *Canadian Family Physician*, 53, 1713-1720.

Sandhu, M. A., Saeed, A. A., Khilji, M. S., Ahmed, A., Latif, M. S. Z., Khalid, N. (2013). Genotoxicity evaluation of chlorpyrifos: a gender related approach in regular toxicity testing. *Journal of Toxicological Sciences*, 38, 237-244.

Sathyanarayana, S., Basso, O., Karr, C. J., Lozano, P., Alavanja, M., Sandler, D. P., Hoppin, J. A. (2010). Maternal Pesticide Use and Birth Weight in the Agricultural Health Study. *Journal of Agromedicine*, 15, 127-136.

Saulsbury, M. D., Heyliger, S. O., Wang, K. Y., Johnson, D. J. (2009). Chlorpyrifos induces oxidative stress in oligodendrocyte progenitor cells. *Toxicology*, 259, 1-9.

Saunders, M., Magnanti, B. L., Carreira, S. C., Yang, A. L., Alamo-Hernandez, U., Riojas-Rodriguez, H., Calamandrei, G., Koppe, J. G., von Krauss, M. K., Keune, H., Bartonova, A. (2012). Chlorpyrifos and neurodevelopmental effects: a literature review and expert elicitation on research and policy. *Environmental Health*, 11.

Savolainen, K. (2001): Understanding the Toxic Actions of Organophosphates. In: R. I. Krieger and W. C. Krieger (Eds.), *Handbook of Pesticide Toxicology* (2nd edition, pp. 1013-1041). San Diego, California: Academic Press/Elsevier.

Searles Nielsen, S., McKean-Cowdin, R., Farin, F. M., Holly, E. A., Preston-Martin, S., Mueller, B. A. (2010). Childhood brain tumors, residential insecticide exposure, and pesticide metabolism genes. *Environ. Health Perspect.,* 118, 144-9.

Scheil, V., Zurn, A., Kohler, H. R., Triebskorn, R. (2010). Embryo Development, Stress Protein (Hsp70) Responses, and Histopathology in Zebrafish (Danio rerio) Following Exposure to Nickel Chloride, Chlorpyrifos, and Binary Mixtures of Them. *Environmental Toxicology*, 25, 83-93.

Slotkin, T. A. (1999). Developmental cholinotoxicants: Nicotine and chlorpyrifos. *Environmental Health Perspectives*, 107, 71-80.

Slotkin, T. A. (2011). Does early-life exposure to organophosphate insecticides lead to prediabetes and obesity? *Reproductive Toxicology*, 31, 297-301.

Slotkin, T. A., Card, J., Seidler, F. J. (2013). Adverse Benzo[a]pyrene Effects on Neurodifferentiation Are Altered by Other Neurotoxicant Coexposures: Interactions with Dexamethasone, Chlorpyrifos, or Nicotine in PC12 Cells. *Environmental Health Perspectives*, 121, 825-831.

Slotkin, T. A., Card, J., Seidler, F. J. (2012). Chlorpyrifos developmental neurotoxicity: Interaction with glucocorticoids in PC12 cells. *Neurotoxicology and Teratology*, 34, 505-512.

Slotkin, T. A., MacKillop, E. A., Ryde, I. T., Seidler, F. J. (2007). Ameliorating the developmental neurotoxicity of chlorpyrifos: a mechanisms-based approach in PC12 cells. *Environmental Health Perspectives,* 115, 1306-1313.

Slotkin, T. A., Seidler, F. J. (2007). Prenatal chlorpyrifos exposure elicits presynaptic serotonergic and dopaminergic hyperactivity at adolescence: Critical periods for regional and sex-selective effects. *Reproductive Toxicology,* 23, 421-427.

Soares, W. L., Porto, M. F. D. (2009). Estimating the social cost of pesticide use: An assessment from acute poisoning in Brazil. *Ecological Economics*, 68, 2721-2728.

Sogorb, M. A., Vilanova, E. (2002). Enzymes involved in the detoxification of organophosphorus, carbamate and pyrethroid insecticides through hydrolysis. *Toxicology Letters*, 128, 215-228.

Solomon, C., Poole, J., Palmer, K. T., Peveler, R., Coggon, D. (2007). Acute symptoms following work with pesticides. *Occupational Medicine-Oxford*, 57, 505-511.

Solomon, K. R., Williams, W. M., Mackay, D., Purdy, J., Giddings, J. M., Giesy, J. P. (2014). Properties and uses of chlorpyrifos in the United States. *Reviews of environmental contamination and toxicology*, 231, 13-34.

Song, X., Violin, J. D., Seidler, F. J., Slotkin, T. A. (1998). Modeling the developmental neurotoxicity of chlorpyrifos in vitro: Macromolecule synthesis in PC12 cells. *Toxicology and Applied Pharmacology*, 151, 182-191.

Staljanssens, D., Van Camp, J., Herregods, G., Dhaenens, M., Deforce, D., Van de Voorde, J., Smagghe, G. (2011). Antihypertensive effect of insect cells: In vitro and in vivo evaluation. *Peptides*, 32, 526-530.

Stoate, C., Boatman, N. D., Borralho, R. J., Carvalho, C. R., de Snoo, G. R., Eden, P. (2001). Ecological impacts of arable intensification in Europe. *Journal of Environmental Management*, 63, 337-365.

Sutiakova, I., Kovalkovicova, N., Legath, J., Pistl, J., Kovac, G., Sutiak, V., Novotny, J., Vaczi, P. (2009). Effect of subchronic exposure to tolylfluanid on the lactate dehydrogenase activity and its isoenzymes in sheep. *Slovak Journal of Animal Sciences, Supplement*, 42, 104-109.

Tang, Q. H., Yang, W., Tan, Y. X., Zhou, W. P. (2007). Sodium butyrate induces rat liver oval cells WB-F344 differentiating into billiary epithelium cells in vitro. *Academic Journal of Second Military Medical University,* 28, 32-35.

Thornley, J. A., Trask, H. W., Ridley, C. J. A., Korc, M., Gui, J., Ringelberg, C. S., Wang, S., Tomlinson, C. R. (2011). Differential regulation of polysome mRNA levels in mouse Hepa-1C1C7 cells exposed to dioxin. *Toxicology in Vitro*, 25, 1457-1467.

Thrasher, J. D., Madison, R., Broughton, A. (1993). Immunologic abnormalities in humans exposed to chlorpyrifos: preliminary observations. *Arch. Environ. Health*, 48, 89-93.

Timchalk, C., Nolan, R. J., Mendrala, A. L., Dittenber, D. A., Brzak, K. A., Mattsson, J. L. (2002). A physiologically based pharmacokinetic and pharmacodynamic (PBPK/PD) model for the organophosphate insecticide chlorpyrifos in rats and humans. *Toxicological Sciences*, 66, 34-53.

Timchalk, C., Poet, T. S., Kousba, A. A. (2006). Age-dependent pharmacokinetic and pharmacodynamic response in preweanling rats following oral exposure to the organophosphorus insecticide chlorpyrifos. *Toxicology*, 220, 13-25.

Tripathi, S., Srivastav, A. K. (2010). Nephrotoxicity induced by long-term oral administration of different doses of chlorpyrifos. *Toxicology and Industrial Health*, 26, 439-447.

US EPA. 2003a. Organophosphate Pesticides. US Environmental Protection Agency. Available: http://www.epa.gov/pesticides/reregistration/status_ op.htm.

US EPA. 2003b. Organophosphorous Pesticides in Food—a Primer on Reassessment of Residue Limits. 735-F-99-014:1999. Washington, DC: US Environmental Protection Agency. Available: http://www.epa.gov/ pesticides/reregistration/status_op.htm.

US EPA. 2002. (US Environmental Protection Agency). 2002. US Environmental Protection Agency Office of Pesticide Programs reregistration eligibility decision for chlorpyrifos. Available: http://www. epa.gov/pesticides/reregistration/REDs/ chlorpyrifos_red.pdf.

US EPA, 2008. Memorandum. November 28, 2008. EFED Registration review – preliminary problem formulation for ecological risk and environmental fate, endangered species and drinking water assessments for chlorpyrifos (PC Code 059101; DP Barcode D355212).

US EPA, 2011. *Chlorpyrifos Preliminary Human Health Assessment for Registration Review.* Available: http://www.epa.gov/oppsrrd1/registration _review/chlorpyrifos/.

Udarbe Zamora, E. M., Liu, J., Pope, C. N. (2008). Effects of chlorpyrifos oxon on M2 muscarinic receptor internalization in different cell types. *Journal of Toxicology and Environmental Health - Part A: Current Issues*, 71, 1440-1447.

Uzun, F. G., Demir, F., Kalender, S., Bas, H., Kalender, Y. (2010). Protective effect of catechin and quercetin on chlorpyrifos-induced lung toxicity in male rats. *Food and Chemical Toxicology*, 48, 1714-1720.

Van den Hazel, P., Zuurbier, M. (2005): *PINCHE project: Final report WP1 exposure assessment*, The Institute of Public Health in Ireland, Arnhem, The Netherlands.

Van der Mark, M., Brouwer, M., Kromhout, H., Nijssen, P., Huss, A., Vermeulen, R. (2012). Is Pesticide Use Related to Parkinson Disease? Some Clues to Heterogeneity in Study Results. *Environmental Health Perspectives,* 120, 340-347.

Viswanath, G., Chatterjee, S., Dabral, S., Nanguneri, S. R., Divya, G., Roy, P. (2010). Anti-androgenic endocrine disrupting activities of chlorpyrifos and piperophos. *Journal of Steroid Biochemistry and Molecular Biology*, 120, 22-29.

Walker, K. R., Ricciardone, M. D., Jensen, J. (2003). Developing an international consensus on DDT: a balance of environmental protection and disease control. *International Journal of Hygiene and Environmental Health,* 206, 423-435.

Wang, J., Yi, C. H., Liu, X. Z., Yuan, Y., Bian, J. C., Liu, Z. P. (2009). Effects of cadmium acetate on oxidative stress in primary cultured rat hepatocytes. *Chinese Veterinary Science/Zhongguo Shouyi Kexue*, 39, 1010-1013.

Wang, J., Zhu, L., Xie, H., Shao, B., Hou, X. (2014). The enzyme toxicity and genotoxicity of chlorpyrifos and its toxic metabolite TCP to zebrafish Danio rerio. *Ecotoxicology*.

Ward, M. H., Colt, J. S., Metayer, C., Gunier, R. B., Lubin, J., Crouse, V., Nishioka, M. G., Reynolds, P., Buffler, P. A. (2009). Residential Exposure to Polychlorinated Biphenyls and Organochlorine Pesticides and Risk of Childhood Leukemia. *Environmental Health Perspectives*, 117, 1007-1013.

Ward, M. H., Lubin, J., Giglierano, J., Colt, J. S., Wolter, C., Bekiroglu, N., Camann, D., Hartge, P., Nuckols, J. R. (2006). Proximity to crops and residential exposure to agricultural herbicides in Iowa. *Environmental Health Perspectives,* 114, 893-897.

Ware, G. W., Whitacre, D. M. (2004). *The pesticide book* (6th edition). Willoughby, Ohio: MeisterPro Information Resources.

Ware, G. W., Whitacre, D. M. (2004): *An Intorduction to insecticide* (4th edition). In: *The pesticide book* (6th edition, pp. Willoughby, Ohio: MeisterPro Information Resources.

Watson, A., Bakshi, K., Opresko, D., Young, R., Houschild, V., King, J. (2006): Cholinesterase inhibitors as chemical warfare agents: community preparedness guidelines. In: R. C. Gupta (Ed.), *Toxicology of Organophosphate and Carbamate Compounds* (1st edition, pp. 47-68). Amsterdam, The Netherlands: Academic Press/Elsevier.

Wen, Y. J., Su, T., Tang, J. W., Zhang, C. Y., Wang, X., Cai, S. Q., Li, X. M. (2006). Cytotoxicity of phenanthrenes extracted from Aristolochia contorta in human proximal tubular epithelial cell line. *Nephron. Exp. Nephrol.*, 103, e95-e102.

WHO (1992): *Our Planet, Our Health; Report of the WHO Commission on Health and Environment.* Geneva, Switzerland: WHO.

Whyatt, R. M., Barr, D. B. (2001). Measurement of organophosphate metabolites in postpartum meconium as a potential biomarker of prenatal exposure: A validation study. *Environmental Health Perspectives*, 109, 417-420.

Whyatt, R. M., Rauh, V., Barr, D. B., Camann, D. E., Andrews, H. F., Garfinkel, R., Hoepner, L. A., Diaz, D., Dietrich, J., Reyes, A., Tang, D. L., Kinney, P. L., Perera, F. P. (2004). Prenatal insecticide exposures and birth weight and length among an urban minority cohort. *Environmental Health Perspectives,* 112, 1125-1132.

Williams, W. M., Giddings, J. M., Purdy, J., Solomon, K. R., Giesy, J. P. (2014). Exposures of aquatic organisms to the organophosphorus insecticide, chlorpyrifos resulting from use in the United States. *Rev. Environ. Contam. Toxicol.*, 231, 77-117.

Wilson, C., Tisdell, C. (2001). Why farmers continue to use pesticides despite environmental, health and sustainability costs. *Ecological Economics*, 39, 449-462.

Wohlfahrt-Veje, C., Main, K. M., Schmidt, I. M., Boas, M., Jensen, T. K., Grandjean, P., Skakkebaek, N. E., Andersen, H. R. (2011). Lower birth weight and increased body fat at school age in children prenatally exposed to modern pesticides: a prospective study. *Environmental Health*, 10.

Wu, J., Daugulis, A. J., Faulkner, P., Goosen, M. F. A. (1992). Correlation of Ldh Activity with Loss of Insect Cell Viability: An Assessment of the Ldh Assay. *Biotechnology Techniques*, 6, 335-340.

Yanagisawa, N. (2014). The nerve agent Sarin: History, clinical manifestations, and treatment. *Brain and Nerve*, 66, 561-569.

Yin, X. H., Zhu, G. N., Li, X. B., Liu, S. Y. (2009). Genotoxicity evaluation of chlorpyrifos to amphibian Chinese toad (Amphibian: Anura) by Comet assay and Micronucleus test. *Mutation Research-Genetic Toxicology and Environmental Mutagenesis*, 680, 2-6.

Young, J. G., Eskenazi, B., Gladstone, E. A., Bradman, A., Pedersen, L., Johnson, C., Barr, D. B., Furlong, C. E., Holland, N. T. (2005). Association between in utero organophosphate pesticide exposure and abnormal reflexes in neonates. *Neurotoxicology*, 26, 199-209.

Young, J. T., Grandjean, M. (1988). Chlorpyrifos: 2-year dietary chronic toxicity oncogenicity study in Fischer-344 rats. TXT K-044793-(79). *Unpublished research report of the Dow Chemical Company.*

Zheng, Q., Olivier, K., Won, Y. K., Pope, C. N. (2000). Comparative cholinergic neurotoxicity of oral chlorpyrifos exposures in preweanling and adult rats. *Toxicol. Sci.*, 55, 124-32.

In: Chlorpyrifos
Editor: Caitlin Mayes

ISBN: 978-1-63482-111-7

Chapter 5

TOXIC EFFECTS OF CHLORPYRIFOS AND ITS METABOLITES ON SOME PHYSIOLOGICALLY IMPORTANT ENZYMES: ATPASES, CHOLINESTERASES, PEROXIDASES

Mirjana B. Čolović, Tamara D. Lazarević-Pašti and Vesna M. Vasić*

Department of Physical Chemistry, Institute of Nuclear Sciences "Vinča", University of Belgrade, Belgrade, Serbia

ABSTRACT

The aim of the paper is to give an overview of the mechanism of interaction between chlorpyrifos and the products of its environmental and metabolic transformations with some enzymes which regulate the key functions in many physiological processes. The metabolites of the parent compound formed due to the metabolic processes as well as chemical transformations after its application in the environment will be identified.

The enzymes from the group of ATPases, cholinesterases and peroxidases were chosen as the targets for evaluation of chlorpyrifos and the products of its metabolic/chemical transformations toxicity. The mechanism of inhibition of the selected enzymes induced by chlorpyrifos and the products of its transformations will be elucidated. The recovery effects on the inhibited activities of the selected enzymes will be analyzed.

The overview of the methods for the detection of the parent compound and the products of its transformations in the environment will be given. The toxic effects of chlorpyrifos and its metabolites towards the selected enzymes will be compared and discussed.

---

* Correspondence: Dr. Vesna M. Vasić, Department of Physical Chemistry, "Vinča" Institute of Nuclear Sciences, University of Belgrade, P.O. Box 522, 11001 Belgrade, Serbia. Tel.: + 381-11-3408 287, Fax: + 381-11-8066 434, E-mail: evasic@vin.bg.ac.rs.

Abbreviations

Ab-CG, antibody-colloidal gold
AChE, acetylcholinesterase
A-esterases, associated esterases
AOP, advanced oxidation process
ATPase(s), adenosine triphosphatase(s)
BuChE, *butyrylcholinesterase*
CA, chromosomal aberrations
CaE, chloroacetate esterase
CarbE, carboxylesterase
CGR, carboxylic graphene
ChE, cholinesterase
CPF, chlorpyrifos
CPO, chlorpyrifos-oxon,
DETP, O,O-diethyl thiophosphate
DNA, deoxyribonucleic acid,
FIA, flow-injection analysis
GCE, glass carbon electrode
GC/MS, gas chromatography/mass spectrometry
GST, glutathione-S-transferase
HDL, high density lipoprotein
HOX, hypohalous acids
HPLC, high performance liquid chromatography
ICA, immunochromatographic assay
LC-ESI-MS, liquid chromatography-electrospray ionization-mass spectrometry
LOD, limit of detection
MN, micronuclei
MO, microwave-assisted oxidation
MPO, myeloperoxidase
NADPH, nicotinamide adenine dinucleotide phosphate reduced
NF, Nafion
OPH, organophosphate hydrolase
OP(s), organophosphate(s)
2-PAM, pyridine-2-aldoximes
PBPK/PD, physiologically based pharmacokinetics and pharmacodynamics model
PON1, paraoxonase
ROS, reactive oxygen species
SAMs, self-assembled monolayers
SCE, sister chromatid exchanges
SPR, surface plasmon resonance
ssDNA, single strand oligonucleotide
SWCNT, single walled carbon nanotubes
TCP, 3,5,6-trichloro-2-pyridinol
UPLC, ultra performance liquid chromatography
UV, ultraviolet

1. Chlorpyrifos

Chlorpyrifos [O,O-diethyl-O-(3,5,6-trichloro-2-pyridyl) phosphorothioate] (CPF) (Scheme 1) is one of the most extensively used broad-spectrum organophosphate (OP) insecticides, acaricides and miticides in agricultural and nonagricultural areas to control foliage and soil-borne insect pests on a variety of food and feed crops.

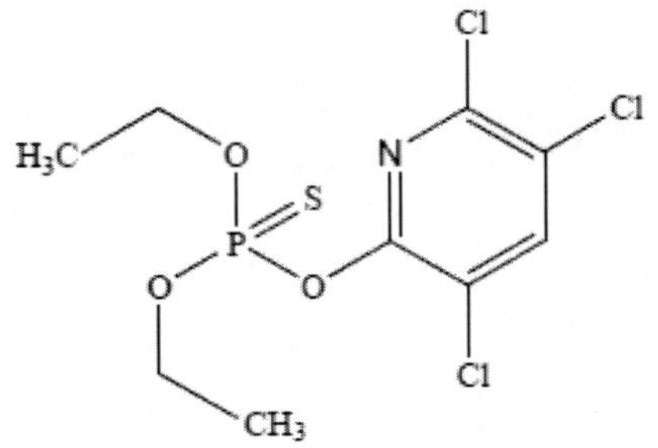

Scheme 1. Structural formula of CPF.

It is available in various formulations under the trade names such as Lorsban, Pyrinex, Spannit, Tricel, Dursban, Piridane, Silrifos, and Talon. Exposure to CPF for long periods results in serious harm to the human nervous system, respiratory tract, and cardiovascular system [Sinha, 2006; Farag, 2003]

CPF is a degradable compound, and a number of metabolic processes and environmental forces may be active in its breakdown. The metabolic and oxidation processes usually lead to the formation of chlorpyrifos-oxon (CPO), the product which is really responsible for CPF toxicity. However, in soil and water, the major pathway of its degradation begins with cleavage of the phosphorus ester bond to yield 3,5,6-trichloro-2-pyridinol (TCP), which is further degraded via microbial activity and photolysis to carbon dioxide and organic matter. This first step is a detoxification, as TCP is toxicologically insignificant.

1.1. Chlorpyrifos Metabolic Transformations

CPF is readily absorbed by laboratory animals or humans following oral exposures and undergoes metabolic transformations mainly by the liver enzymes [Tang, 2001; Estevan, 2014; Xing, 2014]. The metabolism and pharmackokinetics of CPF have been extensively studied using various experimental techniques for identification of metabolites in order to understand the way of its metabolic transformations [Smith, 2012; Wang, 2011; Zou, 2010]. There are several enzymes that play a significant role in the metabolism and toxicity of CPF.

The physiologically based pharmacokinetics and pharmacodynamics model (PBPK/PD) for CPF was developed to describe the time course of absorption, distribution, metabolism and excretion of CPF, CPO and TCP, in humans and rats, as well as the inhibition of target esterases (acetylcholinesterase (AChE), *butyrylcholinesterase* (BuChE) and chloroacetate esterase (CaE)) in blood components and selected tissues following oral and dermal exposure to CPF [Smith, 2012; Timchalk, 2002]. Scheme 2 illustrates the disposition of CPF, CPO and

TCP in blood according the PBPK/PD model after dermal and dietary exposure of CPF. The shaded tissue compartments indicate organs in which B-esterases (AChE, BuChE and CaE) enzyme activity is described. For illustration of pharmacokinetic data, some constants of CPF metabolic transformation are indicated in Table 1.

Table 1. Kinetic parameters of CPF metabolic transformations

CPF transformation path	Rat		Human		Ref.
	K_m (μmol/l)	V_{max} (μmol/h/kg)	K_m (μmol/l)	V_{max} (μmol/h/kg)	
CYP450 CPF-to-CPO (liver)	2.86	80	2.86	80	[Ma, 1994]
CYP450 CPF-to-TCP	24	273	24	273	[Ma, 1994]
A-EST CPO-to-TCP (liver)	240	74421	240	74421	[Mortensen,1996]
A-EST CPO-to-TCP (blood)	250	57003	250	57003	[Mortensen,1996]

K_m - Michaelis constant for saturable process; V_{max} - maximum velocity for saturable process.

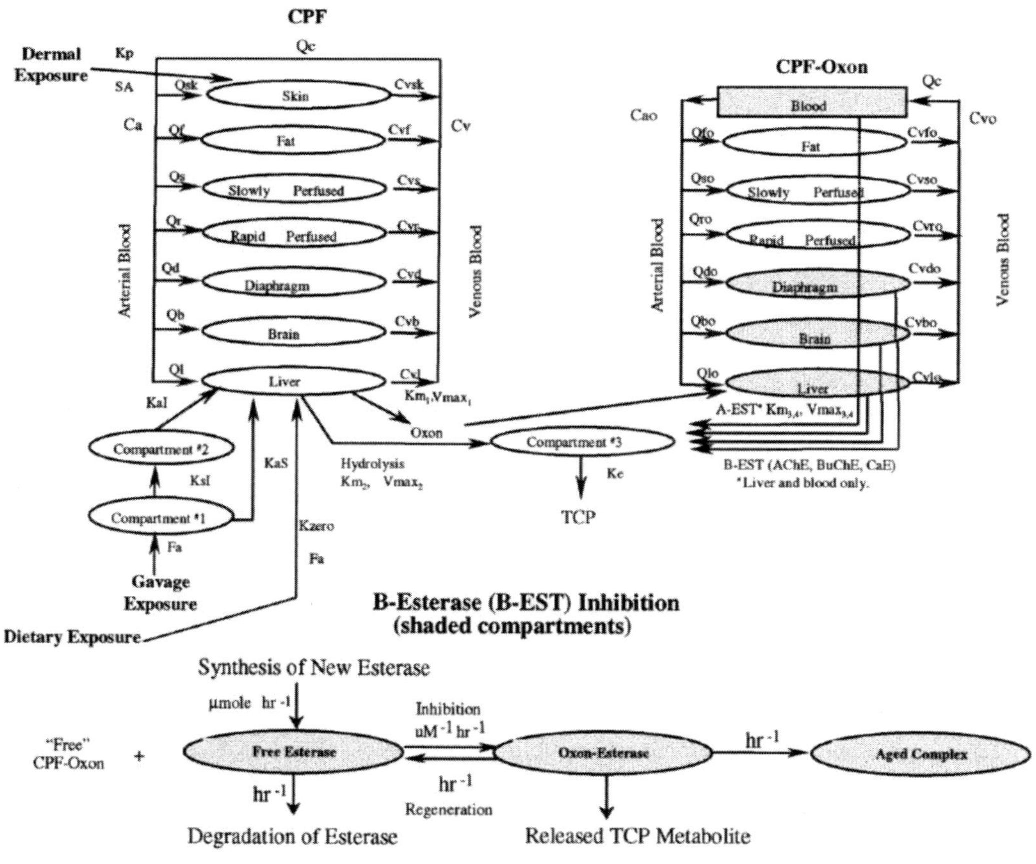

Scheme 2. Physiologically based pharmacokinetic and pharmacodynamic model for the disposition of CPF, CPO and TCP, following oral and dermal exposure to CPF [Timchalk, 2002].

1.1.1. Metabolism of Chlorpyrifos by Human Cytochrome P450 Isoforms

The *in vivo* toxicity of CPF is a result of its bioactivation by cytochrome P450 – mediated monooxygenases to a more potent cholinesterase (ChE) inhibitor, CPO. The initial metabolic attack on the CPF molecule is its desulfuration, which proceeds through a possible phosphooxythiiran intermediate (Scheme 3) and induces its transformation to CPO [Tang, 2001]. This is highly reactive metabolite and either binds to a serine esterase or is rapidly deactivated through hydrolytic cleavage by a process called dearylation, releasing TCP. Simultaneously during the desulfuration process, dearylation is also acting on the parent CPF, leading to the release of TCP, which is further, conjugated to form glycine or glucuronide conjugates and eliminated into the urine. The both CPF and CPO target serine esterases, such as BuChE and AChE by binding to the active site and inactivating the ChE enzyme.

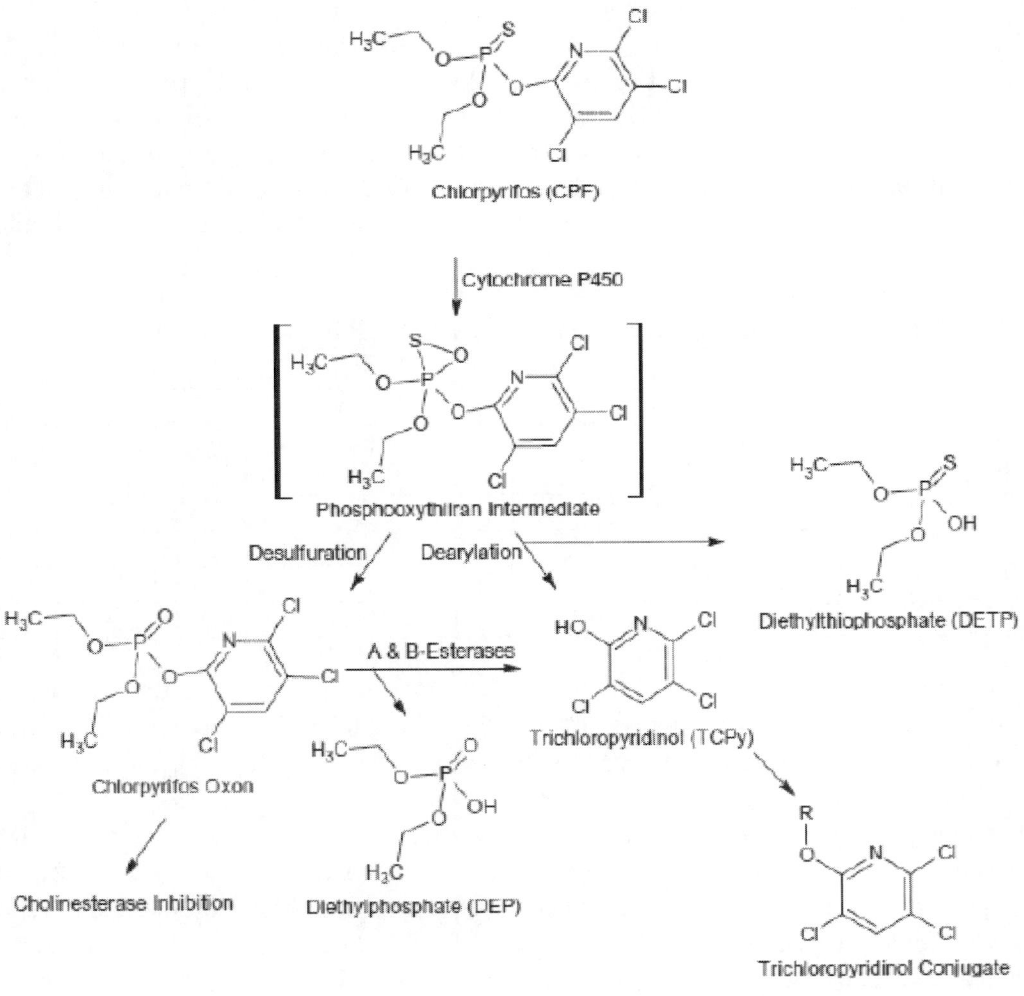

Scheme 3. Enzymatic metabolism of CPF, CPO and TCP [Smith, 2012].

1.1.2. The role of A-esterases in Detoxification of Chlorpyrifos

Another group of important enzymes in the metabolic detoxification of CPF is the high-density lipoprotein (HDL)-associated esterase (A-esterases), which can hydrolyse OPs without being inhibited by them. These calcium activated enzymes are also known as paraoxonase (PON1), and they are distributed in multiple forms in various tissues including the liver, brain and blood [Walker, 1987; Hunter Coombes, 2014]. PON1 protects against the acute toxic effects of several OP anticholinesterases through hydrolysis of the parent compounds or their active metabolites (oxons) [Furlong, 2008]. In routine exposures to CPF, CPO would not be expected to be present at high levels because of the kinetics of formation from CPF by cytochromes P450. The detoxication of CPO by PON1 at low (< 5 mmol/l) concentrations has been studied, confirming the importance of PON1 in detoxifying CPO [Sogorb, 2008; Sogorb, 2010].

A-esterases hydrolyze CPO, which further binds irreversibly to other non-target serine esterases, carboxylesterase (CarbE), which is considered as a scavenger of the CPO removing it from reaching the target AChE enzyme. This binding has been described to be stoichometric. CarbEs are distributed among different tissues (liver, blood (not human), lung, brain) with highest abundance in the liver. The pharmacokinetics and pharmacodynamics studies of CPF and TCP in rat saliva and blood were performed after *in vivo* CPF administration. These studies enabled the development of various sensors for monitoring of pesticide exposure to humans such as agricultural workers, when rapid pesticide exposure assessments are required [Wang, 2011; Zou, 2010]. Table 2 illustrates the elimination rate of CPF, as well the formation and elimination rate of TCP concentrations in rat blood [Smith, 2012].

Table 2. Pharmacokinetics data for CPF and TCP in blood [Smith, 2012]

	CPF	TCP	
Dose (mg/kg)	elimination rate (h^{-1})	formation rate (h^{-1})	elimination rate (h^{-1})
0.5	0.25 ± 0.27	0.96 ± 0.30	0.54 ± 0.45
1	1.08 ± 0.15	5.80 ± 1.54	0.38 ± 0.25
2.5	1.47 ± 0.12	5.11 ± 4.14	0.30 ± 0.10
5	1.35 ± 0.47	11.48 ± 8.49	0.34 ±0.13

The results suggest that CPF is rapidly metabolized after *in vivo* administration, followed by the dose dependent decrease of ChE activity in blood. Formation of TCP from administered CPF at the low dose (0.5 mg/kg) was slower than from higher CPF doses, potentially due to differences in plasma protein binding to CPF.

Some publications report that orally administered CPF is more extensively metabolized than systemic exposures of CPF, and vehicle of administration also has an effect on absorption rates. Thus, equivalent doses via different routes and/or vehicles of administration could potentially lead to different body burdens of CPF, different rates of bioactivation to CPO, and different toxic responses [Smith, 2009; Foxenberg, 2011] However, species differences in susceptibility to organophosphorothioate poisoning of animals and humans are dependent on pharmacokinetics (e.g. tissue distribution), metabolism (e.g. rate of oxon formation), and pharmacodynamics (e.g. AChE inhibition) [Nolan, 1984; Smith, 2009]. TCP was the major metabolite in the blood, while the glucuronide conjugate of TCP was the major metabolite in urine and bile of channel cat fish. The toxic metabolite, CPO was not detected in any samples of blood, tissues, or excreta [Barron, 1991].

1.1.3. The Role of Glutathione S-transferases in the Metabolic Transformations of Chlorpyrifos

The glutathione-S-transferases (GST) may also play an important role in the secondary metabolism of CPF producing water soluble metabolites that are readily excreted into the urine [Fujioka, 2007]. Formation of glutathione (GSH) conjugates was studied *in vitro* with and without metabolic activation of CPF, CPO and TCP in the presence of GSH by human liver microsomes or P450 3A4 with Metabolites were analyzed by liquid chromatography-electrospray ionization-mass spectrometry (LC-ESI-MS).

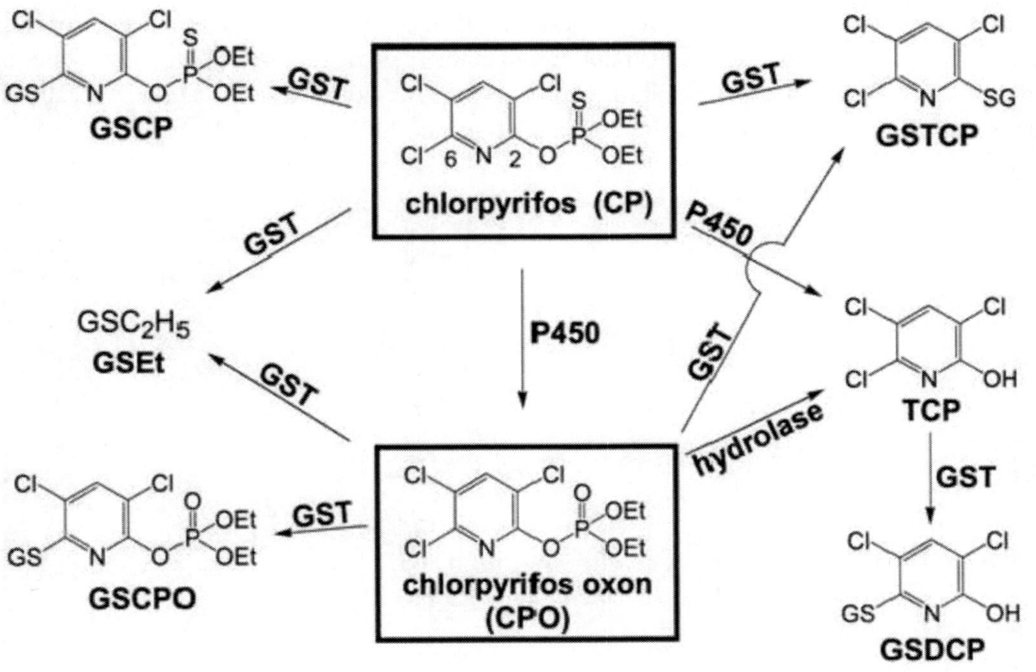

Scheme 4. Metabolic transformations of CPF and CPO catalyzed by GST and P450 [Fujioka, 2007].

In the presence of GSH, CPF and CPO undergo dealkylation, dearilayion, and also dearylation, giving different conjugates by attack at 2- or 6- position of the pyridinil moiety after phosphorothionate oxidation and P-O-pyrinidil and P-O ethyl cleavage. Eight GSH conjugates were identified for CPF and CPO (Scheme 4). In GS-chlorpyrifos (GSCPF) and GS-chlorpyrifos oxon (GSCPO), the 6-chloro substituent is displaced by GSH. Moreover, TCP metabolites formed are *S*-(3,5,6-trichloropyridin-2-yl)glutathione (GSTCP); *S*-(3,5-dichloro-6-hydroxypyridin-2-yl)glutathione; since *S*-ethylglutathione (GSEt) is also one of three major CPF and CPO metabolites [Fujioka, 2007]. CPF and CPO gave also one minor metabolite - *S*-(3,5-dichloro-6-hydroxypyridin-2-yl)glutathione (GSDCP) on incubation with GST and GSH. The formation of these metabolites is completely dependent on added GSH. When human liver microsomes or placental GST were used, CPF gave only nGSCPF. When incubated with human liver microsomes, NADPH, and the equine GST/GSH system, CPF gave GSCPF as the only significant metabolite, presumably formed by an endogenous GST in the microsomal preparation.

1.2. Chlorpyrifos Transformation in the Environment

1.2.1. Biodegradation of Chlorpyrifos and 3,5,6-trichloro-2-pyridinol

The potential of microorganisms as bioremediation agents for CPF and TCP has attracted the attention of many researchers [Li, 2007; Yang, 2005; Chen, 2012; Bhagobaty, 2006]. Evidence from soil degradation studies indicates that cleavage and mineralization of the heterocyclic ring occurs in soil due to the activities of microorganisms. It has been suggested that the accumulated TCP in liquid medium or soil, which has antimicrobial property, prevents the proliferation of microorganisms involving in degrading CPF [Li, 2010].

Recently, the new fungal strain Hu-01 with high CPF degradation activity was isolated [Chen, 2012] and the optimum degradation conditions were determined. The proposed pathway for CPF degradation is illustrated in Scheme 5. As the major degradation product of CPF, TCP has greater water solubility than its parent molecule and causes widespread contamination of soils and aquatic environments [Anwar, 2009].

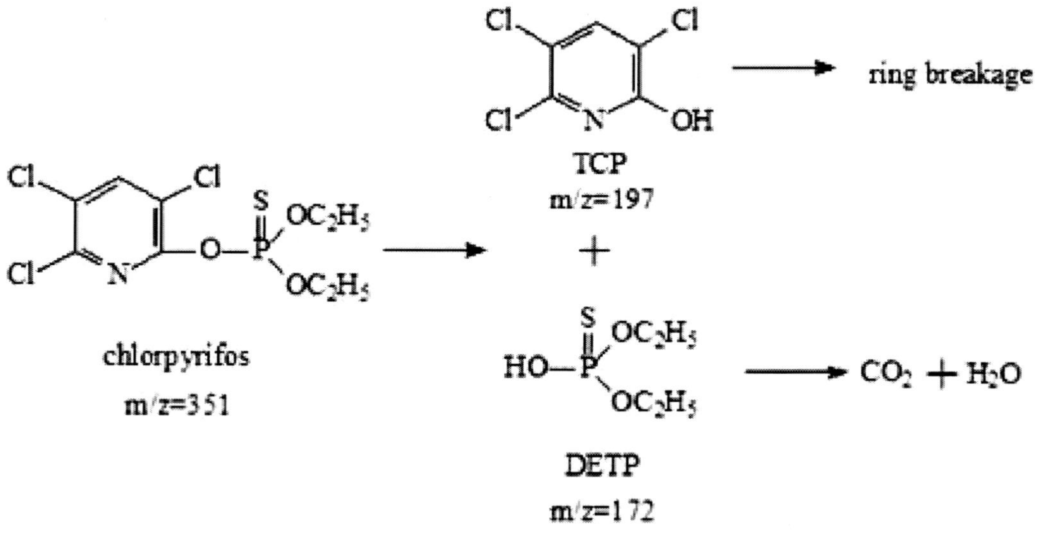

Scheme 5. The proposed pathway for the CPF degradation by strain Hu-01 [Chen, 2012].

Recently, some fungal isolates have also been successfully used for degradation of CPF and TCP with different capabilities [Maya, 2012]. The concentrations of CPF and TPC after biodegradation were measured by HPLC. The results exhibited peaks for TCP, indicating that the fungal isolates have the capacity to produce this main metabolite during bio-degradation of CPF. It was also found that the biodegradation of CPF proceeded through TCP pathway which ends with pyruvic acid prior to the initiation of Krebs cycle and consequent release of carbon dioxide and water by fungi [Singh, 2006a]. In this study, the genetic relatedness of isolate F1 to *Aspergillus sp.*, F2 and F3 to *Penicillium sp.*, F4 to *Eurotium sp.* and F5 to *Emericella sp.* were confirmed. The kinetic parameters of biodegradation of CPF and TCP, obtained using the Michaelis–Menten kinetics model to fit the experimental data, are presented in Table 3.

Table 3. Kinetic parameters for the fungal isolates degrading CPF and TCP [Maya, 2012]

	CPF			TCP		
Isolates	K_s	V_{max}	V_{max}/K_s	K_s	V_{max}	V_{max}/K_s
1	169.9	40.37	0.2382	163.9	40.40	0.2475
2	129.87	20.10	0.1548	95.2	15.66	0.1645
3	112.35	15.90	0.1415	97.08	14.61	0.1504
4	97.10	10.56	0.1088	97.08	12.24	0.1261
5	66.66	6.56	0.0984	53.19	3.41	0.0641

K_s (mg/l), V_{max} (mg/l/d), and V_{max}/K_s (1/d).

1.2.2. Chlorpyrifos Chemical Degradation

The major pathways of chemical transformation of CPF in environment are hydrolysis, oxidation and photolysis [Wu, 2006; Murillo, 2010]. In all systems (soil, water, plants and animals), these pathways begin with cleavage of the phosphorus ester bond to CPO after the replacement of sulphur by oxygen in the thiophosphoryl bond (P=S) [Duirk, 2008]. CPF can also be remediated in the presence of noble metal nanoparticles [Bootharaju, 2012]. The combination of various methods has a synergistic effect in reducing CPF concentration [Pengphol, 2012; Lazarević-Pašti, 2011; Reddy, 2013].

1.2.2.1. Chlorpyrifos Hydrolysis

CPF readily undergoes hydrolysis in basic pH [Faust, 1972; Macalady, 1983; Liu, 2001], whereas it persists in neutral and slightly acidic conditions typical of soils, surface waters, and aquifer sediments. The scheme of CPF hydrolysis is given in Scheme 6.

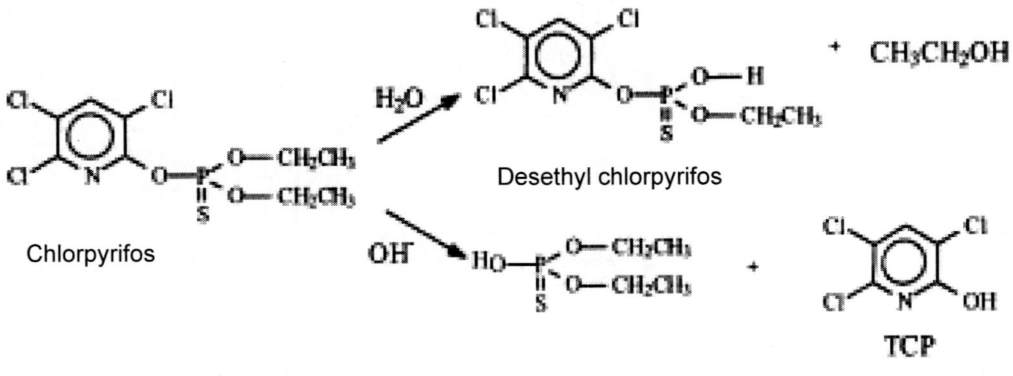

Scheme 6. Possible mechanisms for hydrolysis of CPF [Liu, 2001].

Hydrolysis is strongly dependent on the pH and other solution constituents. CPF possesses two tertiary alkyl ester bonds and one phosphate ester (pyridyl) bond that are candidates for hydrolytic cleavage. The hydrolysis is usually initiated via a nucleophilic or electrophilic attack, which leads to the breakdown of these bonds. As shown in Scheme 6, neutral hydrolysis of CPF involves the nucleophilic attack of water at the less saturated carbon (soft center) with loss of the alkyl leaving groups [Smith, 1978], since the alkaline hydrolysis is initiated by nucleophilic attack of the hydroxide ion at the phosphorus atom (hard center), bringing about loss of the most acidic alcoholic or phenolic leaving group.

However, in natural water samples these processes are up to 16 folds faster than under laboratory conditions [Liu, 2001].

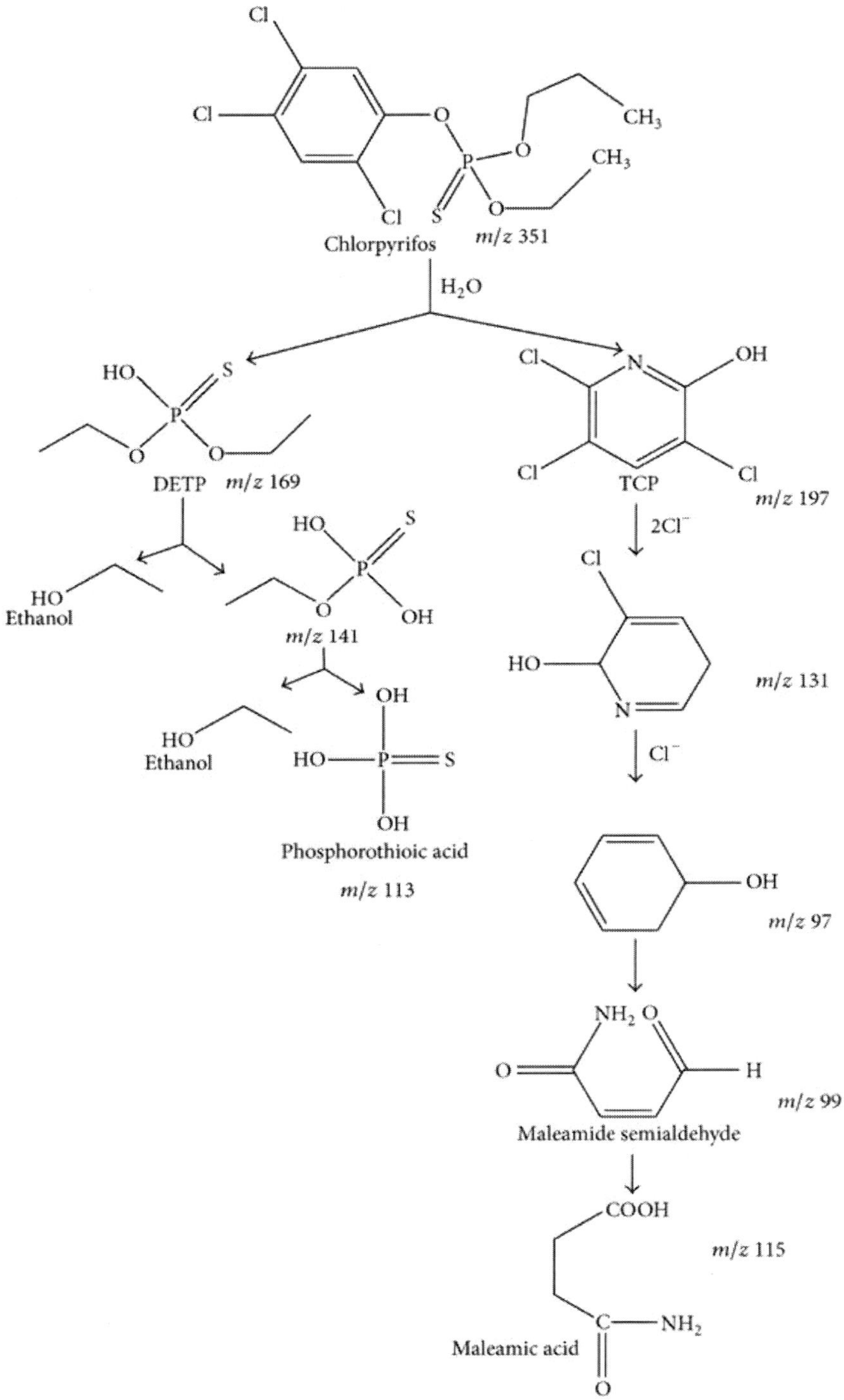

Scheme 7. Proposed pathway for the degradation of CPF by zerovalent nano iron particles [Reddy, 2013].

The nanoparticles are also powerful tool for remediation of CPF [Reddy, 2013; Bootharaju, 2012]. Recently, the zerovalent nano iron particles were used for remediation of CPF contamined soils, since they can be nontoxic, noninvasive and environmentally compatible. Hence, the degradation of CPF with zerovalent nano iron particles had occurred through a stepwise dechlorination mechanism. The scheme of decomposition is presented in Scheme 7, where TCP and O,O-diethyl thiophosphate (DETP) were found as by-products. DETP is further decomposed yielding two molecules of ethanole and phoshorthionic acid. Continous dechlorination of TCP resulted in the formation of maleamic acid.

1.2.2.2. Chlorpyrifos Oxidation

The enzymatic oxidation of CPF leads usually to the formation of more toxic CPO, and can be used in bioanalytical methods for its determination [Lazarević-Pašti, 2011; Čolović, 2013a; Čolović, 2013b]. However, the application of various methods in combination for the remediation of CPF contaminated waters and soils by oxidation can have a synergistic effect in reducing CPF concentration [Samet, 2010; Reddy, 2013; Murillo, 2010]. The combined process of ultrasonication and ozonation offers several advantages over the individual process. The main advantage is the increased mass transfer of ozone from the gas phase to the bulk solution to react with substrate by mechanical effects of ultrasound, followed by the released chloride, nitrate and sulphate anions from the CPF structure, during decomposition [Pengphol, 2012].

Moreover, the gamma irradiation of CPF in aqueous solution contributes also to remediation of CPF in the environment [Ismail, 2013] since it leads to the producing of primary transient species, the oxidizing hydroxyl radicals ($HO^{\bullet}$) and reducing hydrated electrons (e^{-}_{aq}), and also hydrogen atoms ($H^{\bullet}$) that can oxidize or reduce contaminants simultaneously. When using ionizing radiation as an advanced oxidation process (AOP), pollutants can be degraded in most cases to less toxic and in some cases to more biodegradable compounds that do not pose adverse environmental consequences. AOPs that combine microwave radiation (MO), ultraviolet (UV) radiation, titanium dioxide (TiO_2) tablets and hydrogen peroxide (H_2O_2) solutions have proven particularly useful in degrading of CPF. The results showed that the $MO/UV/TiO_2/H_2O_2$ process was the most efficient to remove CPF, with 100% removal reached within 6 minutes reaction time [Barros, 2013].

Experiments measured CPF loss in the presence of free chlorine identified degradation pathways based on mass balances over the pH range of 6.3-11. A total of two transformation products were observed - CPO and TCP obtained via oxidation by hypochlorous acid (HOCl). Both CPF and CPO were observed to hydrolyze to TCP which was found to be a stable end-product, generally less toxic than the parent compounds [Duirk, 2005].

1.2.2.3. Chlorpyrifos Photolysis

Photodegradation is one of the most destructive pathways for pesticides degradation after their release into the environment, where the sunlight plays the crucial role in this process [Amarathunga, 2014].

Direct photodegradation has been shown not to be efficient systems for the degradation of CPF. Several studies have been conducted on CPF photodegradation under artificial solar sources of irradiation or UV light in combination with catalysts, such as TiO_2, metals ions, hydrogen peroxide, and humic substances [Burrows, 2002; Muhamad, 2010; Hossain, 2013].

Furthermore, direct and indirect degradation may cause different transformation products which are more or less toxic, or complete mineralization. In natural water samples, the rate of catalytic photodegradation of CPF was lower than that in distilled water, and following the order: distilled water > ground water > lake water > river water >drinking water [Muhamad, 2010].

2. Influence of Chlorpyrifos on Physiologically Important Enzymes

2.1. Influence of Chlorpyrifos and Its Transformation Products on Acetylcholinesterase

2.1.1. Acetylcholinesterase-Primary Target of Chlorpyrifos Toxic Action

The primary target of CPF action, as well as other OPs, is AChE, and the main mechanism of toxicity in acute OP exposure involves the specific irreversible inhibition of this enzyme activity in the nervous system and blood, manifesting as a cholinergic crisis with excessive glandular secretions and weakness, miosis and fasciculation of muscle, which may lead to death [Rahimi, 2006; Savolainen, 2001; Sharma, 2005].

AChE (EC 3.1.1.7) is a serine protease mainly found at neuromuscular junctions and cholinergic brain synapses. Its principal biological role is termination of impulse transmission at cholinergic synapses by rapid hydrolysis of the neurotransmitter acetylcholine to acetate and choline. AChE has a remarkably high specific catalytic activity, especially for a serine hydrolase - each molecule of AChE degrades about 25000 molecules of acetylcholine per second, approaching the rate of a diffusion-controlled reaction [Quinn, 1987; Taylor, 1994].

The molecule has an ellipsoidal shape with dimensions ~ 45 Å by 60 Å by 65 Å. The enzyme monomer is an α/β protein containing 12 – stranded central mixed β sheet surrounded by 14 α helices. The most remarkable feature of the structure is a deep and narrow gorge, ~ 20 Å long penetrating halfway into the enzyme and widens out close to its base [Manavalan, 1985]. The active site of AChE is located 4 Å from the bottom of the molecule and comprises two subsites - the "anionic" subsite and "esteratic" subsite corresponding to the catalytic machinery and the choline-binding pocket, respectively [Nachmansohn, 1951]. The anionic subsite, uncharged and lipophilic, binds the positive quartenary amine of choline moiety of acetylcholine, as well as both quartenary ligands (edrophonium, N-methylacridinium) acting as competitive inhibitors [Wilson, 1958; Mooser, 1974], and quartenary oximes which effectively reactivate organophosphate-inhibited AChE [Froede, 1971]. The cationic substrates are not bound by a negatively-charged amino acid in the anionic site, but by interaction of 14 aromatic residues that line the gorge leading to the active site [Radić, 1992; Ordentlich, 1995; Ariel, 1998]. All 14 amino acids in the aromatic gorge are highly conserved across different species [Ordentlich, 1993]. Among the aromatic amino acids, tryptophan 84 is critical and its substitution with alanine causes a 3000-fold decrease in the enzyme activity [Tougu, 2001]. The esteratic subsite, where acetylcholine is hydrolyzed to acetate and choline, contains, similar to the catalytical subsites of other serine hydrolases, the catalytic triad of three amino acids: serine 200, histidine 440 and glutamate 327 (Scheme 8).

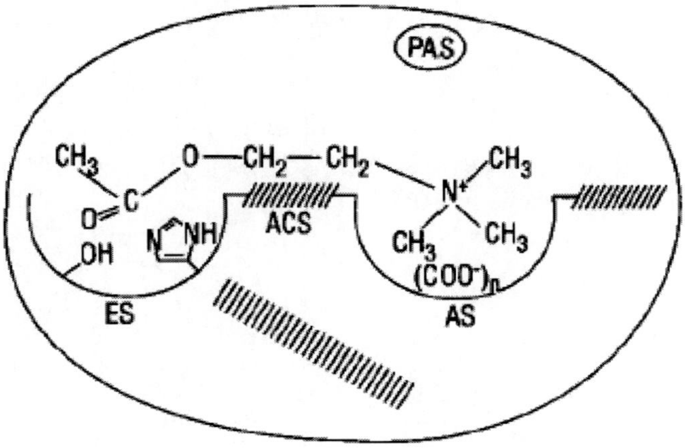

Scheme 8. Schematic presentation of AChE binding sites: ES – esteratic site, AS – anionic site, PAS – peripheral anionic site, ACS – selective aromatic cation binding site.

The hydrolysis reaction of the carboxyl ester leads to the formation of an acyl-enzyme and free choline. Then, the acyl-enzyme undergoes nucleophilic attack by a water molecule, assisted by the histidine 440 group, liberating acetic acid and regenerating the free enzyme (Scheme 9) [Nachmansohn, 1951].

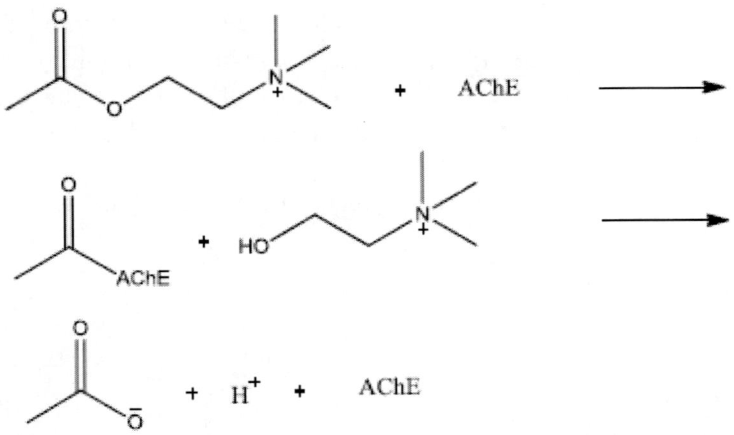

Scheme 9. Mechanism of acetylcholine hydrolysis catalyzed by AChE.

In addition to two subsites of active centre, AChE contains one or more "peripheral" anionic sites distinct from the choline-binding pocket of the active site (Scheme 8). It serves for binding acetylcholine and other quartenary ligands acting as uncompetitive inhibitors that bind at a site clearly distinct from that occupied by the monoquartenary competitive inhibitors [Taylor, 1975]. This site is involved in the substrate inhibition characteristics of AChE [Reiner, 1991].

2.1.2. Irreversible Inhibition of Acetylcholinesterase by Chlorpyrifos and Its Oxo Analog

The acute toxic effects of CPF are related to irreversible inactivation of AChE [WHO, 1986a]. Actually, OPs are substrate analogues to acetylcholine, and like natural substrate enter the active site covalently binding to serine –OH group. As in acetylation, OP is split and the enzyme is phosphorylated (Scheme 10).

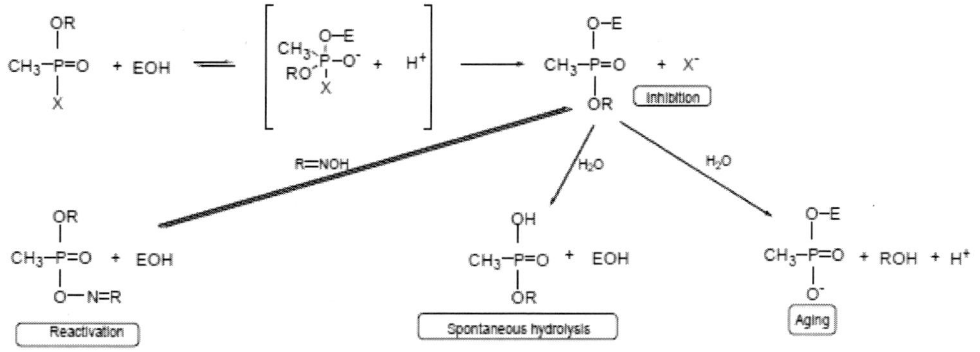

Scheme 10. Mechanism of AChE inhibition induced by OPs; reactivation, spontaneous hydrolysis, and aging of the phosphorylated enzyme.

While the acyl enzyme is quickly hydrolyzed to regenerate the free enzyme, dephosphorylation is very slow (on the order of days), and phosphorylated enzyme cannot hydrolyze the neurotransmitter [Boublik, 2002]. The inhibition of the enzyme leads to accumulation of acetylcholine in the synaptic cleft resulting in over-stimulation of nicotinic and muscarinic ACh receptors and impeded neurotransmission [WHO, 1986a; WHO, 1986b; Hayden, 2010; Costa, 2006].

Mechanism of OPs induced AChE inhibition is presented using the reaction scheme:

$$E + PX \leftrightarrows E^{*}PX \rightarrow EP + X \tag{I}$$

where, E–enzyme, PX–OP, $E^{*}PX$–reversible enzyme-OP complex, EP–phosphorylated enzyme, X–OP leaving group [Aldridge, 1950].

Irreversible inhibition occurs in two steps; the first one is fast, short term reversible enzyme inactivation, and its influence is dominant in the begining of the inhibition. The next step is slow irreversible inhibition producing a very stable enzyme-OP complex (phosphorylated enzyme)-inhibitor is covalently bonded to the enzyme [Knežević-Jugović, 2008]. Time dependent irreversible inhibition can be described by the equation:

$$\ln \frac{E}{E_o} = -\frac{k_3\, t}{1 + K_I/(I)} \tag{1}$$

where, E/Eo–remaining enzyme activity related to initial enzyme activity (control) (E_o), K_I–dissociation constant for enzyme-inhibitor complex $E^{*}PX$, k_3–the first rate constant for the conversion of the reversible enzyme-inhibitor complex to phosphorylated enzyme, EP, (I)–

inhibitor (OP) concentration, t–time interval after the enzyme and inhibitor mixing. If (I) » (E_o), the reciprocal slope value of linear dependence ln(E/Eo) - t (Figure 1a) can be presented in the form:

$$\frac{1}{k_{app}} = \frac{1}{k_3} + \frac{K_I}{k_3} \cdot \frac{1}{(I)} \qquad (2)$$

The values of inhibition parameters, K_I and k_3, are calculated from the slop and intersection of $1/k_{app}$ - 1/(I) linear dependence (Figure 1b) [Krstić, 2008b; Čolović, 2011].

The rate of irreversible inhibition for CPF and its oxo analog, CPO was determined according to Ellman's method [Ellman, 1961]. The enzyme (AChE from electric eel; specific activity 288 IU/mg solid, 408 IU/mg protein) was *in vitro* exposed to CPO (concentration range from 2×10^{-8} to 2×10^{-7} mol/l) and CPF (2×10^{-6} - 2×10^{-5} mol/l) at several preincubation times (enzyme-inhibitor contact time). The results show that the inhibition of AChE activity by CPF increases as a function of exposure time within the investigated time interval (Figure 1a) in accordance with Equation 1. The values of k_{app}, obtained from the slope of the dependence of $\ln E/E_o$ vs. t, were then plotted according to Equation 2 and presented in Figure 1b. Reasonably good straight lines were obtained that intercept the positive y axis, confirming that reversible complexes are formed in the concentration range studied.

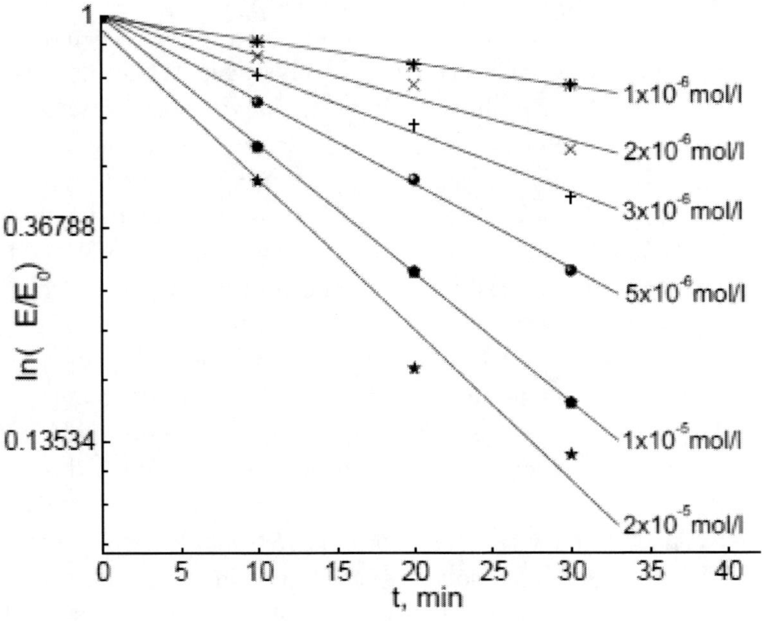

Figure 1a. Progressive development of inhibition produced by reaction of AChE with different concentrations of CPF plotted as semi logarithmic curve in accordance with Equation (1) [Čolović, 2011].

The inhibition parameters K_I and k_3 are calculated from the slope and intercept of Equation 2 [Čolović, 2011] and are given in Table 4.

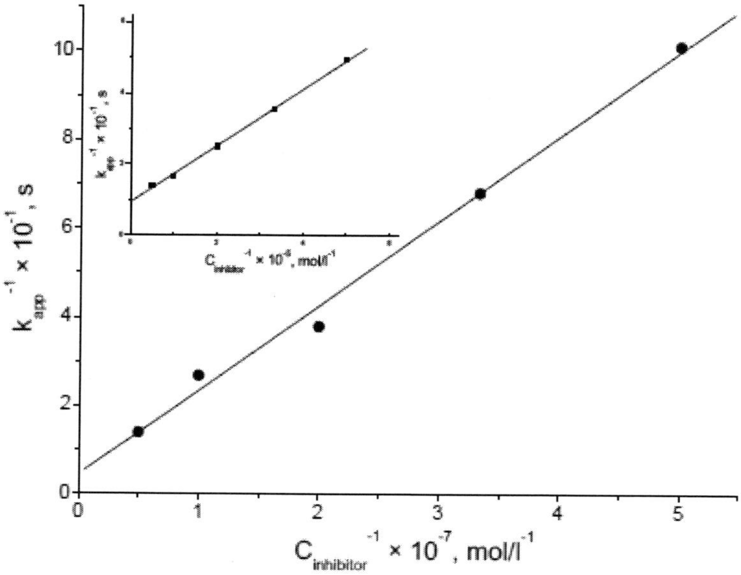

Figure 1b. The dependence of k_{app} upon the concentration of CPO and CPF (inset) plotted as reciprocals in accordance with Equation (2) [Čolović, 2011; Čolović, 2014].

2.1.3. In Vitro Influence of Chlorpyrifos and Its By-products on Acetylcholinesterase Activity

The separate influence of CPF and its oxidation and hydrolysis products-CPO and TCP, respectively, usually formed due to the chemical conversion or photochemical treatment of the selected organophosphates, on AChE activity was investigated in the concentration range from 1×10^{-10} to 1×10^{-2} mol/l. The inhibition of AChE activity was measured using slightly modified Ellman's procedure [Ellman, 1961] in the absence (control) and presence of CPF and its related compounds. The experiments were performed by *in vitro* exposure of 0.02 IU commercial AChE from electric eel to investigated compounds in 50 mmol/l phosphate buffer pH 8.0 (final volume 0.650 ml). The standard medium assays were preincubated for 20 min at 37°C in the absence and presence of desired concentration of the inhibitors. The control tubes contained the corresponding concentration of organic solvents without the organophosphate. Ten μl acetylthiocholine iodide (0.075 mol/l) was used as the enzyme substrate in combination with 20 μl 5,5'-dithio-bis-2- nitrobenzoic acid (0.01 mol/l in 50 mmol/l phosphate buffer pH 7.0) as a chromogenic reagent. The reaction was started by the addition of acetylthiocholine iodide, and allowed to proceed for 5 min until stopped by 65 μl sodium dodecyl sulfate (10%). The product 5-thio-2-nitrobenzoate, released in the reaction of thiocholineiodide (product of enzymatic reaction) and 5,5'-dithio-bis-2- nitrobenzoic acid, was measured at 412 nm (in buffer solution) using Perkin Elmer Lambda 35 UV-VIS spectrophotometer (Shelton, USA). The obtained results show (Figure 2a) that, CPF and CPO inhibit AChE in a concentration-dependent manner but with different inhibitory potencies. Dependence of the AChE activity, expressed as the percent of control value (obtained without inhibitor), on the inhibitor concentration fitted a sigmoidal function in all cases (Figure 2a). The inhibition parameters, the concentrations of investigated compounds with capability to inhibit 50% of the enzyme after given exposure time (IC_{50} values) and Hill coefficient, n_H determined using the Hill method (Figure 2b) by linear regression analysis of log [%

activity/(100 – % activity)] *vs.* log (concentration of inhibitor) plots, are given in Table 4. The differences between values of IC_{50} obtained using the Hill method and by fitting the sigmoidal inhibition curves were in the range of experimental error.

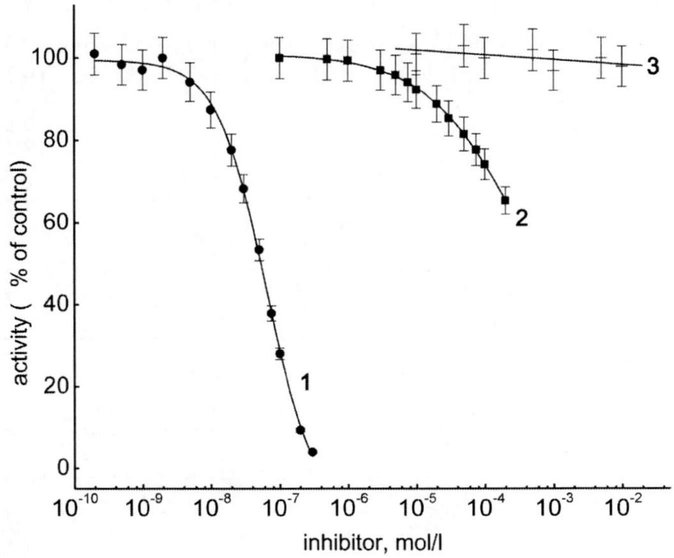

Figure 2a. The concentration-dependent inhibition of electric eel AChE by CPO (1), CPF (2), and TCP (3). The values are expressed as mean ± S.E.M [Čolović, 2011].

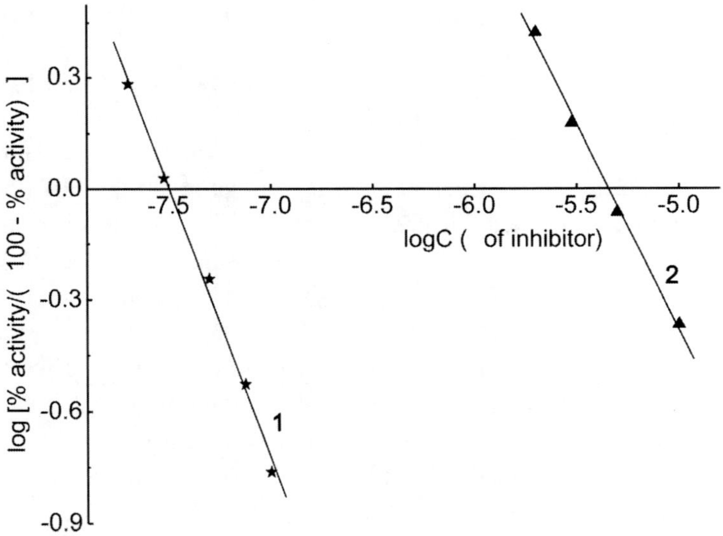

Figure 2b. Hill analysis of inhibition of electric eel AChE activity induced by CPO (1) and CPF (2) [Čolović, 2011].

As can be seen (Figures 2a and 2b, Table 4), inhibitory potencies of CPF and its degradation products are quite different. Concentration of 2×10^{-7} mol/l CPO completely inhibited AChE activity, while the effect of the same CPF concentration on the enzyme activity was negligible. The IC_{50} value of the enzyme activity was achieved at $(4.3 \pm 0.2) \times 10^{-6}$ mol/l of CPF, while the same effect was observed when the concentration of CPO was $(3.0 \pm 0.1) \times 10^{-8}$ mol/l. The hydrolysis product of CPF, TCP (Figures 2a and 2b) did not noticeably affect the enzyme activity at any of the investigated concentrations [Čolović, 2011].

Table 4. The inhibition parameters (IC_{50} values (20 min) and Hill coefficients n_H) of CPF and its decomposition products for AChE obtained by Hill analysis and fitting the experimental data by sigmoidal function [Čolović, 2011]

Compound	Sigmoidal fitting	Hill analysis		Kinetic data		
	IC_{50}, mol/l	IC_{50}, mol/l	n_H	K_I, mol/l	k_3, s^{-1}	k_3/K_I, $(mol/l)^{-1}s^{-1}$
CPF	$(4.3 \pm 0.2) \times 10^{-6}$	4.5×10^{-6}	1.2 ± 0.1	9.6×10^{-6}	0.12 ± 0.06	1.2×10^{4}
CPO	$(3.0 \pm 0.1) \times 10^{-8}$	3.2×10^{-8}	1.5 ± 0.1	4.3×10^{-7}	0.23 ± 0.03	5.3×10^{5}
TCP	Inhibition < 5%	-	-	-	-	-

Although CPF and CPO have similar structures, it is clear that combination of the substituents at the central phosphorous atom is responsible for their different inhibition power. The replacement of sulphur with oxygen in the molecule of CPO causes an increased positive charge density on the central phosphorous atom, facilitating the nucleophilic attack of the –OH group of serine, resulting in higher inhibitory potency of CPO [Krstić, 2007].

2.1.4. Inhibition of Acetylcholinesterase Activity by Simultaneous Exposure to Chlorpyrifos and its Transformation Products

Bavcon Kralj et al. investigated CPF photodegradation during 120 minutes using the xenon light (125 W Cermax Xenon parabolic lamp) [Bavcon Kralj, 2007a]. The analysis of CPF irradiated solutions suggested the presence of the CPO, as the main photodegradation product, and TCP, hydrolysis product of CPF as a result of bond (P–O) cleavage. The oxidation product was detected in the first 3.5 minutes of the irradiation, achieved maximal value at 15 minutes and was observed even in the 120-minute irradiated sample at the end of experiment. Thus, in a toxicity evaluation of photodegraded CPF, it is worthy to study the simultaneous influence of both CPF and its inhibiting oxidation product, CPO present in each of the investigated CPF irradiated sample solutions, on the activity of AChE as their target enzyme. Therefore, synthetic mixtures of CPF and CPO (pure commercial compounds) were composed in various concentration ratios and applied in the AChE assay (20 minutes preincubation). The final concentrations of the parent compound and its oxo analogue in the medium assays were around or lower than their IC_{50} (20 min) values (Table 5). The results of the influence of simultaneous and single (separate) exposure toward the selected CPF/CPO concentrations on AChE activity are presented in Table 5. According to the results of variance analysis (One-way ANOVA), additive inhibition effects were obtained in the presence of lower CPF (mol/l)/CPO (mol/l) concentrations: $3 \times 10^{-7}/3 \times 10^{-9}$, $3 \times 10^{-7}/5 \times 10^{-9}$, $3 \times 10^{-7}/1 \times 10^{-8}$, $3 \times 10^{-7}/2 \times 10^{-8}$, $5 \times 10^{-7}/3 \times 10^{-9}$, $5 \times 10^{-7}/5 \times 10^{-9}$, $5 \times 10^{-7}/1 \times 10^{-8}$, $5 \times 10^{-7}/2 \times 10^{-8}$, $1 \times 10^{-6}/3 \times 10^{-9}$, $1 \times 10^{-6}/5 \times 10^{-9}$, $2 \times 10^{-6}/3 \times 10^{-9}$, $2 \times 10^{-6}/5 \times 10^{-9}$, $3 \times 10^{-6}/3 \times 10^{-9}$. Higher concentrations of the investigated organophosphates induced antagonistic enzyme inhibitions (Table 5). The presence of non-inhibiting (Figure 2a, Table 4) CPF hydrolysis product (identified in the irradiated solutions), TCP in the CPF/CPO mixtures does not alter their inhibition potencies against AChE (data not shown) [Čolović, 2011].

The additive inhibition of AChE activity obtained by simultaneous exposure to low concentrations of CPF/CPO confirms that both inhibitors bind to the same binding site and that there is an excess of inhibitor binding sites. The antagonistic inhibition of electric eel AChE in the presence of inhibitors mixtures at higher concentrations of CPF/CPO $\geq 5 \times 10^{-6}/3 \times 10^{-8}$ mol/l can be explained by competition between inhibitors for a limited number of inhibitor binding sites on the enzyme [Carfagna, 1996].

Table 5. Inhibition of AChE activity induced by simultaneous exposure to CPF and CPO in mixtures. Values in parenthesis represent organophosphate-induced inhibition measured separately (by single exposure) [Čolović, 2011]

Inhibition (%)						
CPF, mol/l / CPO, mol/l	3×10^{-7} (5.5)	5×10^{-7} (11.2)	1×10^{-6} (16.2)	2×10^{-6} (26.0)	3×10^{-6} (44.3)	5×10^{-6} (53.0)
3×10^{-9} (12.5)	15.9	23.5	33.3	44.8	53.0	55.0
5×10^{-9} (20.7)	20.6	27.4	35.7	45.5	54.0	58.1
1×10^{-8} (28.4)	26.9	37.9	37.5	47.5	57.0	60.4
2×10^{-8} (41.0)	38.1	42.9	46.3	50.7	61.0	68.2
3×10^{-8} (50.6)	45.0	49.9	60.9	64.7	71.0	78.0
5×10^{-8} (60.0)	56.9	64.1	71.2	73.6	77.1	84.8

2.2. Influence of Chlorpyrifos on Na^+/K^+-ATPase

2.2.1. Nonspecific Toxic Effects of Organophosphates

The primary target of OP action is AChE, and the main mechanism of toxicity in acute OP exposure involves the specific irreversible inhibition of this enzyme activity in the nervous system and blood, manifesting as a cholinergic crisis with excessive glandular secretions and weakness, miosis and fasciculation of muscle, which may lead to death [Savolainen, 2001; Sharma, 2005]. Additionally, many studies suggest that both acute and chronic intoxication disturb the redox processes changing the activities of antioxidative enzymes and causing enhancement of lipid peroxidation in many organs, and there is little correlation between organ damage and the degree of OP induced AChE inhibition [Soltatinejad, 2009; Buyukokuroglu, 2008; Fortunato, 2006]. Indeed, in acute, and rather subchronic or chronic OP exposition, induction of oxidative stress has been reported as the main mechanism of its toxicity [Ranjbar, 2005]. Oxidative stress is defined as an imbalance between the production of free radicals – reactive oxygen species (ROS) and the antioxidant defense system – enzymatic and non-enzymatic. The ROS may be generated as the result of the metabolism of OPs by cytochrome P450s, monooxygenases that catalyze oxidation by addition of one atom of molecular oxygen into the substrate (OP) by electron transport pathway [Chambers, 2001]. This disorder is manifested by changes in the activity of antioxidative enzymes (catalase, superoxide dismutase, glutathione peroxidase, glutathione reductase), increased malondialdehyde concentration and/or altered levels of non-enzymatic antioxidants (reduced glutathione, vitamin C, vitamin E, beta-carotene), as oxidative stress parameters and the marker of ROS increased level and lipid peroxidation [Akhgari, 2003; Abdollahi, 2004]. There is some evidence that OPs may affect liver, kidney, muscles, immune, and hematological system, causing many human body disorders [Teimouri, 2006; Possamai, 2007; Dandapani, 2003]. Also, some findings indicate oxidative stress as an important pathomechanism of neurological disorders such as Alzheimer's disease and Parkinson's disease, as well as of cardiovascular diseases [Abdollahi, 2004; Petrovitch, 2002; Mates, 1999].

Furthermore, OP induced ROS attack lipids, proteins and DNA, causing oxidation and membrane damage, enzyme inactivation, DNA damage and cell death [Halliwell, 1999; Valko, 2004]. The highly reactive free radicals attack DNA resulting in single and double strand breaks, as well as oxidative damage to sugar and base residues that can later be converted to strand breaks [Wallace, 1988]. On the other hand, phosphorus moiety in the OPs appears to be a good substrate for nucleophilic attack leading to phosphorylation of DNA which is an instance of DNA damage [Das, 2007]. Some reported studies indicate the increase in chromosomal aberrations (CA), micronuclei (MN) and sister chromatid exchanges (SCE), as the markers of cytogenetic damage, in cultured lymphocytes isolated from peripheral blood taken from exposed individuals. Thus, cytogenetic damage in circulating lymphocytes has been widely used as a biomarker of exposure and effects of pesticides [Bolognesi, 2003; Bull, 2006]. It has been reported that AChE non-inhibiting OP decomposition products exert stronger genotoxic potency compared to the parent compound [Čolović, 2010], suggesting that the risk of genotoxicity from some insecticides might be appreciably greater than that predicted from standard toxicity tests [Attia, 2007]. Moreover, DNA damage leads to genomic instability that may result in mutagenesis and carcinogenesis [Turner, 2002], and some epidemiological studies demonstrate cancer risk due to pesticides exposure [Settimi, 2003; Mink, 2008; Band, 2011].

2.2.2. *Influence of Chlorpyrifos on Na^+/K^+-ATPase Activity*

Like other OPs, the main toxic action of CPF is inhibition of AChE by phosphorylation of the serine hydroxyl group in the substrate-binding domain of the enzyme, which results in accumulation of acetylcholine and associated neurotoxicity [Fulton, 2001; Oruc, 2007]. Additionally, it is established that some OPs inhibit different adenosine triphosphatases (ATPases), the group of enzymes playing an important role in biochemical processes [Vasić, 2008] in various tissues [Rahman, 1997]. Na^+/K^+-ATPase (EC 3.6.1.37) is a cell membrane located enzyme that establishes and maintains the high internal K^+ and low internal Na^+ concentrations, which are essential for neurotransmission and represent a convenient driving force for the secondary transport of metabolic substrates such as amino acids and glucose [Köksoy, 2002; Wang, 2012]. Moreover, recent studies show that in addition to pumping ions, Na^+/K^+-ATPase interacts with neighboring membrane proteins and organized cytosolic cascades of signaling proteins to send messages to the intracellular organelles [Xie, 2002; Xie, 2003; Wang, 2012]. Pesticides exert biological effect on the ATPase system by partitioning in the enzyme complex [Kinter, 1972], and it is also reported that an allosteric change results in a decrease of ATPase activity [Reddy, 1992]. The proper functioning of these enzymes is critical for cellular viability because they control many essential cellular functions, and any inhibition of the activity may be considered as a sensitive indicator of toxicity [Čolović, 2010; Oruc, 2007]. Besides, pesticides may also induce oxidative stress leading to generation of free radicals and lipid peroxidation [Banerjee, 1999; Goel, 2005; Kehrer, 1993]. As Na^+/K^+-ATPase is a membrane bound enzyme, its activity depends on the phospholipids environment of the membrane. Therefore, any change in the lipid component of the membrane directly affects the activity of this enzyme [Sahoo, 1999].

In vitro influence of CPF on Na^+/K^+-ATPase activity was investigated using rat erythrocyte membranes [Singh, 2006b]. Erythrocytes are convenient model systems for the study of interactions of xenobiotics with biomembranes, because they lead to change in membrane permeability [Singh, 2004]. Generally, OPs are lipophilic in nature, and thus have strong affinity for interaction with lipid rich biomembranes and living organisms [Costa, 1990]. Membrane preparation from male albino *Wistar strain* rat blood was exposed to 0.2 mg/l CPF. Results of obtained Na^+/K^+-ATPase activity in treated erythrocyte membranes indicated 75% decrease in the enzyme activity related to control value ($P \leq 0.001$). Concerning the significant effect on erythrocyte Na^+/K^+-ATPase, this membrane bound enzyme may be used for CPF toxicological evaluation.

In vivo influence of CPF on Na^+/K^+-ATPase was studied in fore-, mid and hind brain regions of albino rats (*Wistar strain*) [Mehta, 2005]. Adult male rats were treated with CPF dissolved in 0.3 ml dimethylsulfoxide intramuscularly in thigh muscles for three days at doses of 20, 50 and 100 mg/kg body wt (1/10, 1/4 and 1/2 LD_{50}, respectively). After animal decapitations, brains were excised out and separated in different regions. Na^+/K^+-ATPase activity was determined in previously prepared synaptosomal fractions [Mehrotra, 1989; Jorgensen, 1974]. CPF caused decrease in Na^+/K^+-ATPase activity in dose-dependent and region-specific manner. The activity of Na^+/K^+-ATPase was decreased to 71, 75 and 88% in fore-brain, 58, 91 and 92% in mid-brain and 88 and 90% in hind-brain of rats, receiving 20, 50 and 100 mg/kg body wt, respectively, for three days (Table 6) [Mehta, 2005].

Differences in susceptibilities of different brain regions to CPF may be due to differences in capabilities of different regions to combat oxidative stress. Actually, the distribution of enzymatic and non-enzymatic antioxidants is not uniform throughout the brain [Mehta, 2005; Somani, 1999].

Table 6. Effect of CPF treatment on Na^+/K^+-ATPase activity in different regions of rat brain [Mehta, 2005]

	Na^+/K^+-ATPase activity, nmol Pi/min/mg protein		
Dose	Fore-brain	Mid-brain	Hind-brain
Control	177.00 ± 15.00	194.50 ± 14.74	162.00 ± 7.88
20 mg	51.30 ± 5.15	81.40 ± 3.95	19.69 ± 5.26
50 mg	44.00 ± 2.00	18.00 ± 0.92	15.01 ± 3.32
100 mg	21.56 ± 0.42	15.00 ± 1.19	not determined

Recently, Narra [Narra, 2014] investigated *in vivo* short-term effect of CPF on Na^+/K^+-ATPase activity in gills, the principal organ for respiration of freshwater crabs, *Barytelphusa guerini.* Crabs were exposed to a sublethal concentration of CPF (0.07 mg/l) for a total of 8 days (at intervals of 1, 2, 4, and 8 days) in clean water. After the exposure period, crabs were released into freshwater and kept in the same medium for 8 days for the recovery study. Na^+/K^+-ATPase activity was estimated in the supernatant of dissected and homogenized gills [Shiosaka, 1971]. Gill Na^+/K^+-ATPase activity was inhibited (62 %) after 8 days of exposure and recovered in a time-dependent manner. At the end of recovery period (8 days) the enzyme activity recovered completely and achieved the control value ($p > 0.05$, insignificant difference compared with control) (Table 7) [Narra, 2014].

Table 7. Na^+/K^+-ATPase activity in gills of crab exposed to a sublethal dose (0.07 mg/l) of CPF for 8 days and after 8 days of recovery [Narra, 2014]

Na^+/K^+-ATPase activity, μmol Pi/mg protein			
Control	9.72 ± 0.7		
Exposure period (days)			
1	2	4	8
29.25 ± 6.8	28.36 ± 8.3	21.52 ± 9.3	17.85 ± 9.8
Recovery period (days)			
1	2	4	8
24.85 ± 10.2	31.78 ± 8.8	38.22 ± 9.3	46.02 ± 8.1

The enzyme activity recovery after 8 days indicates that crab has the capacity to recover from toxicant stress after CPF intoxication, and successfully adapt for survival in a pesticide-extreme environment. Detoxification mechanisms are enzymatic adaptations developed by animals to protect against toxic xenobiotics produced by living organisms. Anthropogenic xenobiotics, which are carried into the organism at portals (gills, gastrointestinal tract, integument and excretory system), may not be efficiently metabolized [Narra, 2014].

The *in vivo* effect of various sublethal CPF concentrations on Na^+/K^+-ATPase activity in the immune organs, spleen and head kidney of common carp (*Cyprinus carpio L.)* was studied [Wang, 2013]. Carps were exposed to 1.16, 11.6 and 116 μg/l CPF for 40 days, followed by 20 days of recovery in pesticide-free water. The obtained results are presented in Table 8. In general, Na^+/K^+-ATPase activity in both spleens and head kidneys was reduced proportional to increasing concentrations of CPF. Na^+/K^+-ATPase activity in the spleen, for all dose groups of CPF, was significantly lower ($P < 0.05$) compared to the control group. In the head kidney, Na^+/K^+-ATPase activity, in the presence of 1.16 μg/l CPF, was significantly lower ($P < 0.05$) than the activity from the control group. After recovery for 20 days, the spleen Na^+/K^+-ATPase activity of 1.16 μg/l CPF was significantly altered compared with that of the corresponding exposure-only group. In the head kidney, Na^+/K^+-ATPase activity for recovery groups after 11.6 and 116 μg/l CPF treatment was significantly higher ($P < 0.05$) than in the corresponding exposure-only groups (Table 8).

Table 8. Na^+/K^+-ATPase activity in the spleen and head kidney tissues from the common carp after 40-day exposure to various CPF concentrations, and 20-day recovery period [Wang, 2013]

	Na^+/K^+-ATPase activity, μmol Pi/mg protein			
	Spleen tissue		Head kidney tissue	
CPF, μg/l	Exposure	Recovery	Exposure	Recovery
Control, 0	9.76 ± 0.54	7.38 ± 0.76	6.07 ± 0.87	6.17 ± 0.37
1.16	7.44 ± 0.47	9.97 ± 0.33	4.89 ± 0.78	5.25 ± 0.42
11.6	5.82 ± 0.39	4.14 ± 0.66	2.88 ± 0.59	5.56 ± 0.36
116	4.69 ± 0.75	4.66 ± 0.47	1.91 ± 0.56	3.87 ± 0.44

Na^+/K^+-ATPase exhibited sensitivity to CPF, suggesting that its activity may act as a potential biomarker for assessing the environmental CPF risk for carp.

2.3. Influence of Chlorpyrifos on Human Myeloperoxidase

Myeloperoxidase (MPO) is an important member of the haem peroxidase-cyclooxygenase superfamily. This enzyme is physiologically expressed in circulating neutrophils, monocytes and some tissue macrophages including microglia [Agner, 1941]. MPO plays an essential role in the antimicrobial and antiviral system of humans. The microbicidal activity of MPO exists due to its capability to oxidize halide and pseudohalide ions (Cl^-, Br^-, I^- and SCN^-) by H_2O_2, producing in that way respective hypohalous acids (HOX) [Klebanoff, 1968]. During the phagocytosis of pathogens, azurophilic granules release their content together with MPO into phagolysosomes. On the other hand, MPO can be discharged outside the phagocytes. By this means, tissue damage during inflammation is greatly promoted by MPO-derived oxidants [Labro, 2000]. Regarding its activity, MPO is a key factor in a great number of conditions within the group of cardiovascular diseases [Daugherty, 1994; Malle, 2006], inflammatory diseases, neurodegenerative diseases [Green, 2004; Maki, 2009; Choi, 2005], kidney diseases [Malle, 2003; Malle, 1997] and immune-mediated diseases [Edwards, 1988; Nurcombe, 1991; Schiller, 1996]. Therefore, MPO and its downstream inflammatory pathways might be attractive targets for both prognostic and therapeutic intervention in the prophylaxis of all mentioned illnesses. Nowadays, structure and reaction mechanism of MPO are known, which enables rational strategy in development of specific MPO inhibitors that still preserve MPO activity during host defense from bacteria, but hinder pathophysiologically persistent activation of MPO.

The influence of CPF and CPO on MPO peroxidase activity was investigated for OPs concentrations ranging from 1×10^{-11} to 1×10^{-4} mol/l in phosphate buffer at pH 6

[Lazarević-Pašti, 2013]. The effect was monitored through the oxidation of *o*-dianisidine in the presence of H_2O_2. The inhibition of 7 nmol/l MPO peroxidase activity induced by CPF and respective oxon was evaluated using optimal experimental conditions for the inhibition test, determined earlier [Lazarević-Pašti, 2013]. The incubation time of enzyme with pesticides was 30 minutes. The obtained results are shown in Figure 3.

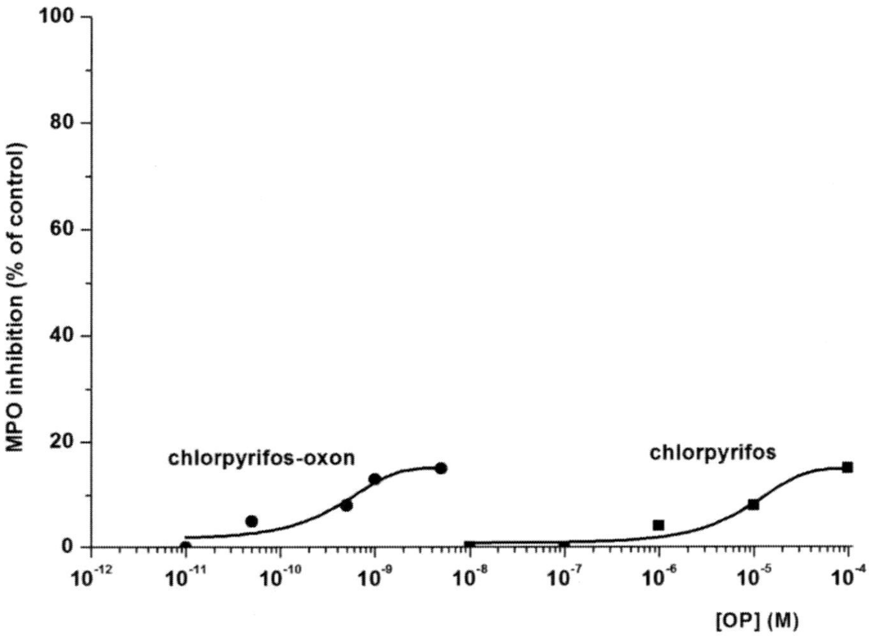

Figure 3. The concentration dependent inhibition of MPO peroxidase activity induced by chlorpyrifos (CPF) and chlorpyrifos oxon (CPO) [Lazarević-Pašti, 2013].

As can be seen from the inhibition curves (Figure 3), peroxidase activity of MPO depends on OPs concentration. It is obvious that CPO is more powerful inhibitor of MPO peroxidase activity than respective thio-form. It is also clear that CPFand his oxo-analogue did not exert significant inhibitory power under the given experimental conditions, hence it was not possible to determine IC_{50} values. Compared to the other OPs examined under the same conditions, CPF is a poor MPO inhibitor [Lazarević-Pašti, 2013]. As the study showed, the lowest activities are observed in the cases of malathion and phorate, which have the simplest structures. MPO activity further increases, i.e. the degree of inhibition decreases, when going from diazinon to CPF (having one aromatic ring within the structure) to azynphos-methyl (two rings). The same approximate trend is also followed in the case of oxo-analogs, leading to the conclusion that inhibition trend is rendered by molecular structure which invokes steric hindrance for OPs interaction with MPO's active center responsible for peroxidase activity or allosteric site [Lazarević-Pašti, 2013].

The influence of CPF and CPO on MPO chlorination activity was also examined [Lazarević-Pašti, 2013]. Neither of the investigated OPs inhibited MPO chlorination activity under specified experimental conditions. It was possible to conclude that, unlike peroxidase activity of MPO, chlorination activity is not affected by CPF and its oxo-analogue under given experimental conditions. This is of particular importance having in mind that the chlorination activity is the physiological function of MPO [Klebanoff, 1991]. The results also confirmed that the inhibition of MPO peroxidase activity by a certain compound does not necessarily indicate that the compound would also inhibit the chlorination activity of the enzyme [Kettle, 1995]. On the other hand, majority of the known MPO inhibitors do affect both of the enzyme's activities [Kettle, 1995; Momić, 2008].

As overviewed previously by Arnhold and Flemming [Arnhold, 2010] active MPO was observed in atheromatous plaques, brain tissue samples with both Alzheimer-type alterations and Parkinson's disease, multiple sclerosis lesions, synovial fluid of patients with rheumatoid arthritis, glomerular basement membrane at membranous glomerulonephritis, and polymorphonuclear neutrophil-mediated liver diseases. Moreover, enhanced plasma levels of MPO are considered to be a risk factor for coronary artery disease [Arnhold, 2010; Baldus, 2003; Nicholls, 2005; Zhang, 2001]. Literature survey unambiguously shows that MPO is one of the main targets to look at, from different angles, in various chronic inflammatory diseases [Savenkova, 1994]. Considering all mentioned, MPO appears to be one of the decisive pharmacological targets. Numerous studies have been focused on its inhibition. A great number of nonsteroidal and anti-inflammatory drugs, anilines, and phenols inhibit MPO reversibly [Momić, 2008]. Previous studies have demonstrated that some anti-inflammatory drugs may inhibit MPO activity. In addition to anti-inflammatory drugs with inhibition ability toward MPO, numerous phenolic acids derivatives as quercetin, curcumin, ferulic, caffeic and gallic have demonstrated a strong MPO inhibition [Kato, 2003]. Though it is difficult to envisage pharmaceutical usage of organophosphorous compounds, information about its inhibitory effects on MPO is valuable for discovery of new inhibitors through the accumulation of knowledge about structure-activity relationship.

3. Methods for Identification of Chlorpyrifos and Its Transformation/Degradation Products

The potential for exposure to CPF is present in several occupational fields, including agriculture, manufacturing, animal health technicians, pesticide applicators, and others. Humans and animals may be exposed to metabolites of CPF through dietary sources and from background levels found in the environment. The metabolites excreted by humans and animals are in the same family of chemicals as products of degradation that form when CPF is broken down in the environment [CDC, 2005; Timchalk, 2007]. CPF is also stable in soils with reported half-lives ranging between 7 and 120 days [EPA, 1999]. Studies have found this pesticide in fields for over one year after application [Kamrin, 1997]. Soil persistence may depend on the formulation, rate of application, soil type, climate and other conditions [Roberts, 1999]. CPF binds strongly to soils, is relatively immobile, and has low water solubility [EPA, 1999; Kamrin, 1997]. Due to its wide application and high resistivity, it is very important to have an optimal method for the identification and detection of CPF and its transformation/degradation products.

3.1. Conventional Instrumental Analytical Techniques for Detection of Chlorpyrifos

Since the OP pesticides are often used in agriculture and control of insects, their detection is very important. In the past, analytical procedures for the determination of OPs were very time consuming and complicated because of the limitations of available methods. It was difficult to identify the whole spectrum of various compounds and quantification was possible only in the case of high concentrations. The only methods of choice were paper chromatography and some colorimetric methods.

Development of gas chromatography as analytical method and microculometric and electron-capture detectors set new goals and made easier the analysis of halogenated compounds. However, it was not a good solution for compounds with electronegative functional groups. Later came detectors for gas chromatography specific for individual elements, among them the flame photometer for the determination of phosphorus and sulfur. Microculometric detector for halogenated compounds was used at that time, but his defect was insufficient sensitivity and robustness for routine analysis. Over time, methods with adequate sensitivity have been developed, but the duration and the simplicity of the analysis is still a problem.

Separation of pesticides with chromatographic methods before determining the individual components is now routine. Gas chromatography and liquid chromatography are commonly used for this purpose. Thin-layer chromatography was once very popular, but the new chromatographic techniques completely suppressed it at one point. However, with development of new detection systems, thin-layer chromatography is returned to use. Lately, capillary zone electrophoresis became popular, since it was proved to be complementary to gas and liquid chromatography.

Gas chromatography has been the method of choice for the separation of the first generation of pesticides for a long time. They were very hydrophobic, volatile and stable at high temperatures. The problem became apparent when polar and labile pesticides came into use. Then the liquid chromatography gained in importance in this area. The detection limit of this method is about 1 nmol/l [Betowski, 1988; Cabanillas, 1991; Guardino, 1998]. The usage of new types of detectors, such as a diode array detector, fluorescence and mass detector was of particular importance.

Gas chromatograph with a mass spectrometer is an instrument that is nowadays commonly used for analysis of the OP, among them CPF [Brzak, 1998; Thompson, 1996]. The reason is the possibility of determining microgram quantities of pesticides, which can be achieved thanks to the great power of the separation and connection with a mass spectrometer, which has an extremely high sensitivity (0.05 nmol/l) [Guardino, 1998; Liapis, 2003]. However, the sensitivity is not high enough to reach the limit of quantification prescribed by the Commission of the European Communities for individual pesticides, and pre-concentration of samples is always necessary. CPF in waterways is regulated as a hazardous substance under section 311(b)(2)(A) of the Federal Water Pollution Control Act and falls under the Clean Water Act Amendments of 1977 and 1978 [CEC, 2014]. The regulation is inclusive of all CPF isomers and hydrates in any solution or mixture. EPA has not yet set a drinking water regulatory standard for CPF but has established a federal drinking water guideline of 2 ug/L [GCF, 2008].

Although conventional instrumental techniques are used daily to detect pesticides and have low detection limits, they are expensive and complicated. It is therefore necessary to develop other methods that will be able to quickly, reliably and cost-effectively analyze CPF and other OP pesticides.

3.2. Bioanalytical Enzyme Tests for Detection of Chlorpyrifos

Bioanalytical methods are often more cost-effective and more sensitive than instrumental procedures. Recent developments in bioanalytical methods may provide more convenient applications for their use in pesticide residues analysis.

CPF and other OPs being an inhibitors or substrates of some enzymes is the key fact that can be used in their detection. Enzymes that are commonly used for that purpose are organophosphate hydrolase (OPH) and cholinesterases (AChE and BuChE).

In the catalytic process of OPs by OPH, two protons are released during each hydrolysis reaction, and direct detection of CPF and other OPs is thus possible *via* measurement of the pH change associated with enzyme activity (Scheme 11). The stoichiometric production of hydrogen ions offers an opportunity to detect the activity of the enzyme on a substrate by the change in pH [Paliwal, 2008].

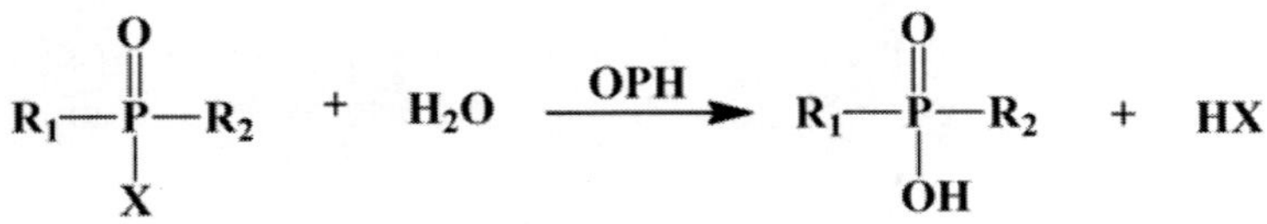

Scheme 11. Catalytic process of OPH.

In the case of cholinesterases, CPF and other OPs are inhibitors of these enzymes. As already mentioned, AChE is a key enzyme of the nervous system of animals. The main biological role of AChE is termination of the transmission of impulses in the cholinergic synapses by rapid hydrolysis of the neurotransmitter acetylcholine. Unlike AChE, BuChE is produced in liver and found primarily in blood plasma. OPs are irreversible inhibitors of cholinesterases and covalently bind to the active site of the enzymes.

Ellman's procedure [Ellman, 1961] is commonly used for the determination of cholinesterase activity and also for monitoring of the acetylcholine hydrolysis by AChE or BuChE *in vitro.* In these experiments, acetylthiocholine is disposed instead of acetylcholine because its similar qualitative kinetic behaviour provides on-line monitoring of the hydrolysis by the Ellman's method. Ellman's procedure can be applied also for the determination of cholinesterase activity of tissue extracts, homogenates, cell suspensions etc. but the procedure is disturbed by high hemoglobin absorption at 412 nm. Therefore, modified methods were developed [Worek, 1999; Willig, 1996]. The original Ellman's colorimetric procedure is based on the reaction of thiocholine (one of the products of enzymatic hydrolysis of acetylthiocholine with 5,5-dithiobis-2-nitrobenzoic acid (DTNB, Ellman's reagent) forming a yellow product (5-mercapto-2-nitrobenzoic acid (TNB) and its dissociated forms) at pH 8 (Scheme 12). The maximum absorption coefficient was found at 412 nm. As the substrate is hydrolyzed, the increase of absorbance at 412 nm is recorded, and the activity of the enzyme calculated as the reaction rate from the slope of the linear part of the time-dependence of the absorbance. These are the basic principles that underlie most of the methods for the detection of CPF.

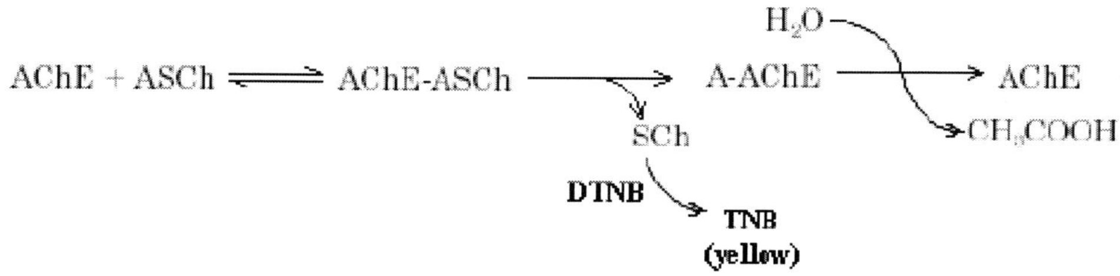

Scheme 12. Ellman's procedure: Acetylthiocholine (ASCh) is hydrolysed by AChE, producing acetic acid (CH_3COOH) and thiocholine (SCh), which reacts with 5,5″-dithio-bis-2-nitrobenzoic acid (DTNB), producing intensively yellow coloured compound 5-thio-2-nitrobenzoate (TNB). The intensity of yellow colour is the measure of the enzyme activity, and proportional to the concentration of the target analyte.

As already mentioned, the toxic effects of CPF and other OPs are primarily associated with the irreversible inhibition of the AChE [Van Dyk, 2011]. In the case of thio-OPs, toxic potential is even more increased upon chemical transformations after metabolic oxidation during which sulfur atom from the thionate group is replaced by oxygen atom, transforming OPs to their oxo-forms (oxons) [Zhang, 1999]. The reason for increased toxicity of mentioned oxo-forms is sought in higher inhibitory power towards AChE, compared to parent compounds.

3.2.1. Application of Chlorpyrifos Oxidation in Methods for its Detection Based on Acetylcholinesterase Inhibition

As discussed earlier, well known conventional techniques such as gas chromatography [Kouloumbos, 2003] and liquid chromatography [Perez-Ruiz, 2005; Abu-Qare, 2001; Kaur, 1997] are widely used for the analysis of CPF and other OPs, but time-consuming and complicated. Therefore, various cholinesterase based bioanalytical methods for OPs detection were developed [Arduini, 2006; Mulchandani, 1998; Trettnak, 1993; Pogačnik, 1999; Pogačnik, 2003]. The principle of these methods is based on the fact that the AChE activity is inhibited by OPs and that the degree of AChE inhibition is dependent on the concentration of OPs. Since the oxidized forms of OPs show stronger inhibitory effects towards AChE than unoxidized, thio forms [Krstić, 2008], efficient oxidation of samples is a crucial pre-step for lowering detection limit of the methods [Lee, 2002; Hernandez, 1998; Lazarević-Pašti, 2012].

Numerous compounds were used for the oxidation of CPF, and other OPs as well. Most of them were inorganic: bromine [Kim, 2000; Kumaran, 1992; Kumaran, 1995], ozone [Skowronska, 1983], dinitrogen tetraoxide and nitric acid [Michalski, 1970], peroxide [Stec, 1976]. Frequently, after oxidation of OPs by inorganic compounds, formation of products of hydrolysis was observed [Zhang, 1999], being undesirable for their efficient detection. Oxidation can be carried out by some enzymes, too. Chloroperoxidase, lignin peroxidase, horseradish peroxidase and cytochrome c were assayed for biocatalytic oxidation of different OPs, while the chloroperoxidase gave the best results and was able to transform 7 out of 10 tested pesticides [Hernandez, 1998].

Besides inorganic and enzymatic oxidation methods, the transformation of OPs from thio to oxo forms can be achieved by electrochemical methods, which gave very good and reliable results. Evtugyn et al. reported electrochemical "activation" of the OPs-containing samples by electrolysis in the presence of 0.1mol/l NaCl [Evtugyn, 1997]. After the electrochemical treatment, detection limit of OPs was lowered by a factor of 2 using potentiometric detection.

3.2.1.1. Application of Human Myeloperoxidase for Oxidation of Chlorpyrifos

The oxidation of CPF by H_2O_2 catalyzed by enzyme human MPO was investigated in details [Lazarević-Pašti, 2011]. The reaction medium was exposed to the enzyme during 10 min. The products obtained due to the reaction between CPF and MPO were analyzed by ultra performance liquid chromatography (UPLC) measurements. The results indicated only one major reaction product which was identified as the oxon derivative of the parent compound – CPO. The reaction scheme for CPF oxidation is presented in Scheme 13. This experiment confirmed that MPO was able to replace the sulfur atom by an oxygen atom in the molecule of CPF, transforming the phosphorothionate group to the more toxic oxon derivative.

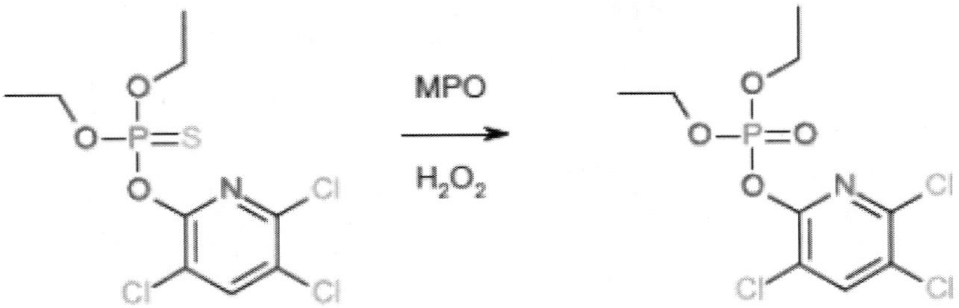

Scheme 13. Reaction of CPF with MPO in the presence of H_2O_2.

UPLC chromatograms of CPF and the products of his MPO induced oxidation were recorded during 60 min after the oxidation was stopped by catalase. The chromatograms did not change during this period, indicating no concentration change of CPO formed (Figure 4) [Lazarević-Pašti, 2011]. These results also confirmed that there were no other products of oxidation, i.e. the further cleavage of oxo-form was not observed.

The AChE inhibition induced by CPF before and after oxidation together with respective oxo-form in the concentration range from 1×10^{-3} to 1×10^{-9} mol/l was evaluated [Lazarević-Pašti, 2011]. Results obtained are presented in Figure 5. As can be seen from the inhibition studies, CPO exerted for about two to three orders of magnitude stronger inhibitory power, compared to the parent compound under the same experimental conditions (Figure 5, curves 1 and 2). Further, CPF in the same concentration range was oxidized by MPO under the optimal conditions (10 min incubation with 100 nmol/l MPO in the presence of 50 μmol/l H_2O_2, at room temperature and in 50 mmol/l phosphate buffer pH 6). The inhibition of AChE activity obtained using solution of oxidized CPF is presented in Figure 5 (curve 3), for comparison. It is obvious that after the incubation of CPF with MPO, the inhibition curve was shifted to the lower concentration range, compared to the inhibition curve of parent compound due to the formation of CPO. Oxidation is an important step of the analytical methods for determination of OPs.

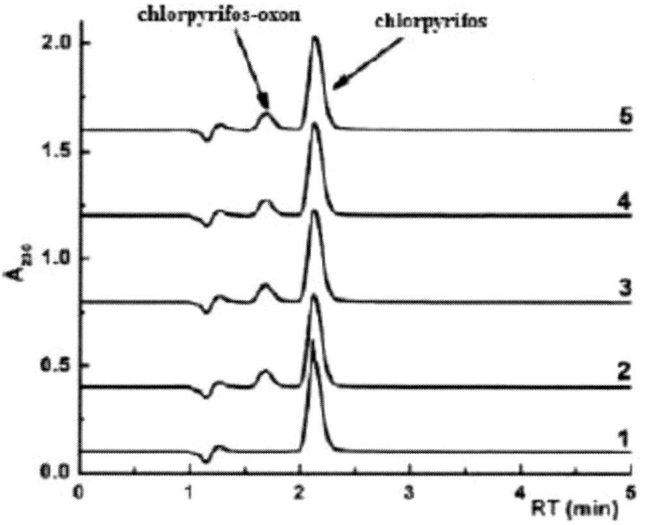

Figure 4. UPLC chromatograms of 1×10^{-6} mol/l chlorpyrifos before (1) and after (2–5) 10 min of incubation with 100 nmol/l MPO. Oxidized samples were followed within 60 min. The time interval between two consecutive chromatograms was 20 min [Lazarević-Pašti, 2011].

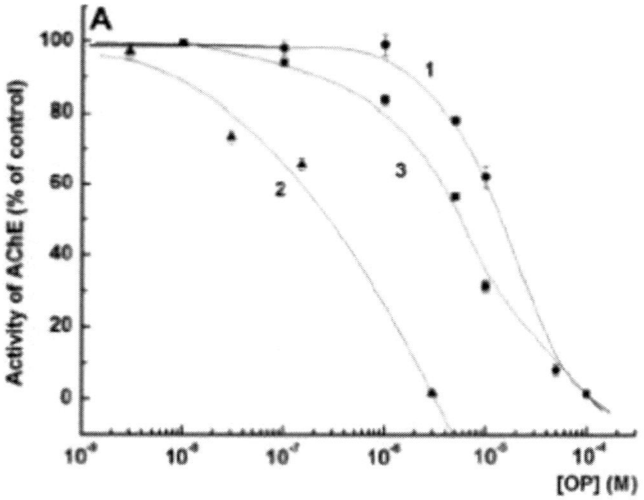

Figure 5. The concentration dependent inhibition of AChE activity induced by CPF before (1) and after (3) exposure to 100 nmol/l MPO during 10 min and CPO (2) [Lazarević-Pašti, 2011].

Method for the oxidation of CPF using MPO is considered to be rapid, robust, environmental-friendly and simply to perform. From the data presented in Figure 5, it can be seen that the oxidation of about 0.5 μmol/l for CPF induces at least 10% of AChE inhibition under the experimental conditions described above. This concentration presents the detection limit for detection of CPF by single exposure to AChE using the presented method of oxidation. The obtained results are also in consistency with previously reported ones for the cases of CPF [Bavcon Kralj, 2006; Ivanov, 2003] and azinphosmethyl [Bavcon Kralj, 2006]. These results are also comparable with those obtained in case of CPF oxidation by chloroperoxidase, where limit of detection was 10 μg/l [Roepcke, 2010]. On the other hand, limit of detection for HPLC analysis of this pesticide is around 1 nmol/l [Guardino, 1998; Cabanillas, 1991; Betowski, 1988] and for GC/MS analysis is around 0.05 nmol/l [Guardino, 1998; Liapis, 2003]. Although sensitivity of conventional techniques is higher, these methods are complicated, expensive, time consuming and cannot be used in the field. New analytical approach which comes with the oxidation of CPF by MPO is simple to perform, robust, cheap and mobile, which gives it great advantage over conventional techniques. Chemicals used in this analytical method are harmless, which makes oxidation procedure environmental-friendly, simpler to perform and gives it advantage over the oxidation with other oxidants, especially when it comes to field testing.

3.2.1.2. Indirect Electrochemical Oxidation of Chlorpyrifos via Electrogenerated Halogens

It was already mentioned that CPF is more toxic when it is converted into his oxo-form, CPO. It is due to the fact that oxo-forms of organophosphate pesticides are stronger inhibitors of AChE. At first, it was shown for the case of CPF [Duirk, 2006] that the presence of bromide, naturally occurring aqueous constituent, can boost the rate of oxidation by formation of more reactive halogen species.

CPF has been indirectly electrochemically oxidized in aqueous media using anodically evolved Cl_2, Br_2 or I_2 as a pre-step for their detection *via* AChE-based test [Lazarević-Pašti, 2012]. Electrochemical oxidation of CPF was performed in an indirect manner *via* anodically electrogenerated halogens in 0.1 mol/l solution containing X^- anions (X = Cl, Br or I) and a given concentration of pesticide in a total volume of 20 ml. As a source of halide ions and supporting electrolytes either NaCl, KBr or KI were used. Halogen species were generated under galvanostatic conditions at a constant current density of j = 2 mA cm_2, with pH of the solution adjusted to 6.5 using 50 mmol/l phosphate buffer. Duration of anodic oxidation was 15 min. The presence of single oxidation product, corresponding oxo-form-CPO, was confirmed by UPLC analysis, as well as its stability with respect to hydrolysis. Consecutive recording of UPLC chromatograms during 2 h after anodic treatment showed that there were no changes in the concentration of formed oxon derivatives, i.e. the further cleavage of CPF was not observed. Similar results were also obtained when chloroperoxidase was used for biocatalytic oxidation of some OPs [Hernandez, 1998; Boškin, 2009]. Formation of OPs hydrolysis products was mostly observed due to cytochromes P450 bioactivation [Kulkarni, 1980; Jauregui, 2003], UV treatment of OPs [Shemer, 2006; Bavcon Kralj, 2007], and oxidation of OPs by some inorganic oxidants [Zhang, 1999; Skowronska, 1983; Marty, 1995]. These results led to a conclusion that indirect electrochemical oxidation using evolved halogens presents an elegant way to selectively oxidize CPF to CPO without further degradation of the oxidation product, and can be used as a pre-step for its detection.

Effect of duration of anodic treatment was investigated by indirect oxidation of CPF [Lazarević-Pašti, 2012]. Oxidation was followed for up to 15 min. From the obtained data, it was obvious that AChE activity decreased with the increase of the time of anodic treatment with all tested oxidants, which implies that the concentration of CPO increased. Also, for the same time of indirect oxidation of CPF with electrochemically generated halogens, the lowest AChE activity was observed when Br_2 is used as oxidant, which corresponds to the highest degree of conversion of CPF to corresponding oxo-form. Comparing different halogens (Figure 6), the best results were obtained using Br_2 as the oxidant due to high reactivity of

HOBr, which is formed upon chemical reaction of anodically formed Br_2 with water. Limit of detection of CPF was lowered upon indirect electrochemical oxidation with Br_2 for two orders of magnitude or more, comparing to unoxidized parent thio-form. In fact, the lowest possible detection limit for CPF using proposed analytical procedure was achieved, as being determined by detection limits of corresponding oxo-form.

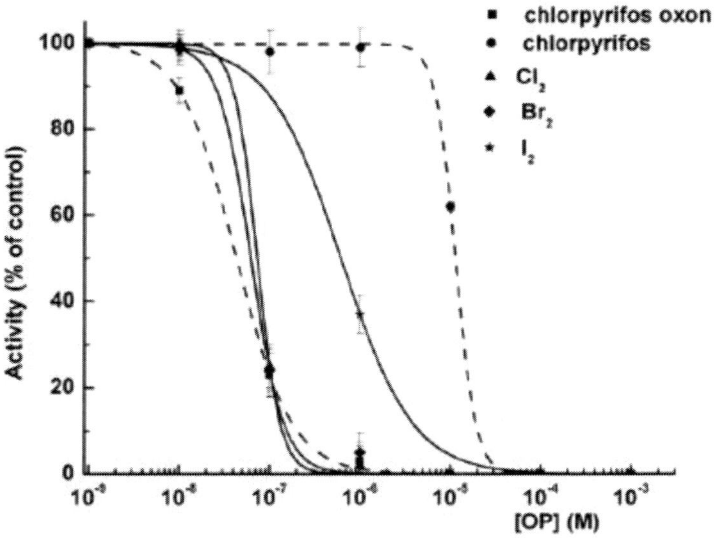

Figure 6. The concentration-dependent inhibition of AChE activity induced by chlorpyrifos (CPF) before and after his oxidation with electrochemically generated Cl_2, Br_2 or I_2 for 15 min. Inhibition curve of chlorpyrifos oxon (CPO) is also enclosed [Lazarević-Pašti, 2012].

3.3. Biosensors for Detection of Chlorpyrifos and Its Metabolites

The term sensor is defined as a device or system that responds to the physical and chemical quantity by producing an output signal that is a measure of this quantity. These are usually the electrodes or optical fiber, which detect or respond to specific interactions with chemicals in the environment. Function of the biosensor is based on the coupling of the immobilized biological sensor element (a biologically active component, e.g. an antibody, an enzyme or cells) with the instrument (physicochemical transducer), which transforms the specific recognition to an electrical, optical or thermal signal, and the results are obtained immediately [Byfield, 1994]. The measurement of the target analyte is accomplished by transduction of selective reactions of biomolecule and analyte into a measurable signal. The most important features of biosensor are the specificity and high sensitivity to the target analyte, which is achieved as a result of an optimized molecular recognition of the bioactive component and its natural substrate with a greater affinity than the recognition with the other components in the system. One of the advantages of biosensors over other analytical techniques is the fact that biosensors, because they are biologically based, can measure toxicity, whereas analytical techniques measure only concentration.

Since many pesticides belong to the group of organophosporous compounds, which are known as inhibitors of cholinesterases activity, the pesticide biosensors are usually based on the inhibition of these enzymes by OPs, particularly AChE. That is also the case with CPF and its metabolites. Principles of signal detection are the same as in plain enzyme tests for OPs detection, based on Ellman's procedure [Ellman, 1961].

It was shown that one enzyme may be the target of a large number of toxic compounds with various characteristics [Tran-Minh, 1993.]. Therefore, biosensors which are based on the inhibition of AChE measure the sum of toxic effects of all pesticides in a sample, but do not measure their individual concentrations [Marty, 1995]. In combination with a variety of transducers, such as potentiometric [Ivnitskii, 1994; Ristori, 1996; Stein, 1993; Rogers, 1995; Dumschat, 1991; Gyurcsányi, 1999], amperometric [Skládal, 1997; La Rosa, 1995; Cremisini, 1995; Besombes, 1995; Günther, 1995; Martorell, 1994], optical [Roda, 1994; Moris, 1995; García de María, 1994], piezoelectric [Abad, 1998] and conductometric [Dzydevich, 1994], they give satisfactory results.

The most commonly used optical transducers are spectrophotometric and fluorimetric. However, spectrophotometry is often not sufficiently sensitive for detection of pesticides in concentrations which are found in real samples (e.g. drinking water, vegetables, fruits, soil). Therefore, it is necessary to apply a more sensitive technique for measuring the absorbance of the product of reaction between a substrate and AChE. On the other hand, fluorimetry is a very sensitive method, but requires substrates that generate fluorescent products. Recently, optical methods are enhanced using laser [Pogačnik, 2001a]. Despite this, the majority of biosensors based on the inhibition of AChE have an electrochemical transducer. Majority of them are amperometric and potentiometric transducers. Potentiometric devices are the simplest to construct and use.

Although sensitive and useful as disposable sensors for monitoring pollution in the environment, these biosensors have some limitations. Firstly, they have long and tedious protocols that require long incubation with inhibitors prior to analysis in order to obtain good sensitivity. Also, incubation with pyridine-2-aldoximes (2-PAM) is required in order to regenerate AChE activity after inhibition with the OPs [La Rosa, 1995; Dzydevich, 1994; Marty, 1992]. Additionally, AChE may be inhibited with various neurotoxins, which includes a large number of OPs, carbamates and many other compounds. This is the reason why these biosensors are not selective and cannot be used for quantification of individual pesticides without coupling with some other method.

In order to improve the sensitivity of assays toward AChE inhibitors, an effective method was developed for the conversion of CPF and some other OPs (diazinon, malathion, azinphos-methyl and phorate) into more toxic inhibitors. This was accomplished by converting them from the thio form into their oxo form using MPO. Then, the oxo forms, which are the only products of conversion, were determined by AChE bioassays, using either the free enzyme, or a flow injection analysis manifold with immobilized AChE and spectrophotometric detection, which presents the biosensor construction. The method is considered to represent an excellent means for improving the sensitivity of assays for determination of OPs, since usage of immobilized AChE lowered the limit of detection for about three orders of magnitude [Lazarević-Pašti, 2010].

A methodology for the determination of CPF in water-supply samples and in milk from dairy cattle was developed [Catalina Rodríguez, 2013]. An amperometric biosensor was used to inhibit AChE immobilized by the cross-linking method (crosslink between the enzyme and the sensor). The concentration of CPF was determined in the range of 1×10^{-6} mol/l to 5×10^{-2} mol/l with a detection limit of 5×10^{-6} mol/l. Spiked water samples showed high recoveries

(91.32 and 93.98% for low and high CPF levels, respectively), while milk samples exhibited a matrix effect with recoveries of 82.81 and 79.77% for high and low CPF levels, respectively. The average concentration of CPF in the water supply samples (5.11×10^{-6} mol/l), determined using the biosensor, was compared using gas chromatography and gave an average value of 3.04×10^{-6} mol/l. The results allow it to be concluded that although chromatographic methods are still more exact, biosensors are promising tools for the determination of analytes in the field, as they have a low cost, a reduced analysis time and good reproducibility in the data.

Mentioned biosensors have real application in various situations. The efficiency of CPF degradation was controlled using a highly sensitive biosensor allowing the detection of OP concentration as low as 0.004 μg/l. Under optimal conditions, the method was shown to be adapted to the decontamination of real samples of pesticides with concentrations up to 20 μg/l [Istamboulie, 2010].

A novel, low potential and highly sensitive AChE biosensor was developed based on 1-butyl-3-methylimidazolium tetrafluoroborate/multiwalled carbon nanotube composite gel thiocholine sensor. Composite gel promoted electron transfer reaction at a lower potential (+50 mV) and catalyzed electrochemical oxidation of thiocholine with high sensitivity. AChE was immobilized in sol-gel matrix that provides a good support for enzyme without any inhibition effect from the ionic liquid. The amount of immobilized enzyme and incubation time with CPF were optimized. CPF could be determined in the range of 10^{-8}–10^{-6} mol/l with a detection limit of 4 nmol/l. Fast and efficient enzyme reactivation was obtained at low obidoxime concentration (0.1 mmol/l). Moreover, the biosensor exhibited a good stability and reproducibility and could be use for multiple determinations of pesticide with no loss of the enzyme activity [Rotariu, 2011].

AChE biosensor based on Ag NPs, carboxylic graphene (CGR) and Nafion (NF) hybrid modified glass carbon electrode (GCE) has been successfully developed [Liu, 2014]. Ag NPs-CGR-NF possessed predominant conductivity, catalysis and biocompatibility and provided a hydrophilic surface for AChE adhesion. Chitosan was used to immobilize AChE on the surface of Ag NPs-CGR-NF/GCE to keep the AChE activities. The AChE biosensor showed favorable affinity to acetylthiocholine chloride and could catalyze the hydrolysis of acetylthiocholine chloride with an apparent Michaelis-Menten constant value of 133 μmol/l, which was then oxidized to produce a detectable and fast response. Under optimum conditions, the biosensor detected CPF at concentrations ranging from 1×10^{-13} to 1×10^{-8} mol/l. The detection limits for CPF was 5.3×10^{-14} mol/l. The developed biosensor exhibited good sensitivity, stability, reproducibility and low cost, thus providing a promising tool for analysis of enzyme inhibitors.

Another electrochemical biosensor for the determination of CPF has been described [Viswanathan, 2009]. The self-assembled monolayers (SAMs) of single walled carbon nanotubes (SWCNT) wrapped by thiol terminated single strand oligonucleotide (ssDNA) on gold was utilized to prepare nano size polyaniline matrix for AChE immobilization. The key step of this biosensor was AChE-acetylcholine enzymatic reaction which causes the small changes of local pH in the vicinity of an electrode surface. The dynamic range for the determination of CPF was found to be between 1×10^{-11} and 1×10^{-6} mol/l with good reproducibility and stability. The detection limit of the biosensor was found to be 1×10^{-12} mol/l. The biosensor has been applied for the determination of CPF in spiked river water samples.

OPH is an enzyme that hydrolyzes broad spectrum of organophosphate esters, pesticides such as CPF, parathion, coumaphos and acephate, as well as chemical agents soman, sarin, and tabun [Dave, 1993; Dumas, 1989; Dumas, 1990; Lai, 1994]. The hydrolysis of each molecule of CPF gives two protons, and the correlation of the number of protons and the concentration of the molecules based is on potentiometric measurement. In this way a simple, more direct and faster measurements of only pesticides that are structurally organophosphates are possible. In addition, the biosensors which are based on OPH could potentially be used for quantification of the individual OPs when used as the detector in conjunction with the HPLC system [Mulchandani, 1998; Simonian, 2005].

3.3.1. Flow-injection Analysis System Based on Immobilized Acetylcholinesterase for Detection and Determination of Organophosphates

The inhibitory power of CPF, its oxo analog-CPO and hydrolysis product, TCP was tested on flow-injection analysis (FIA) system containing immobilized AChE. FIA-AChE system consisted of a HPLC pump (Dionex AMP-1), an injection valve (Waters U6K) and a bioanalytical column (40 × 4 mm) filled with 0.55 mg of beads with immobilized AChE. The carrier buffer (flow rate: 0.2 ml/min) was pumped through a flow-through cell (250 μl). The initial (control) enzyme activity (a_0) of the immobilized enzyme in the bioanalytical column was determined by injecting a mixture of the substrate (acetylthiocholine iodide) and a chromogenic reagent (5,5'-dithio-bis-2-nitrobenzoic acid). The absorbance at 412 nm was measured spectrophotometrically. After measurement of initial activity a_0, the sample containing the selected organophospate was injected. The remaining enzyme activity (a_r) was determined by another injection of a mixture of acetylthiocholine iodide and 5,5'-dithio-bis-2-nitrobenzoic acid (Figure 7). After each injection of solution containing OP, the activity of the enzyme was reactivated to its initial value by five consecutive injections of a reactivator (4 mmol/l 2-PAM) [Krstić, 2008a; Krstić, 2008b]. All measurements were made in triplicate.

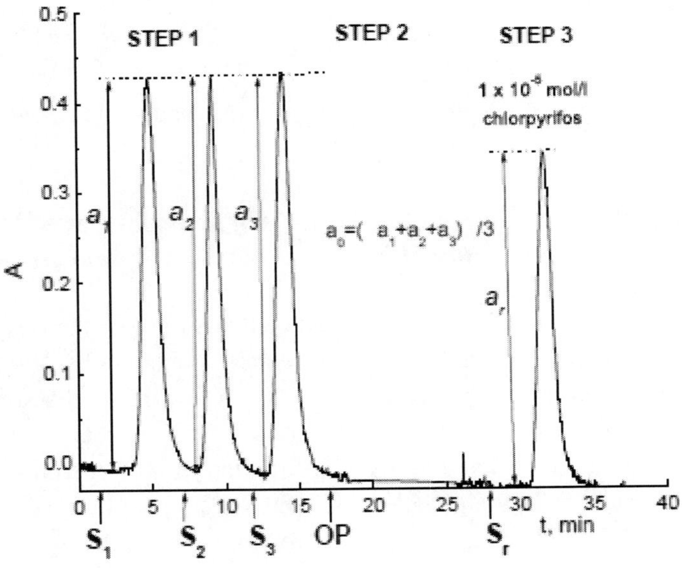

Figure 7. FIA signals obtained in the absence (control-a_1, a_2, a_3) and presence of 1.5×10^{-5} mol/l chlorpyrifos (a_r); S-substrate, OP-organophosphate [Čolović, 2014].

3.3.1.1. Acetylcholinesterase Immobilization

0.55 mg AChE was immobilized on 0.15 g activated controlled pore glass [Pogačnik, 2001b]. The glass particles were boiled in 5% nitric acid. The activation step with 3-aminopropyl triethoxysilan was repeated three times; afterwards, the glass beads were washed with deionised water and dried. Alkylamino glass was cross-linked with 2.5% glutaraldehyde solution in a stream of nitrogen and washed with deionised water. The enzyme was dissolved in 3 ml cold phosphate buffer and added to the pre-treated glass in a stream of nitrogen. The mixture was then vacuum-filtered. The filtrate was collected in a small graduated cylinder in order to measure its exact volume. The volumes of the filtrates were 95–98% of those of the original AChE solution. The immobilized enzyme was washed with phosphate buffer and then with cold water to ensure the removal of any unlinked enzyme. The immobilized beads were stored at 48 °C in phosphate buffer (pH 6.0) until use. The protein content before and after the immobilization was measured by the Lowry method using BSA as a standard at 750 nm [Lowry, 1961]. The free AChE activities in the initial AChE solution and in the filtrate were measured by Ellman's procedure [Ellman, 1961]. The amount of the protein adsorbed on the glass (P_g, mg g^{-1}) and the immobilization efficiency (%) were determined from the following equations:

$$P_g = (c_oV_o - c_fV_f)/w = 1 \text{ mg g}^{-1} \qquad (3)$$

$$\text{Immobilization efficiency (\%)} = [(E_oV_o - E_fV_f)/E_oV_o] \times 100 = 99\% \qquad (4)$$

where c_o is the protein concentration in the initial AChE solution (mg/ml), E_o is its AChE activity (IU/ml), V_o is its volume (ml); c_f is the protein concentration in the filtrate (mg/ml); E_f is its AChE activity (IU/ml); V_f is its volume (ml); and w is the weight of the glass used (g).

The degree of adsorption was calculated as:

$$\alpha = P_g / P_{g,m} = 1 \qquad (5)$$

where $P_{g,m}$ (mg/g) is the maximal amount of protein adsorbed on the controlled pore glass [Krstić, 2008b].

3.3.1.2. Inhibitory Potency and Detection of Chlorpyrifos and Its Degradation Products Determined Using Flow-injection Analysis System with Immobilized Acetylcholinesterase

The response of bioanalytical FIA system that contains immobilized AChE from bovine erythrocytes in the presence of CPF and its related compounds-CPO and TCP was investigated in the concentration range from 1×10^{-9} to 1×10^{-2} mol/l. FIA signals obtained in the presence of 1.5×10^{-5} mol/l CPF are presented in Figure 8a. The dependence of immobilized AChE activity in FIA system on concentration of CPF and CPO fitted sigmoidal function (Figure 8a), while TCP altered AChE activity up to 10%. Hill analysis of the inhibition curves for CPF and CPO is given in Figure 8b) [Čolović, 2014].

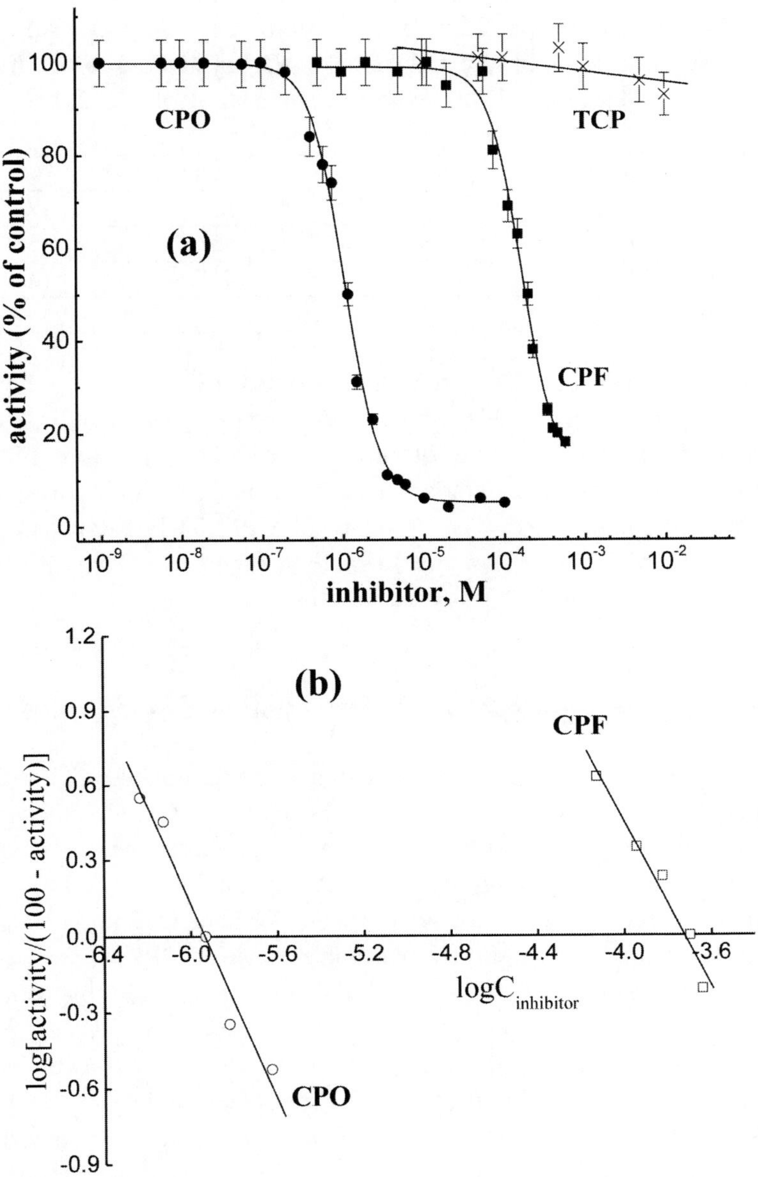

Figure 8. a) Concentration-dependent inhibition of immobilized AChE in FIA system by CPF, CPO and TCP; b) Hill analysis of the inhibition sigmoidal curves presented in figure a) [Čolović, 2014].

Concentrations decreasing immobilized AChE activity in FIA system, e.i. the obtained signal as the response of FIA system to injection of an OP, 50% (IC_{50}) related to the control enzyme activity (obtained before inhibitor injection) are presented in Table 9. IC_{50} values are calculated from sigmoidal function obtained by fitting experimental data and Hill analysis. Table 9 also contains LOD (limit of detection) values of FIA-AChE system for CPF and its related compounds. LOD value is determined as concentration of investigated compound decreasing immobilized AChE activity 10% related to control enzyme activity [Čolović, 2014].

Table 9. Inhibition parameters (IC_{50} and Hill coefficient) and LOD values of CPF and its related compounds for immobilized AChE in FIA system, obtained by Hill analysis and sigmoidal fitting experimental data [Čolović, 2014]

	Sigmoidal fitting		Hill analysis	
Compound	IC_{50}, mol/l	IC_{50}, mol/l	n_H	LOD, mol/l
CPF	$(1.8 \pm 0.2) \times 10^{-4}$	1.9×10^{-4}	1.6 ± 0.2	$(5.9 \pm 0.3) \times 10^{-5}$
CPO	$(1.1 \pm 0.1) \times 10^{-6}$	1.2×10^{-6}	1.9 ± 0.2	$(3.8 \pm 0.2) \times 10^{-7}$
TCP	Inhibition < 10%	Inhibition < 10%	-	-

IC_{50} i LOD values (Table 9) indicate that FIA-AChE bioassay is the most sensitive toward CPO (IC_{50} = 1,2×10^{-6} mol/l), while its thio analog, CPF (IC_{50} = 1,9×10^{-4} mol/l) is about a hundred times weaker inhibitor of immobilized AChE. Therefore, FIA-AChE system can detect 100 times lower concentrations of oxon compared to its parent thio compound (Table 9). The investigated hydrolysis product of CPF, TCP did not cause statistically significant decrease in FIA signal (< 10%) (Table 9, Figure 8a). Thus, it is not possible to detect its existence using FIA-AChE bioassay [Čolović, 2014].

3.4. Application of Immunoassays for Detection of Chlorpyrifos

Nowadays, it is very widespread to use the immune-based methods for the detection of different pesticides, among them CPF and its metabolites. These methods are numerous, efficient and very convenient. Nanoparticles are often used in this methods, which makes this field interesting, actual and tempting for further development.

Continuous monitoring of the organophosphate pesticide CPF in real water samples has been performed using a portable surface plasmon resonance (SPR) immunosensor [Mauriz, 2006]. The immunoassay was based on a binding inhibition test by using the covalent immobilization of an analyte derivative onto the gold-coated sensing surface. Sensor reusability was assured through the formation of alkanethiol self-assembled monolayers (SAMs). In this format, the biosensor surface remained stable over 190 measurements including regeneration cycles. The assay provided a highly sensitive detection of CPF at nanogram per litre levels (55 ng/l) in laboratory buffer. No significant matrix effects were found for ground, river and drinking water samples, where the LODs ranged from 45 to 64 ng/l and the recovery values were between 80 and 120%. The validation of CPF immunoassay was performed using conventional chromatographic methods (GC/MS). Rapid analysis of samples, without previous extraction or clean-up, was accomplished in only 20 min, including the regeneration cycle. The high sensitivity, low-time of response and portability of the β-SPR biosensor platform, would allow its application as a field analytical method for environmental waters.

Also, an immunochromatographic assay (ICA) based on competitive antigen-coated format using colloidal gold as the label was developed for the detection of CPF [Kim, 2011]. The ICA test strip consisted of a membrane with a detection zone, a sample pad and an absorbent pad. The membrane was separately coated with CPF Hapten-OVA conjugate (test line) and anti-mouse IgG (control line). Based on the fact that the competition is between the migrating analyte in the sample and the analyte hapten immobilized on the test strip for the binding sites of the antibody-colloidal gold (Ab-CG) conjugate migrating on the test strip, this study suggests that the relative migration speed between the two migrating substances is a critically important factor for the sensitive detection by competitive ICA. This criterion was utilized for the confirmation of appropriateness of a nitrocellulose membrane for CPF ICA. The LOD of the ICA for CPF standard and CPF spiked into agricultural samples were 10 and 50 ng/ml, respectively. The assay time for the ICA test was less than 10 min, suitable for rapid on-site testing of CPF.

ACKNOWLEDGMENTS

The authors would like to thank to the Ministry of Education, Science and Technological Development of the Republic of Serbia for their financial support (Project no. 172023).

REFERENCES

Abad, JM; Pariente, F; Hernandez, L; Abruna, HD; Lorenzo, E. Determination of Organophosphorus and Carbamate Pesticides Using a Piezoelectric Biosensor. *Anal Chem*, 1998, 70, 2848–2855.

Abdollahi, M; Rainba, A; Shadnia, S; Nikfar, S; Rezaie, A. Pesticide and oxidative stress: a review. *Med Sci Monitor*, 2004, 10, RA141–RA147.

Abu-Qare, A; Abou-Donia, M. Simultaneous determination of malathion, permethrin, DEET (N, N-diethyl-m-toluamide), and their metabolites in rat plasma and urine using high performance liquid chromatography. *J Pharm Biomed Anal*, 2001, 26, 291−299.

Agner, K. Verdoperoxidase. A ferment isolated from leukocytes. *Acta Physiol Scand*, 1941, 2, 1–62.

Akhgari, M; Abdollahi, M; Kebryaeezadeh, A; Hosseini, R; Sabzevari, O. Biochemical evidence for free radical induced lipid peroxidation as a mechanism for subchronic toxicity of malathion in blood and liver of rats. *Hum Exp Toxicol*, 2003, 22, 205–208.

Aldridge, WN. Some properties of specific cholinesterase with particular reference to the mechanism of inhibition by diethyl *p*-nitrophenyl thiophosphate and analogs (E605). *Biochem J*, 1950, 46, 451–460.

Amarathunga, AAD; Kazama, F. Photodegradation of chlorpyrifos with humic acid-bound suspended matter. *J Hazard Mater*, 2014, 280, 671–677.

Anwar, S; Liaquat, F; Khan, QM; Khalid, ZM; Iqbal, S. Biodegradation of chlorpyrifos and its hydrolysis product 3,5,6-trichloro-2-pyridinol by Bacillus pumilus strain C2A1. *J Hazard Mater*, 2009, 168, 400–405.

Arduini, F; Ricci, F; Tuta, CS; Moscone, D; Amine, A; Palleschi, G. Detection of carbamic and organophosphorous pesticides in water samples using a cholinesterase biosensor based on Prussian Blue-modified screen-printed electrode. *Anal Chim Acta*, 2006, 580, 155−162.

Ariel, N; Ordentlich, A; Barak, D; Bino, T; Velan, B; Shafferman, A. The 'aromatic patch' of three proximal residues in the human acetylcholinesterase active centre allows for versatile interaction modes with inhibitors. *Biochem J,* 1998, 335, 95–102.

Arnhold, J; Flemmig, J. Human myeloperoxidase in innate and acquired immunity. *Arch Biochem Biophys*, 2010, 500, 92–106.

Attia, SM. Chromosomal composition of micronuclei in mouse bone marrow treated with rifampicin and nicotine, analyzed by multicolor fluorescence in situ hybridization with pancentromeric DNA probe. *Toxicology*, 2007, 235, 112–118.

Baldus, S; Heeschen, C; Meinertz, T; Zeiher, AM; Eiserich, JP; Munzel, T; Simoons, ML; Hamm, CW. Myeloperoxidase serum levels predict risk in patients with acute coronary syndromes. *Circulation*, 2003, 108, 1440–1445.

Band, PR; Abanto, Z; Bert, J; Lang, B; Fang, R; Gallagher, RP; Le, ND. Prostate cancer risk and exposure to pesticides in British Columbia Farmers. *Prostate*, 2011, 71, 168–183.

Banerjee, BD; Seth, V; Bhattacharya, A; Pasha, ST; Chakraborty, AK. Biochemical effects of some pesticides on lipid peroxidation and free-radical scavengers. *Toxicol Lett*, 1999, 107, 33–47.

Barron, MG; Plakas, SM; Wilga, PC. Chlorpyrifos pharmacokinetics and metabolism following intravascular and dietary administration in channel cat fish. *Toxicol Appl Pharmacal*, 1991, 108, 474–482.

Barros, FCF; Barros, AL; Silva, MAA; do Nascimento, RF. Use of Microwave-Assisted Oxidation for Removal of the Pesticide Chlorpyrifos from Aqueous Media. *IJCEE-IJENS*, 2013, 13, 16–27.

Bavcon Kralj, M; Trebše, P; Franko, M. Oxidation as a pre-step in determination of organophosphorus compounds by the AChE-TLS bioassay. *Acta Chim Slov*, 2006, 53, 43–51.

Bavcon Kralj, M; Franko, M; Trebše, P. Photodegradation of organophosphorus insecticides – investigations of products and their toxicity using gas chromatography mass-spectrometry and AChE-thermal lens spectrometric bioassay. *Chemosphere*, 2007a, 67, 99–107.

Bavcon Kralj, M; Černigoj, U; Franko, M; Trebše, P. Comparison of photocatalysis and photolysis of malathion, isomalathion, malaoxon, and commercial malathion-products and toxicity studies. *Water Res*, 2007b, 41, 4504–4514.

Besombes, J-L; Cosnier, S; Labbé, P; Reverdy, G. A biosensor as warning device for the detection of cyanide, chlorophenols, atrazine and carbamate pesticides. *Anal Chim Acta*, 1995, 311, 255–263.

Betowski, LD; Jones, TL. The analysis of organophosphorus pesticide samples by high-performance liquid chromatography/mass spectrometry and high-performance liquid chromatography/mass spectrometry/mass spectrometry. *Environ Sci Technol*, 1988, 22, 1430–1434.

Bhagobaty, RK; Joshi, SR; Malick, A. Microbial degradation of organophosphorous pesticide: Chlorpyrifos (Mini-Review). *Inter J Microbiol*, 2006, 4, 1–12.

Bolognesi, C. Genotoxicity of pesticides: a review of human biomonitoring studies. *Mutat Res-Rev Mutat*, 2003, 543, 251–272.

Bootharaju, MS; Pradeep, T. Understanding the Degradation Pathway of the Pesticide, Chlorpyrifos by Noble Metal Nanoparticles. *Langmuir*, 2012, 28, 2671–2679.

Boškin, A; Tran, C; Franko, M. Oxidation of organophosphorus pesticides with chloroperoxidase enzyme in the presence of an ionic liquid as co-solvent. *Environ Chem Lett*, 2009, 7, 267–270.

Boublik, Y; Saint-Aguet, P; Lougarre, A; Arnaud, M; Villatte, F; Estrada-Mondaca, S; Fournier, D. Acetylcholinesterase engineering for detection of insecticide residues. *Protein Eng Des Sel,* 2002, 15, 43–50.

Brzak, KA; Harms, DW; Bartels, MJ; Nolan, RJ. Determination of chlorpyrifos, chlorpyrifos oxon, and 3,5,6-trichloro-2-pyridinol in rat and human blood. *J Anal Toxicol*, 1998, 22, 203–210.

Bull, S; Fletcher, K; Boobis, AR; Battershill, JM. Evidence for genotoxicity of pesticides in pesticide applicators: a review. *Mutagenesis*, 2006, 21, 93–103.

Burrows, HD; Canle, M; Santaballa, JA; Steenken, S. Reaction pathways and mechanisms of photodegradation of pesticides. *J Photoch Photobio B*, 2002, 67, 71–108.

Buyukokuroglu, M; Cemek, M; Yurumez, Y; Yavuz, Y; Aslan, A. Antioxidative role of melatonin in organophosphate toxicity in rats. *Cell Biol Toxicol*, 2008, 24, 151–158.

Byfield, MP; Abuknesha, RA. Biochemical aspects of biosensors. *Biosens Bioelectron,* 1994, 9, 373–399.

Cabanillas, CG; Bushway, R. Analysis of Phosmet and Azinphos-Methyl in Apples by High-Performance Liquid Chromatography. *J Liq Chromatogr*, 1991, 14, 3603–3613.

Carfagna, MA; Ponsler, GD; Muhoberac, BB. Inhibition of ATPase activity in rat synaptic plasma membranes by simultaneous exposure to metals. *Chem-Biol Interact*, 1996, 100, 53–65.

Catalina Rodríguez, D; Carvajal, S; Peñuela, G. Effect of chlorpyrifos on the inhibition of the enzyme acetylcholinesterase by cross-linking in water-supply samples and milk from dairy cattle. *Talanta*, 2013, 111, 1–7.

CDC. Third National Report on Human Exposure to Environmental Chemicals; U.S. Department of Health and Human Services, Centers for Disease Control and Prevention, Atlanta, 2005, pp. 349–377.

CEC, Designation of Hazardous Substances [40 CFR 116], Protection of Environment [40 CFR]. Retrieved 2014-07-23.

Chambers, JE; Carr, RL; Boone, S; Chambers, HW. The metabolism of organophosphorus insecticides. In: Handbook of Pesticide Toxicology, Academic Press, USA, 2001, pp. 919–927.

Chen, S; Liu, C; Peng, C; Liu, H; Hu, M; Zhong, G. Biodegradation of Chlorpyrifos and Its Hydrolysis Product 3,5,6-Trichloro-2-Pyridinol by a New Fungal Strain Cladosporium cladosporioides Hu-01. *PLoS ONE*, 2012, 7, e47205.

Choi, DK; Pennathur, S; Perier, C; Tieu, K; Teismann, P; Wu, DC; Jackson-Lewis, V; Vila, M; Vonsattel, JP; Heinecke, JW; Przedborski, S. Ablation of the inflammatory enzyme myeloperoxidase mitigates features of Parkinson's disease in mice. *J Neurosci*, 2005, 25, 6594–6600.

Čolović, M; Krstić, D; Petrović, S; Leskovac, A; Joksić, G; Savić, J; Franko, M; Trebše, P; Vasić, V. Toxic effects of diazinon and its photodegradation products. *Toxicol Lett*, 2010, 193, 9–18.

Čolović, MB; Krstić, DZ; Uščumlić, S; Vasić, VM. Single and simultaneous exposure of acetylcholinesterase to diazinon, chlorpyrifos and their photodegradation products. *Pestic Biochem Phys,* 2011, 100, 16–22.

Čolović, MB; Krstić, DZ; Vasić, VM; Bondžić, AM; Uśćumlić, GS; Petrović, SD. Organophosphorus insecticides: toxic effects and bioanalytical tests for evaluating toxicity during degradation processes. *Hem Ind*, 2013a, 67, 217–230.

Čolović, MB; Krstić, DZ; Lazarević-Pašti, TD; Bondžić, AM; Vasić, VM. Acetylcholinesterase Inhibitors: Pharmacology and Toxicology. *Curr Neuropharmacol*, 2013b, 11, 315–335.

Čolović, M. Bioanalytical methods for detection and toxicity evaluation of organo-thiophosphorous insecticides and their degradation products. Faculty of Technology and Metallurgy, University of Belgrade, Belgrade, 2014.

Costa, LG. Current issues in organophosphate toxicology. *Clin Chim Acta*, 2006, 366, 1–13.

Costa, LG. Toxicology of pesticides: Experimental, Clinical and Regulatory Perspective. Nato ASI Series, Springer-Verlag, Berlin, 1990, pp. 77–91.

Cremisini, C; Di Sario, S; Mela, J; Pilloton, R; Palleschi, G. Evaluation of the use of free and immobilised acetylcholinesterase for paraoxon detection with an amperometric choline oxidase based biosensor. *Anal Chim Acta*, 1995, 311, 273–280.

Dandapani, M; Zachariah, A; Kavitha, MR; Jeyaseelan, L; Oommen, A. Oxidative damage in intermediate syndrome of acute organophosphorus poisoning. *Indian J Med Res,* 2003, 117, 253–259.

Das, P; Shaik, A; Jamil, K. Genotoxicity induced by pesticide mixtures: in-vitro studies on human peripheral blood lymphocytes. *Toxicol Ind Health*, 2007, 23, 449–458.

Daugherty, A; Dunn, JL; Rateri, DL; Heinecke, JW. Myeloperoxidase, a catalyst for lipoprotein oxidation, is expressed in human atherosclerotic lesions. *J Clin Investig*, 1994, 94, 437–444.

Dave, KI; Miller, CE; Wild, JR. Characterization of organophosphorus hydrolases and the genetic manipulation of the phosphotriesterase from Pseudomonas diminuta. *Chem Biol Interact*, 1993, 87, 55–68.

Duirk, SE; Collette, TW. Organophosphate Pesticide Degradation Under Drinking Water Treatment Conditions. 2005, pp. EPA/600/R-05/103.

Duirk, SE; Cherney, DP; Tarr, IJ; Collette, TW. Determining Active Oxidant Species Reacting with Organophosphate Pesticides in Chlorinated Drinking Water. in: U.S.E P.A. Agency (Ed.), Athens, GA, 2006, vol. EPA/600/R-06/103, pp. EPA/600/R-06/103.

Duirk, SE; Tarr, JC; Collette, TW. Chlorpyrifos Transformation by Aqueous Chlorine in the Presence of Bromide and Natural Organic Matter. *J Agric Food Chem*, 2008, 56, 1328−1335.

Dumas, DP; Caldwell, SR; Wild, JR; Raushel, FM. Purification and properties of the phosphotriesterase from Pseudomonas diminuta. *J Biol Chem*, 1989, 264 19659–19665.

Dumas, DP; Durst, HD; Landis, WG; Raushel, FM; Wild, JR. Inactivation of organophosphorus nerve agents by the phosphotriesterase from Pseudomonas diminuta. *Arch Biochem Biophys*, 1990, 277, 155–159.

Dumschat, C; Müller, H; Stein, K; Schwedt, G. Pesticide-sensitive ISFET based on enzyme inhibition. *Anal Chim Acta*, 1991, 252, 7–9.

Dzydevich, SV; Shulga, AA; Soldatkin, AP; Hendji, AMN; Jaffrezic-Renault, N; Martelet, C. Conductometric biosensors based on cholinesterases for sensitive detection of pesticides. *Electroanal*, 1994, 6, 752–758.

Edwards, SW; Hughes, V; Barlow, J; Bucknall, R. Immunological detection of myeloperoxidase in synovial fluid from patients with rheumatoid arthritis. *Biochem J*, 1988, 250, 81–85.

Ellman, GL; Courtney, KD; Andreas, V; Featherstone, RM. A new and rapid colorimetric determination of acetylcholinesterase activity. *Biochem Pharmacol*, 1961, 7, 88–90.

EPA, Reregistration Eligibility Science Chapter for Chlorpyrifos Fate and Environmental Risk Assessment Chapter. U.S. Environmental Protection Agency, Office of Prevention, Pesticides and Toxic Substances, Office of Pesticide Programs, Environmental Fate and Effects Division, U.S. Government Printing Office, Washington, DC, 1999.

Estevan, C; Fuster, E; del Río, E; Pamies, D; Vilanova, E; Sogorb, MA. Organophosphorus pesticide chlorpyrifos and its metabolites alter the expression of biomarker genes of differentiation in D3 mouse embryonic stem cells in a comparable way to other model neurodevelopmental toxicants. *Chem Res Toxicol,* 2014, 27, 1487-1495.

Evtugyn, GA; Rizaeva, EP; Stoikova, EE; Latipova, VZ; Budnikov, HC. The Application of Cholinesterase Potentiometric Biosensor for Preliminary Screening of the Toxicity of Waste Waters. *Electroanal*, 1997, 9, 1124–1128.

Farag, AT; El-Okazy, AM; El-Aswed, AF. Developmental toxicity of chlorpyrifos in rats. *Reprod Toxicol*, 2003, 17, 203−208.

Faust, SD; Gomaa, HM. Chemical hydrolysis of some organic phosphorous and carbamate pesticides in aquatic environments. *Environ Lett*, 1972, 3, 171−201.

Fortunato, J; Agostinho, F; Reus, G; Petronilho, F; Dal-Pizzol, F; Quevedo, J. Lipid peroxidative damage on malathion exposure in rats. *Neurotox Res*, 2006, 9, 23–28.

Foxenberg, RJ; Ellison, CA; Knaak, JB; Ma, C; Olson, JR. Cytochrome P450-specific human PBPK/PD models for the organophosphorus pesticides: Chlorpyrifos and parathion. 2011, 285, 57–66.

Froede, HC; Wilson, IB. Acetycholinesterase. In: Boyer, PD. (Ed.), The Enzymes, Academic Press, New York, 1971, vol. 5, pp. 87–114.

Fujioka, K; Casida, JE. Glutathione S-transferase conjugation of organophosphorus pesticides yields S-phospho-, S-aryl, and S-alkylglutathione derivatives. *Chem Res Toxicol*, 2007, 20, 1211–1217.

Fulton, MH; Key, PB. Acetylcholinesterase inhibition in estuarine fish and invertebrates as an indicator of organophosphorus insecticide exposure and effects. *Environ Toxicol Chem,* 2001, 20, 37–45.

Furlong, CE. Paraoxonases: an historical perspective. In: Mackness, B; Mackness, M; Aviram, M; Paragh, G. (Eds.), The Paraoxonases: Their Role in Disease Development and Xenobiotic Metabolism, Springer, The Netherlands, 2008, pp. 3–32.

García de María, C; Manzano Muñoz, T; Townshend, A. Reactivation of an immobilized enzyme reactor for the determination of acetylcholinesterase inhibitors. Flow injection determination of paraoxon. *Anal Chim Acta*, 1994, 295, 287–296.

GCF (2008-01-10), Hazardous Substances Data Bank: Chlorpyrifos. U.S. National Library of Medicine TOXNET (Reference Resources).

Goel, A; Dani, V; Dhawan, DK. Protective effects of zinc on lipid peroxidation, antioxidant enzymes and hepatic histoarchitecture in chlorpyrifos-induced toxicity. *Chem-Biol Interact*, 2005, 156, 131–140.

Green, PS; Mendez, AJ; Jacob, JS; Crowley, JR; Growdon, W; Hyman, BT; Heinecke, JW. Neuronal expression of myeloperoxidase is increased in Alzheimer's disease. *J Neurochem*, 2004, 90, 724–733.

Guardino, X; Obiols, J; Rosell, MG; Farran, A; Serra, C. Determination of chlorpyrifos in air, leaves and soil from a greenhouse by gas-chromatography with nitrogen-phosphorus detection, high-performance liquid chromatography and capillary electrophoresis. *J Chromatogr A*, 1998, 823, 91–96.

Günther, A; Bilitewski, U. Characterisation of inhibitors of acetylcholinesterase by an automated amperometric flow-injection system. *Anal Chim Acta*, 1995, 300, 117–125.

Gyurcsányi, RE; Vágföldi, Z; Tóth, K; Nagy, G. Fast Response Potentiometric Acetylcholine Biosensor. *Electroanal*, 1999, 11, 712–718.

Halliwell, B; Gutteridge, JMC. Oxidative stress: Adaptation, damage, repair and death. In: Free Radicals in Biology and Medicine, Oxford University Press, Oxford, 1999, pp. 246–349.

Hayden, KM; Norton, MC; Darcey, D; Ostbye, T; Zandi, PP; Breitner, JCS; Welsh-Bohmer, KA. Occupational exposure to pesticides increases the risk of incident AD: the Cache County study. *Neurology,* 2010, 74, 1524–1530.

Hernandez, J; Robledo, NR; Velasco, L; Quintero, R; Pickard, MA; Vazquez-Duhalt, R. Chloroperoxidase-mediated oxidation of organophosphorus pesticides. *Pestic Biochem Phys*, 1998, 61, 87–94.

Hossain, MS; Fakhruddin, ANM; Alamgir Zaman Chowdhury, M; Khorshed Alam, M. Degradation of chlorpyrifos, an organophosphorus insecticide in aqueous solution with gamma irradiation and natural sunlight. *J Environ Chem Eng*, 2013, 1, 270–274.

Hunter Coombes, R; Meek, EC; Chambers, JE. Human paraoxonase 1 hydrolysis of nanomolar chlorpyrifos-oxon concentrations is unaffected by phenotype or Q192R genotype. *Toxicol Lett*, 2014, 230, 57–61.

Ismail, M; Khan, HM; Sayed, M; Cooper, WJ. Advanced oxidation for the treatment of chlorpyrifos in aqueous solution. *Chemosphere*, 2013, 93, 645–651.

Istamboulie, G; Durbiano, R; Fournier, D; Marty, JL; Noguer, T. Biosensor-controlled degradation of chlorpyrifos and chlorfenvinfos using a phosphotriesterase-based detoxification column. *Chemosphere*, 2010, 78, 1–6.

Ivanov, A; Evtugyn, G; Budnikov, H; Ricci, F; Moscone, D; Palleschi, G. Cholinesterase sensors based on screen-printed electrodes for detection of organophosphorus and carbamic pesticides. *Anal Bioanal Chem*, 2003, 377, 624–631.

Ivnitskii, DM; Rishpon, J. A potentiometric biosensor for pesticides based on the thiocholine hexacyanoferrate (III) reaction. *Biosens Bioelectron,* 1994, 9, 569–576.

Jauregui, J; Valderrama, B; Albores, A; Vazquez-Duhalt, R. Microsomal transformation of organophosphorus pesticides by white rot fungi. *Biodegradation*, 2003, 14, 397–406.

Jorgensen, PL. Isolation of (Na^{+},K^{+})-ATPase. *Methods Enzymol*, 1974, 32, 277–290.

Kamrin, MA. Pesticide Profiles Toxicity, Environmental Impact, and Fate. Lewis Publishers, Boca Raton, FL, 1997, pp. 147–152.

Kato, Y; Nagao, A; Terao, J; Osawa, T. Inhibition of myeloperoxidase-catalyzed tyrosylation by phenolic antioxidants *in vitro*. *Biosci Biotechnol Biochem*, 2003, 67, 1136–1139.

Kaur, I; Mathur, RP; Tandon, SN. Parameters affecting the decay of some organophosphorus pesticides: a study by high-performance liquid chromatography. *Biomed Chromatogr*, 1997, 11, 22–24.

Kehrer, JP; Free radicals as mediators of tissue injury and disease. *Crit Rev Toxicol*, 1993, 23, 21–48.

Kettle, AJ; Gedye, CA; Hampton, MB; Winterbourn, CC. Inhibition of myeloperoxidase by benzoic acid hydrazides. *Biochem J*, 1995, 308, 559–563.

Kim, YA; Lee, S; Park, YC; Lee, YT. A convenient method for oxidation of organophosphorus pesticides in organic solvents. *Environ Res*, 2000, 84, 303−309.

Kim, YA; Lee, EH; Kim, KO; Lee, YT; Hammock, BD; Lee, HS. Competitive immunochromatographic assay for the detection of the organophosphorus pesticide chlorpyrifos. *Anal Chim Acta*, 2011, 693, 106–113.

Kinter, WB; Merkens, LS; Janicki, RH; Guarino, AM. Studies on the mechanism of toxicity of DDT and polychlorinated biphenyls (PCB's): disruption of osmoregulation in marine fish. *Environ Health Persp*, 1972, 1, 169–173.

Klebanoff, SJ. Myeloperoxidase-halide-hydrogen peroxide antibacterial system. *J Bacteriol*, 1968, 95, 2131–2138.

Klebanoff, SJ. Myeloperoxidase: occurrence and biological function. In: Everse, J; Everse, KE; Grisham, MB. (Eds.), Peroxidase in Chemistry and Biology, CRC Press, Boca Raton, 1991, pp. 1–35.

Knežević-Jugović, ZD. Enzimsko inženjerstvo. Tehnološko-metalurški fakultet, Beograd, 2008.

Köksoy, AA. Na^+, K^+-ATPase: a review. *J Ankara Med Sch*, 2002, 24, 73–82.

Kouloumbos, V; Tsipi, D; Hiskia, A; Nikolic, D; van Breemen, R. Identification of photocatalytic degradation products of diazinon in TiO_2 aqueous suspensions using GC/MS/MS and LC/MS with quadrupole time-of-flight mass spectrometry. *J Am Soc Mass Spectrom*, 2003, 14, 803−817.

Krstić, D; Čolović, M; Krinulović, K; Djurić, D; Vasić, V. Inhibition of AChE by single and simultaneous exposure to malathion and its degradation products. *Gen Physiol Biophys*, 2007, 26, 247–253.

Krstić, D; Čolović, M; Bavcon Kralj, M; Trebše, P; Krinulović, K; Vasić, V. The influence of malathion and its decomposition products on free and immobilized acetylcholinesterase. *Russ J Phys Chem A*, 2008a, 82, 663–668.

Krstić, D; Čolović, M; Bavcon Kralj, M; Franko, M; Krinulović, K; Trebše, P; Vasić, V. Inhibition of AChE by malathion and some structurally similar compounds. *J Enz Inh Med Chem*, 2008b, 23, 562–573.

Kulkarni, AP; Hodgson, E. Metabolism of insecticides by mixed function oxidase systems. *Pharmacol Ther*, 1980, 8, 379–475.

Kumaran, S; Tran-Minh, C. Determination of organophosphorous and carbamate insecticides by flow injection analysis. *Anal Biochem*, 1992, 200, 187−194.

Kumaran, S; Morita, M. Application of a cholinesterase biosensor to screen for organophosphorus pesticides extracted from soil. *Talanta*, 1995, 42, 649−655.

La Rosa, C; Pariente, F; Hernández, L; Lorenzo, E. Amperometric flow-through biosensor for the determination of pesticides. *Anal Chim Acta*, 1995, 308, 129–136.

Labro, M-T. Interference of Antibacterial Agents with Phagocyte Functions: Immunomodulation or "Immuno-Fairy Tales"? *Clin Microbiol Rev*, 2000, 13, 615–650.

Lai, K; Dave, KI; Wild, JR. Bimetallic binding motifs in organophosphorus hydrolase are important for catalysis and structural organization. *J Biol Chem*, 1994, 269, 16579–16584.

Lazarević-Pašti, T; Momić, T; Onija, A; Vujisić, Lj; Vasić, V. Myeloperoxidase-mediated oxidation of organophosphorus pesticides as a pre-step in their determination by AChE based bioanalytical methods. *Microchim Acta*, 2010, 170, 289–297.

Lazarević-Pašti, T; Nastasijević, B; Vasić, V. Oxidation of chlorpyrifos, azinphos-methyl and phorate by myeloperoxidase. *Pestic Biochem Phys*, 2011, 101, 220–226.

Lazarević-Pašti, TD; Bondžić, AM; Pašti, IA; Vasić, VM. Indirect electrochemical oxidation of organophosphorous pesticides for efficient detection *via* acetylcholinesterase test. *Pestic Biochem Phys*, 2012, 104, 236–242.

Lazarević-Pašti, T; Momić, T; Radojević, MM; Vasić, V. Influence of organophosphorus pesticides on peroxidase and chlorination activity of human myeloperoxidase. *Pestic Biochem Phys,* 2013, 107, 55–60.

Lee, HS; Kim, YA; Cho, YA; Lee, YT. Oxidation of organophosphorus pesticides for the sensitive detection by a cholinesterase-based biosensor. *Chemosphere*, 2002, 46, 571–576.

Li, JQ; Liu, J; Shen, WJ; Zhao, XL; Hou, Y; Cao, H; Cui, ZL. Isolation and characterization of 3,5,6-trichloro-2-pyridinol-degrading Ralstonia sp. strain T6. *Bioresource Technol*, 2010, 101, 7479–7483.

Li, X; He, J; Li, S. Isolation of a chlorpyrifos-degrading bacterium, Sphingomonas sp. strain Dsp-2, and cloning of the mpd gene. *Res Microbiol*, 2007, 158, 143–149.

Liapis, KS; Aplada-Sarlis, P; Kyriakidis, NV. Rapid multi-residue method for the determination of azinphos methyl, bromopropylate, chlorpyrifos, dimethoate, parathion methyl and phosalone in apricots and peaches by using negative chemical ionization ion trap technology. *J Chromatogr A*, 2003, 996, 181–187.

Liu, B; McConnell, LL; Torrents, A. Hydrolysis of chlorpyrifos in natural waters of the Chesapeake Bay. *Chemosphere*, 2001, 44, 1315–1323.

Liu, Y; Wang, G; Li, C; Zhou, Q; Wang, M; Yang, L. A novel acetylcholinesterase biosensor based on carboxylic graphene coated with silver nanoparticles for pesticide detection. *Mater Sci Eng C Mater Biol Appl*, 2014, 35, 253–258.

Ma, T; Chambers, JE. Kinetic parameters of desulphuration and dearylation of parathion and chlorpyrifos by rat liver microsomes. Food Chem Toxicol, 1994, 32, 763–767.

Macalady, DL; Wolfe, NL. New perspectives on the hydrolytic degradation of the organophosphorothioate insecticide chlorpyrifos. *J Agric Food Chem*, 1983, 31, 1139–1147.

Maki, RA; Tyurin, VA; Lyon, RC; Hamilton, RL; Dekosky, ST; Kagan, VE; Reynolds, WF. Aberrant expression of myeloperoxidase in astrocytes promotes phospholipid oxidation and memory deficits in a mouse model of Alzheimer disease. *J Biol Chem*, 2009, 284, 3158–3169.

Malle, E; Woenckhaus, C; Waeg, G; Esterbauer, H; Grone, EF; Grone, HJ. Immunological evidence for hypochlorite-modified proteins in human kidney. *Am J Pathol*, 1997, 150, 603–615.

Malle, E; Buch, T; Grone, HJ. Myeloperoxidase in kidney disease. *Kidney Int*, 2003, 64, 1956–1967.

Malle, E; Marsche, G; Panzenboeck, U; Sattler, W. Myeloperoxidase-mediated oxidation of high-density lipoproteins: Fingerprints of newly recognized potential proatherogenic lipoproteins. *Arch Biochem Biophys*, 2006, 445, 245–255.

Manavalan, P; Taylor, P; Johnson Jr, WC. Circular dichroism studies of acetylcholine sterase conformation. Comparison of the 11 S and 5.6 S species and the differences induced by inhibitory ligands. *BBA-Protein Struct M*, 1985, 829, 365–370.

Martorell, D; Céspedes, F; Martínez-Fàbregas, E; Alegret, S. Amperometric determination of pesticides using a biosensor based on a polishable graphie-epoxy biocomposite. *Anal Chim Acta*, 1994, 290, 343–348.

Marty, JL; Sode, K; Karube, I. Biosensor for detection of organophosphate and carbamate insecticides. *Electroanal*, 1992, 4, 249–252.

Marty, JL; Mionetto, N; Lacorte, S; Barcelo, D. Validation of an enzymatic biosensor with various liquid chromatographic techniques for determining organophosphorus pesticides and carbaryl in freeze-dried waters. *Anal Chim Acta*, 1995, 311, 265–271.

Mates, JM; Perez-Gomez, C; De Castro, I. Antioxidant enzymes and human diseases. *Clin Biochem*, 1999, 32, 595–603.

Mauriz, E; Calle, A; Lechuga, LM; Quintana, J; Montoya, A; Manclús, JJ. Real-time detection of chlorpyrifos at part per trillion levels in ground, surface and drinking water samples by a portable surface plasmon resonance immunosensor. *Anal Chim Acta*, 2006, 561, 40–47.

Maya, K; Upadhyay, SN; Singh, RS;. Dubey, SK. Degradation kinetics of chlorpyrifos and 3, 5, 6-trichloro-2-pyridinol (TCP) by fungal communities. *Bioresource Technol*, 2012, 126, 216–223.

Mehrotra, BD; Moorthy, KS; Reddy, SR; Desaiah, D. Effects of cyclodiene compounds on calcium pump activity in rat brain and heart. *Toxicol*, 1989, 54, 17–29.

Mehta, A; Verma R; Srivastava, N. Chlorpyrifos-induced alterations in rat brain acetylcholinesterase, lipid peroxidation and ATPases. *Indian J Biochem Biophys,* 2005, 42, 54–58.

Michalski, J; Okruszek, A; Stec, W. Stereochemistry of oxidation of organophosphorus thiono-compounds and PIII compounds by nitric acid and dinitrogen tetroxide. *J Chem Soc, Chem Commun*, 1970, 1495–1497.

Mink, PJ; Adami, H-O; Trichopoulos, D; Britton, NL; Mandel, JS. Pesticides and prostate cancer: a review of epidemiologic studies with specific agricultural exposure information. *Eur J Cancer Prev*, 2008, 17, 97–110.

Momić, T; Vujčić, Z; Vasić, V. Kinetics of inhibition of peroxidase activity of myeloperoxidase by quercetin. *Int J Chem Kinet*, 2008, 40, 384–394.

Mooser, G; Sigman, DS. Ligand binding properties of acetylcholinesterase determined with fluorescent probes. *Biochemistry*, 1974, 13, 2299–2307.

Moris, P; Alexandre, I; Roger, M; Remacle, J. Chemiluminescence assays of organophosphorus and carbamate pesticides. *Anal Chim Acta*, 1995, 302, 53–59.

Mortensen, SR; Chanda, SM; Hooper, MJ; Padilla, S. Maturational differences in chlorpyrifos-oxonase activity may contribute to age-related sensitivity to chlorpyrifos. J Biochem Toxicol, 1996, 11, 279–287.

Muhamad, SG. Kinetic studies of catalytic photodegradation of chlorpyrifos insecticide in various natural waters. *Arab J Chem*, 2010, 3, 127–133.

Mulchandani, A; Mulchandani, P; Chen, W. Enzyme biosensor for determination of organophosphates. *Field Anal Chem Technol*, 1998, 2, 363–369.

Murillo, R; Sarasa, J; Lanao, M; Ovelleiro, JL. Degradation of chlorpyriphos in water by advanced oxidation processes. *Water Sci Tech-W Sup*, 2010, 10, 1–6.

Nachmansohn, D; Wilson, IB. The enzymic hydrolysis and synthesis of acetylcholine. *Adv Enzymol*, 1951, 12, 259–339.

Narra, M R. Tissue-Specific Recovery of Oxidative and Antioxidant Effects of Chlorpyrifos in the Freshwater Crab, Barytelphusa guerini. *Arch Environ Contam Toxicol*, 2014, 67, 158–166.

Nicholls, SJ; Hazen, SL. Myeloperoxidase and cardiovascular disease. *Arterioscler Thromb Vasc Biol*, 2005, 25, 1102–1111.

Nolan, RJ; Rick, DL; Freshour, NL; Saunders, JH. Chlorpyrifos: pharmacokinetics in human volunteers. *Toxicol Appl Pharmacol*, 1984, 73, 8–15.

Nurcombe, HL; Bucknall, RC; Edwards, SW. Activation of the neutrophil myeloperoxidase-H_2O_2 system by synovial fluid isolated from patients with rheumatoid arthritis. *Ann Rheum Dis*, 1991, 50, 237–242.

Ordentlich, A; Barak, D; Kronman, C; Flashner, Y; Leitner, M; Segall, Y; Ariel, N; Cohen, S; Velan, B; Shafferman, A. Dissection of the human acetylcholinesterase active center determinants of substrate specificity. Identification of residues constituting the anionic site, the hydrophobic site, and the acyl pocket. *J Biol Chem*, 1993, 268, 17083–17095.

Ordentlich, A; Barak, D; Kronman, C; Ariel, N; Segall, Y; Velan, B; Shafferman, A. Contribution of Aromatic Moieties of Tyrosine 133 and of the Anionic Subsite Tryptophan 86 to Catalytic Efficiency and Allosteric Modulation of Acetylcholinesterase. *J Biol Chem*, 1995, 270, 2082–2091.

Oruc, EO; Usta, D. Evaluation of oxidative stress responses and neurotoxicity potential of diazinon in different tissues of Cyprinus carpio. *Environ Toxicol Phar,* 2007, 23, 48–55.

Paliwal, S. Development of Enzyme-based Biosensors for the Detection of Organophosphate Neurotoxins. ProQuest, 2008.

Pengphol, S; Uthaibutra, J; Arquero, O; Nomura, N; Whangchai, K. Oxidative Degradation and Detoxification of Chlorpyrifos by Ultrasonic and Ozone Treatments. *J Agr Sci*, 2012, 4, 164−172.

Perez-Ruiz, T; Martinez-Lozano, C; Tomas, V; Martin, J. High-performance liquid chromatographic assay of phosphate and organophosphorus pesticides using a post-column photochemical reaction and fluorimetric detection. *Anal Chim Acta*, 2005, 540, 383−391.

Petrovitch, H; Ross, GW; Abbott, RD; Sanderson, WT; Sharp, DS; Tanner, CM; Masaki, KH; Blanchette, PL; Popper, JS; Foley, D; Launer, L; White, LR. Plantation Work and Risk of Parkinson Disease in a Population-Based Longitudinal Study. *Arch Neurol*, 2002, 59, 1787–1792.

Pogačnik, L; Franko, M. Determination of organophosphate and carbamate pesticides in spiked samples of tap water and fruit juices by a biosensor with photothermal detection. *Biosens Bioelectron*, 1999, 14, 569−578.

Pogačnik, L; Development and application of photothermal biosensor for detection of organophosphate and carbamate pesticides. Polytehnic School of Environmental Sciences, University of Nova Gorica, Nova Gorica, 2001a.

Pogačnik, L; Franko, M. Optimisation of FIA system for detection of organophosphorous and carbamate pesticides based on cholinesterase inhibition. *Talanta*, 2001b, 54, 631–641.

Pogačnik, L; Franko, M. Detection of organophosphate and carbamate pesticides in vegetable samples by a photothermal biosensor. *Biosens Bioelectron*, 2003, 18, 1−9.

Possamai, FP; Fortunato, JJ; Feier, G; Agostinho, FR; Quevedo, J; Wilhelm Filho, D; Dal-Pizzol, F. Oxidative stress after acute and sub-chronic malathion intoxication in Wistar rats. *Environ Toxicol Pharmacol,* 2007, 23, 198–204.

Quinn, DM. Acetylcholinesterase: enzyme structure, reaction dynamics, and virtual transition states. *Chem Rev*, 1987, 87, 955–979.

Radić, Z; Gibney, G; Kawamoto, S; MacPhee-Quigley, K; Bongiorno, C; Taylor, P. Expression of recombinant acetylcholinesterase in a baculovirus system: kinetic properties of glutamate 199 mutants. *Biochemistry*, 1992, 31, 9760–9767.

Rahimi, R; Nikfar, S; Abdollahi, M. Increased morbidity and mortality in acute human organophosphate-poisoned patients treated by oximes: a meta-analysis of clinical trials. *Hum Exp Toxicol*, 2006, 25, 157–162.

Rahman, MF; Siddiqui, MKJ; Mustafa, M. Effects of a newphosphorothionate (RPR-V) on ATPases and acetylcholinesterase in rat brain by subchronic dosing. *J Appl Toxicol*, 1997, 17, 273–278.

Ranjbar, A; Solhi, H; Mashayekhi, FJ; Susanabdi, A; Rezaie, A; Abdollahi, M. Oxidative stress in acute human poisoning with organophosphorus insecticides; a case control study. *Environ Toxicol Phar*, 2005, 20, 88–91.

Reddy, AN; Venugopal, NBRK; Reddy, SLN. Effect of endosulfan 35 EC on ATPases in the tissues of a freshwater field crab Barytelphusa guerini. *B Environ Contam Tox*, 1992, 48, 216–222.

Reiner, E; Aldridge, N; Simeon, V; Radic, Z; Taylor, P. In: Massoulie, J. (Ed.), Cholinesterases: structure, function, mechanism, genetics, and cell biology, American Chemical Society, Washington, 1991, pp. 227–234.

Ristori, C; Del Carlo, C; Martini, M; Barbaro, A; Ancarani, A. Potentiometric detection of pesticides in water samples. *Anal Chim Acta*, 1996, 325, 151–160.

Roberts, TR; Hutson, DH. Metabolic Pathways of Agrochemicals - Part 2: Insecticides and Fungicides. The Royal Society of Chemistry, Cambridge, UK, 1999, pp. 235–242.

Roda, A; Rauch, P; Ferri, E; Girotti, S; Ghini, S; Carrea, G; Bovara, R. Chemiluminescent flow sensor for the determination of Paraoxon and Aldicarb pesticides. *Anal Chim Acta*, 1994, 294, 35–42.

Roepcke, CBS; Muench, SB; Schulze, H; Bachmann, T; Schmid, RD; Hauer, B. Analysis of phosphorothionate pesticides using a chloroperoxidase pretreatment and acetylcholinesterase biosensor detection. *J Agric Food Chem*, 2010, 58, 8748–8756.

Rogers, KR; Williams, LR. Biosensors for environmental monitoring: a regulatory perspective. *TrAC - Trend Anal Chem*, 1995, 14, 289–294.

Rotariu, L; Zamfir, LG; Bala, C. A novel, sensitive, reusable and low potential acetylcholinesterase biosensor for chlorpyrifos based on 1-butyl-3-methylimidazolium tetrafluoroborate/multiwalled carbon nanotubes gel. *Biosens Bioelectron*, 2011, 26, 3692–3695.

Sahoo, A; Samanta, L; Das, A; Patra, KS; Chainy, BNG. Hexachlorocyclohexane-induced behavioural and neurochemical changes in rat. *J Appl Toxicol*, 1999, 19, 13–18.

Samet, Y; Agengui, L; Abdelhédi, R. Electrochemical degradation of chlorpyrifos pesticide in aqueous solutions by anodic oxidation at boron-doped diamond electrodes. *Chem Eng J*, 2010, 161, 167–172.

Savenkova, ML; Mueller, DM; Heinecke, JW. Tyrosyl radical generated by myeloperoxidase is a physiological catalyst for the initiation of lipid peroxidation in low density lipoprotein. *J Biol Chem*, 1994, 269, 20394–20400.

Savolainen, K. Understanding the toxic action of organophosphates. In: Handbook of Pesticide Toxicology, Academic Press, USA, 2001, pp. 1013–1043.

Schiller, J; Arnhold, J; Sonntag, K; Arnold, K. NMR studies on human, pathologically changed synovial fluids: Role of hypochlorous acid. *Magn Reson Med*, 1996, 35, 848–853.

Settimi, L; Masina, A; Andrion, A; Axelson, O. Prostate cancer and exposure to pesticides in agricultural settings. *Int J Cancer*, 2003, 104, 458–461.

Sharma, Y; Bashir, S; Irshad, M; Gupta, SD; Dogra, TD. Effects of acute dimethoate administration on antioxidant status of liver and brain of experimental rats. *Toxicology*, 2005, 206, 49–57.

Shemer, H; Linden, KG. Degradation and by-product formation of diazinon in water during UV and UV/H_2O_2 treatment. *J Hazard Mater*, 2006, 136, 553–559.

Shiosaka, T; Okuda H; Fungi, S. Mechanisms of phosphorylation of thymidine by the culture filtrate of Clostridium perfringens and rat liver extract. *Biochem Biophys Acta*, 1971, 246, 171–183.

Simonian, AL; Good, TA; Wang, SS; Wild, JR. Nanoparticle-based optical biosensors for the direct detection of organophosphate chemical warfare agents and pesticides. *Anal Chim Acta*, 2005, 534, 69–77.

Singh, M; Sandhir, R; Kiran R. In vitro effects of organophosphate pesticides on rat erythrocytes. *Indian J Exp Biol*, 2004, 42, 292–296.

Singh, BK; Walker, A; Wright, DJ. Bioremedial potential of fenamiphos and chlorpyrifos degrading isolates: Influence of different environmental conditions. *Soil Biol Biochem*, 2006a, 38, 2682–2693.

Singh, M; Sandhir, R; Kiran R. Erythrocyte antioxidant enzymes in toxicological evaluation of commonly used organophosphate pesticides. *Indian J Exp Biol*, 2006b, 44, 580–583.

Sinha, SN; Pal, R; Dewan, A; Mansuri, MM; Saiyed, HN. Effect of Dissociation Energy on Ion Formation and Sensitivity of An Analytical Method for Determination of Chlorpyrifos in Human Blood, Using Gas Chro-matography-Mass Spectrometer (GC-MS in MS/MS). *Int J Mass Spectrom*, 2006, 253, 48−57.

Skládal, P; Nunes, GS; Yamanaka, H; Ribeiro, ML. Detection of carbamate pesticides in vegetable samples using cholinesterase-based biosensors. *Electroanal*, 1997, 9, 1083–1087.

Skowronska, A; Krawczyk, E. A general method for the conversion of thiophosphoryl and selenophosphoryl groups into phosphoryl groups by ozone oxidation. *Synthesis*, 1983, 6, 509−510.

Smith, JH; Mabey, WR; Bohonos, N; Holt, BR; Lee, SS; Chou, TW; Bomberger, DC; Mill, T. Environmental pathways of selected chemicals in freshwater systems: Part II. Laboratory studies. U.S. Environmental Protection Agency, Athens, GA, 1978, EPA-600/7-78-074.

Smith, JN; Campbell, JA; Busby-Hjerpe, AL; Lee, S; Poet, TS; Barr, DB; Timchalk, C. Comparative chlorpyrifos pharmacokinetics via multiple routes of exposure and vehicles of administration in the adult rat. *Toxicology*, 2009, 261, 47–58.

Smith, JN; Wang, J; Lin, Y; Klohe, EM; Timchalk, C. Pharmacokinetics and Pharmacodynamics of Chlorpyrifos and 3,5,6-Trichloro-2-pyridinol in Rat Saliva After Chlorpyrifos Administration. *Toxicol Sci*, 2012, 130, 245-256.

Sogorb, MA. Garcia-Arguelles, S; Carrera, V; Vilanova, E. Serum albumin is as efficient as paraxonase in the detoxication of paraoxon at toxicologically relevant concentrations. *Chem Res Toxicol*, 2008, 21, 1524–1529.

Sogorb, MA; Vilanova, E. Serum albumins and detoxication of anti-cholinesterase agents. *Chem Biol Interact*, 2010, 187, 325–329.

Soltatinejad, K; Abdollahi, M. Current opinion on the science of organophosphate pesticides and toxic stress: a systematic review. *Med Sci Monit*, 2009, 15, 75–90.

Somani, SM; Husain, K. Exercise and Oxygen Toxicity: A Handbook. Elsevier Sciences, pp. 713-751.

Stec, WJ; Okruszek, A; Michalski, J. Organophosphorus compounds of sulfur and selenium. Stereochemistry of oxidation of thiono- and selenophosphoryl compounds with hydrogen peroxide. *J Org Chem*, 1976, 41, 233–238.

Stein, K; Schwedt, G. Comparison of immobilization methods for the development of an acetylcholinesterase biosensor. *Anal Chim Acta*, 1983, 272, 73–81.

Tang, J; Cao Y; Rose, RL; Brimfield, AA; Dai, D. Goldstein, JA; Hodgson, E. Metabolism of chlorpyrifos by human cytochrome P450 isoforms and human, mouse, and rat liver microsomes. *Drug Metab Dispos*, 2001, 29, 1201–1204.

Taylor, P; Lappi, S. Interaction of fluorescence probes with acetylcholinesterase. Site and specificity of propidium binding. *Biochemistry*, 1975, 14, 1989–1997.

Taylor, P; Radic, Z. The cholinesterases: from genes to proteins. *Annu Rev Pharmacol,* 1994, 34, 281–320.

Teimouri, F; Amirkabirian, N; Esmaily, H; Mohammadirad, A; Aliahmadi, A; Abdollah, M. Alteration of hepatic cells glucose metabolism as a non-holinergic detoxication mechanism in counteracting diazinon-induced oxidative stress. *Hum Exp Toxicol*, 2006, 25, 697–703.

Thompson, TS; Treble, RG. Solid Phase Extraction and GC-MS Techniques for the Confirmation of Chlorpyrifos Contamination of Surface Water Supplies. *B Environ Contam Tox*, 1996, 57, 525–531.

Timchalk, RC; Nolan, J; Mendrala, AL; Dittenber, DA; Brzak, KA; Mattsson, JL. A physiologically based pharmacokinetic and pharmacodynamic (PBPK/PD) model for the organophosphate insecticide chlorpyrifos in rats and humans. *Toxicol Sci*, 2002, 66, 34–53.

Timchalk, C; Busby, A; Campbell, JA; Needham, LL; Barr, DB. Comparative pharmacokinetics of the organophosphorus insecticide chlorpyrifos and its major metabolites diethylphosphate, diethylthiophosphate and 3,5,6-trichloro-2-pyridinol in the rat. *Toxicology*, 2007, 237, 145–157.

Tougu, V. Acetylcholinesterase: Mechanism of Catalysis and Inhibition. *Curr Med Chem-CNS Agents,* 2001, 1, 155–170.

Tran-Minh, C. Biosensors. Chapman & Hall, London, 1993.

Trettnak, W; Reininger, F; Zinterl, E; Wolfbeis, OS. Fiber-optic remote detection of pesticides and related inhibitors of the enzyme acetylcholine esterase. *Sens Actuators B*, 1993, 11, 87–93.

Turner, ND; Braby, LA; Ford, J; Lupton, JR. Opportunities for nutritional amelioration of radiation-induced cellular damage. *Nutrition*, 2002, 18, 904–912.

Valko, M; Izakovic, M; Mazur, M; Rhodes, CJ; Telser, J. Role of oxygen radicals in DNA damage and cancer incidence. *Mol Cell Biochem*, 2004, 266, 37–56.

Van Dyk, JS; Pletschke, B. Review on the use of enzymes for the detection of organochlorine, organophosphate and carbamate pesticides in the environment. *Chemosphere*, 2011, 82, 291–307.

Vasić, MV; Momić, GT; Petković, M; Krstić, ZD. Na^+,K^+-ATPase as the target enzyme for organic and inorganic compounds. *Sensors,* 2008, 8, 8321–8360.

Vijaya Bhaskar Reddy, A; Madhavi, V; Gangadhara Reddy, K; Madhavi, G. Remediation of Chlorpyrifos-Contaminated Soils by Laboratory-Synthesized Zero-Valent Nano Iron Particles: Effect of pH and Aluminium Salts. *J Chem*, 2013, Article ID 521045, 1−7.

Viswanathan, S; Radecka, H; Radecki, J. Electrochemical biosensor for pesticides based on acetylcholinesterase immobilized on polyaniline deposited on vertically assembled carbon nanotubes wrapped with ssDNA. *Biosens Bioelectron*, 2009, 24, 2772–2777.

Walker, CH; Mackness, MI. "A" esterases and their role in regulating the toxicity of organophosphates. *Arch Toxicol*, 1987, 60, 30–33.

Wallace, SS. Detection and repair of DNA base damages produced by ionizing radiation. *Environ Mol Mutagen*, 1988, 12, 431–477.

Wang, L; Lu, D; Wang, J; Du, D; Zou, Z; Wang, H; Smith, JN; Timchalk, C; Liu, F; Lin, Y. A novel immunochromatographic electrochemical biosensor for highly sensitive and selective detection of trichloropyridinol, a biomarker of exposure to chlorpyrifos. *Biosens Bioelectron,* 2011, 26, 2835–2840.

Wang, HY; O'Doherty, GA. Modulators of Na/K-ATPase: a patent review. *Expert Opin Ther Pat*, 2012, 22, 587–605.

Wang, X; Xing, H; Jiang, Y; Wu, H; Sun, G; Xu, Q; Xu, S. Accumulation, histopathological effects and response of biochemical markers in the spleens and head kidneys of common carp exposed to atrazine and chlorpyrifos. *Food Chem Toxicol*, 2013, 62, 148–158.

WHO, Metabolism and mode of action. In: Organophosphorus Insecticides: a General Introduction, World Health Organization, Geneva, 1986a, pp. 39–48.

WHO, Effects of organophosphorus insecticides on the nervous system. In: Organophosphorus Insecticides: a General Introduction, World Health Organization, Geneva, 1986b, pp. 58–69.

Willig, S; Hunter, DL; Dass, PD; Padilla, S. Validation of the use of 6,6-dithiodinicotinic acid as a chromogen in the Ellman method for cholinesterase determinations. *Vet Hum Toxicol*, 1996, 38, 249–253.

Wilson, IB; Quan, C. Acetylcholinesterase studies on molecular complementariness. *Arch Biochem Biophys*, 1958, 73, 131–143.

Worek, F; Mast, U; Kiderlen, D; Diepold, C; Eyer, P. Improved determination of acetylcholinesterase activity in human whole blood. *Clin Chim Acta*, 1999, 288, 73–90.

Wu, T; Jans, U. Nucleophilic substitution reactions of chlorpyrifos-methyl with sulfur species. *Environ Sci Technol*, 2006, 40, 784−790.

Xie, Z; Askari, A. Na(+)/K(+)-ATPase as a signal transducer. *Eur J Biochem*, 2002, 269, 2434–2439.

Xie, Z; Cai, T. Na^+-K^+--ATPase-mediated signal transduction: from protein interaction to cellular function. *Mol Interv*, 2003, 3, 157–168.

Xing, H; Zhang, Z; Yao, H; Liu, T; Wang, L; Xu, S; Li, S. Effects of atrazine and chlorpyrifos on cytochrome P450 in common carp liver. *Chemosphere*, 2014, 104, 244–250.

Yang, L; Zhao, YH; Zhang, BX; Yang, CH; Zhang, X. Isolation and characterization of a chlorpyrifos and 3,5,6-trichloro-2-pyridinol degrading bacterium. *FEMS Microbiol Lett*, 2005, 251, 67–73.

Zhang, Q; Pehkonen, SO. Oxidation of Diazinon by Aqueous Chlorine: Kinetics, Mechanisms, and Product Studies. *J Agric Food Chem*, 1999, 47, 1760−1766.

Zhang, R; Brennan, ML; Fu, X; Aviles, RJ; Pearce, GL; Penn, MS; Topol, EJ; Sprecher, DL; Hazen, SL. Association between myeloperoxidase levels and risk of coronary artery disease. *J Amer Med Assoc*, 2001, 286, 2136–2142.

Zou, Z; Du, D; Wang, J; Smith, JN; Timchalk, C; Li, Y; Lin, Y. Quantum dot-based immunochromatographic fluorescent biosensor for biomonitoring trichloropyridinol, a biomarker of exposure to chlorpyrifos. *Anal Chem*, 2010, 82, 5125–5133.

In: Chlorpyrifos
Editor: Caitlin Mayes
ISBN: 978-1-63482-111-7

Chapter 6

DEVELOPMENTAL NEUROTOXICITY OF CHLORPYRIFOS: MECHANISTIC INFORMATION AND PREDICTIVE SAFETY ASSESSMENT APPLICATIONS

***Magda Sachana*[*,1] and Carmen Estevan Martinez*[2]**
[1]Laboratory of Biochemistry and Toxicology,
Faculty of Veterinary Medicine,
Aristotle University of Thessaloniki, Thessaloniki, Greece
[2]ECHA, Helsinki, Finland

ABSTRACT

Over the last decade a considerable body of evidence has emerged indicating that chlorpyrifos (CPF) targets important processes during nervous system development causing developmental neurotoxicity. These processes include neuronal proliferation, differentiation, axonogenesis, synaptogenesis, myelination and apoptosis. This book chapter attempts to collate mechanistic information derived from in vivo and in vitro studies investigating the neurotoxic properties of CPF in developing nervous system. Furthermore, this existing mechanistic knowledge will be organized accordingly and linked to distinct adverse outcomes that CPF can cause in developing organisms such as motor, sensory and cognitive deficits. These neurodevelopmental effects vary in severity, duration and time of onset and can be reversible or can have permanent, life-long consequences. Thus, the developmental neurotoxicity induced by CPF is complex and here there will be also presented a chemical-specific predictive risk assessment approach that will encompass all the available mechanistic as well as complementary toxicokinetic information. Additionally, this book chapter will revise all the information available from several risk assessment reports performed by different international organizations regarding the neurodevelopmental effects of CPF. An analysis of the various outcomes and the differences in diverse aspects of these assessments will be provided. Finally, the

[*] Email: msachana@vet.auth.gr

implications of these evaluations across different regulatory frameworks will be also described.

INTRODUCTION

CPF is an organophosphorus compound that has been widely used to control insect pests and has been applied for around 50 years in both agricultural and residential areas until recently. Residential use of CPF has been banned in the United States (US) since 2001, and has been phased out in the European Union (EU). It is included since 2005 (European Commission, 2005) in Annex I of the Directive 91/414/EEC for plant protection products (Council of the European Communities, 1991), which has been replaced by the Regulation (EC) No 1107/2009 of the European Parliament and of the Council of 21 October 2009. Currently, nonagricultural uses of CPF are limited and restricted to insect control for public health purposes and on golf courses. However, in other countries outside of the US and the EU it cannot be ruled out the use of CPF in residential areas.

CPF targets primarily the central and peripheral nervous system, due to the ability of its oxon metabolite named chlorpyrifos-oxon (CPO) to inhibit the enzymatic activity of acetylcholinesterase (AChE) in brain, which is involved in cholinergic neurotransmission. Toxicological effects of CPF in tissues other than the nervous system are limited. However, CPF's involvement in teratogenicity and developmental toxicity has not been elucidated yet. There is no consistency regarding the evidence for CPF's ability to induce teratogenicity in doses well below than those that cause maternal toxicity (Eaton et al., 2008). Additionally, there is contradictory data on the capability of CPF to cause neurodevelopmental toxicity since many in vivo studies have reported such effects after exposure to CPF at doses that elicit decrease in brain AChE, while other reports have indicated similar effects at dose levels which do not cause any apparent AChE inhibition (Eaton et al., 2008). Indeed, Slotkin et al. (2006) have suggested that there is a complete dichotomy between the systemic toxicity of OPs, including CPF that is attributed to AChE inhibition and their capability to cause developmental neurotoxicity. Grandjean and Landrigan (2014) place CPF among the emerging developmental neurotoxicants using recent epidemiological studies. These remarks render important the need to assess the risks of exposure to CPF during development by especially considering that exposure occurs in fetuses, newborns and children that live not only in agricultural but also in residential areas. An epidemiological study has shown that CPF displayed a prevalence of 11% in meconium of human placenta at a mean concentration of 53 ppm, suggesting that CPF exposure to fetuses does occur (Ostrea et al., 2002).

Besides neurobehavioral animal and human epidemiological studies, mechanistic information derived from in vivo and in vitro experimental approaches can facilitate the elucidation of CPF's developmental neurotoxicological properties. Throughout the years a significant amount of scientific data has been generated by exploring the effects of CPF on different neurotransmission systems rather than only investigating CPF-induced damage of the cholinergic pathway. Furthermore, CPF's influence on specific processes during nervous system development has also been extensively studied. This accumulating mechanistic knowledge can be organized accordingly and potentially can improve regulatory decisions. At the same time, computational methods and integrated approaches of testing assessment

(IATAs) can be developed incorporating this mechanistic information and facilitating the prediction of developmental neurotoxicity of untested chemical agents.

NEUROBEHAVIORAL ALTERATIONS INDUCED BY CPF AND THE IMPLICATION OF DIFFERENT NEUROTRANSMITTER PATHWAYS

Two main neurotransmitter pathways have been identified to be involved in altered neurobehavior identified in rodents after developmental exposure to CPF. Since CPF is an organophosphorothionate inhibits AChE resulting to excessive accumulation of acetycholine (ACh) in cholinergic receptors of the nervous system and affects primarily animal behaviors related to the cholinergic pathway. The cholinergic system plays a pivotal role in development of the nervous system that begins from the embryonic period and continues up to postnatal years (Lauder and Schambra, 1999). However, the neurobehavioral effects of CPF have also been assessed under noncholinergic concentrations to resemble the situation that most commonly has been reported in fetuses and children. These neurobehavioral studies using noncholinergic doses revealed serotonergic pathway as another potential neurotransmitter pathway that is disturbed causing serotonin-associated behaviors.

The neurobehavioral effects of CPF on developing animals have been extensively studied by a number of research teams. These studies assessed the effects of CPF in the presence (Chakraborti et al., 1993; Stanton et al., 1994; Chanda and Pope, 1996; Moser and Padilla, 1998; Maurissen et al., 2000) or absence of significant inhibition of AChE activity (Dam et al., 2000; Carr et al., 2001; Jett et al., 2001; Ricceri et al., 2006).

The neurobehavioral studies performed until now cover a wide window of brain development including both prenatal (Chanda and Pope, 1996; Maurissen et al., 2000; Levin et al., 2002; Icenogle et al., 2004; Ricceri et al., 2006; Billauer-Haimovitch et al., 2009; Venerosi et al., 2009; Haviland et al., 2010) and early or late postnatal CPF-treatment (Stanton et al., 1994; Moser and Padilla, 1998; Dam et al., 2000; Carr et al., 2001; Jett et al., 2001; Levin et al., 2001; Ricceri et al., 2006; Johnson et al., 2009). The administration of CPF to laboratory animals was mainly oral (Braquenier et al., 2010; Carr et al., 2001; Johnson et al., 2009; Maurissen et al., 2000; Venerosi et al., 2006; 2009; 2010), dermal (Abou-Donia et al., 2006) or subcutaneous (Dam et al., 2000; Jett et al., 2001; Levin et al., 2001; 2002; Ricceri et al., 2003; Icenogle et al., 2004; Aldridge et al., 2005a; Ricceri et al., 2006; Venerosi et al., 2008; Billauer-Haimovitch et al., 2009; Haviland et al., 2010).

Motor activity measurements (Moser and Padilla, 1998; Dam et al., 2000; Maurissen et al., 2000; Carr et al., 2001; Levin et al., 2001; 2002; Phang, 2002; Icenogle et al., 2004; Abou-Donia et al., 2006; Ricceri et al., 2006; Braquenier et al., 2010; Venerosi et al., 2009; Cole et al., 2012; Levin et al., 2014), developmental landmarks (Maurissen et al., 2000; Billauer-Haimovitch et al., 2009; Johnson et al., 2009), neurodevelopmental reflex parameters (Chanda and Pope, 1996; Dam et al., 2000; Billauer-Haimovitch et al., 2009; Johnson et al., 2009; Venerosi et al., 2009; Cole et al., 2012), novelty-induced activity (Ricceri et al., 2003; Icenogle et al., 2004; Aldridge et al., 2005a; Ricceri et al., 2006; Braquenier et al., 2010; De Felice et al., 2014; Levin et al., 2014) or cognitive function endpoints (Stanton et al., 1994; Maurissen et al., 2000; Jett et al., 2001; Levin et al., 2001; 2002; Icenogle et al., 2004; Ricceri et al., 2006; Billauer-Haimovitch et al., 2009; Johnson et al., 2009; Haviland et al., 2010; Cole et al., 2012) have been performed in an attempt to elucidate the various neurobehavioral alterations induced by CPF in developing rodents.

Studies on motor activity alterations induced by CPF exposure during development provided conflicted results (reviewed in Li et al., 2012). Some reports revealed increased motor activity (Ricceri et al., 2003; Ricceri et al., 2006), whereas others measured reduced intersession habituation and decreased linear trend of the habituation curve (Maurissen et al., 2000; Carr et al., 2001; Levin et al., 2002; Icenogle et al., 2004) or no change in motor activity (Venerosi et al., 2009; Braquenier et al., 2010). These mixed results in motor activity have been attributed to the use of different species, the type of activity test and route of CPF exposure.

Impairment of cognitive function has been suggested to be the most pronounced and consistent finding in CPF-treated developing animals (Stanton et al., 1994; Jett et al., 2001; Levin et al., 2001; 2002; Icenogle et al., 2004; Aldridge et al., 2005a; Ricceri et al., 2003; 2006; Billauer-Haimovitch et al., 2009; Johnson et al., 2009; Haviland et al., 2010), although there is a report showing differently and involved oral treatment of dams (Maurissen et al., 2000). Some studies support sex-dependant neurobehavioral alterations induced by CPF in developing rodents (Dam et al., 2000; Levin et al., 2001; 2002; Aldridge et al., 2005a; Ricceri et al., 2006; Haviland et al., 2010). However, other studies failed to confirm sex-selective neurobehavioral deficits after CPF treatment of developing organisms (Moser and Padilla, 1998; Carr et al., 2001; Jett et al., 2001; Icenogle et al., 2004).

CPF-induced neurobehavioral alterations have been associated with disruption of hippocampal cholinergic synaptic function (Levin et al., 2001; Icenogle et al., 2004) or changed expression of muscarinic receptors (Stanton et al., 1994; Chanda and Pope, 1996; Moser and Padilla, 1998; Jett et al., 2001, Yanai et al., 2010). Additionally, CPF has been found to target the developing serotonergic system at noncholinergic doses and to trigger serotonin-related behaviors. Serotonin (5HT) is a neurotransmitter that influences various processes, including autonomic function, motor activity, hormone secretion, cognition as well as complex behaviors associated with affection, emotion and reward. However, only a battery of emotional tests closely related to 5HT function has been used to assess the effects of CPF on 5HT-associated behaviors in adolescence and adulthood after postnatal exposure of rats to CPF, revealing disturbance on the majority of emotional responses tested (Roegge et al., 2008). Venerosi et al. (2010) showed that after gestational treatment of mice with CPF, the administration of a 5HT uptake inhibitor (fluvoxamine) increases swimming duration or frequency during the conduction of a forced swimming test.

The above data indicate that developmental exposure to CPF induces a range of neurobehavioral alterations and reveal that multiple mechanisms mediate the developmental neurotoxicity of this organophosphorothionate. These adverse neurobehavioral effects have been related in a causative manner with the disruption of cholinergic synaptic function in hippocampus, the levels of muscarinic receptors and the disturbance of serotonin pathway. However, accumulating evidence suggest that the cholinergic and serotonergic systems are not the only systems that are perturbed after developmental exposure to CPF. Glutamatergic system and more specifically the genes encoding glutamate receptors have been found to be altered in in vitro studies, implicating oxidative and excitatory mechanisms in developmental neurotoxicity induced by CPF (Slotkin and Seidler, 2009; Slotkin et al., 2010). Endocannabinoid system, which plays pivotal role in nervous system development, has also been found to be affected by CPF (Carr et al., 2011; 2013; 2014). Finally, emerging evidence suggest that CPF can also induce neurobehavioral alterations in developing organisms indirectly by affecting neurohypophyseal hormones (Tait et al., 2009) and thyroid hormone levels (De Angelis et al., 2009; Slotkin et al., 2013) that play by their own means vital role in mammalian neurodevelopment.

The Interference of CPF with Neurodevelopmental Processes

As soon as the developmental neurotoxicity properties of CPF started to emerge, multiple studies investigated its effects on specific processes during nervous system development. These processes involve neuronal proliferation, migration and differentiation from implantation to mid-gestation and synaptogenesis, myelination and apoptosis during late gestation and early neonatal period. It is known that delay, absence or disruption of any of these developmental events can lead to clinical symptoms. It is also well documented that the timing of exposure during neurodevelopment is critical and that different adversities can be recorded because different ontogenetic events can be targeted by toxicological insults.

Effects of Developmental Exposure of CPF on Neuronal Proliferation and the Implication of Cell Signaling Pathways and Transcription Factors

The effects of CPF on the process of neuronal proliferation have been explored indirectly by determining the total brain cell number based on measurement of DNA synthesis. CPF (2 mg/kg) inhibits both DNA and protein synthesis as early as 4 h after acute exposure of neonatal rats (Whitney et al., 1995). This study also reveals that CPF interferes with cell replication by inhibiting DNA synthesis in developing brain cells of rats of postnatal day (PND) 1 through a non-cholinergic mechanism (Whitney et al., 1995). In rats of the same age, when CPF (1 or 5 mg/kg) is administrated for longer (4 days) DNA, RNA and protein levels are affected, indicating significant brain loss (Campbell et al., 1997; Johnson et al., 1998). In contrast, the lower dose (1 mg/kg) induces a decrease in DNA synthesis, but no change is noted in brain RNA and protein synthesis (Dam et al., 1998). Older pups (PND 11) treated with 4 daily doses of CPF (5 or 25 mg/kg) present changes in both DNA and protein levels after 5 days (Campbell et al., 1997). Interestingly, CPF does not target only brain areas rich of cholinergic neurons but also affects other regions such as cerebellum that lacks cholinergic innervation (Campbell et al., 1997).

Besides postnatal, prenatal administration of CPF, at doses that do not affect AChE activity in brain, also causes decrease on the number of brain cells estimated by reduced DNA concentration (Qiao et al., 2002). This effect is less severe compared to postnatal CPF treatments described above (Whitney et al., 1995; Campbell et al., 1997). Furthermore, increased cell size due to elevated protein/DNA ratio has been recorded in rat pups that lasted up to PND 60 (Qiao et al., 2003).

In vitro studies further confirmed the above mentioned in vivo findings, permitting the precise distinction of CPF antiproliferative effects on specific brain cell population (neuronal or glial cells) rather than assessing brain homogenates that contain a mixed population of cells. CPF, at doses that do not alter cell viability, inhibits DNA synthesis in proliferating rat pheochromocytoma PC12 cells within 1 hour and without affecting the activity of AChE (Qiao et al., 2001). In undifferentiated PC12 cells, CPF (5 and 30 μM) inhibits cell replication measured by DNA synthesis that lasts up to 4-6 days (Slotkin et al., 2007a). After 6 days CPF causes significant reduction in cell number determined by DNA content analysis (Slotkin et al., 2007a) whereas, cell size does not change after calculating protein/DNA ratio. These findings related to antimitotic effects of CPF have been further supported by identifying transcriptional changes in 40% of all genes related to cell cycle in both undifferentiated and differentiated PC12 cells and by successfully predicting 60-70% of the genes affected by CPF in vivo (Slotkin and Seidler, 2012). CPO can also interfere with the proliferation of human (ReN CX) and mouse (mCNS) neuroprogenitor cell lines and decreases proliferation by half compared to controls based on BrdU incorporation assay adapted in a high-content screening approach (Culbreth et al., 2012). Finally, the parent compound CPF can affect the proliferation of type-2 progenitor cell populations in the hippocampal neurogenesis of rat offspring (Ohishi et al., 2013).

Several studies have been designed to explore the mechanisms involved in CPF-interference with the processes of cell replication in an attempt to elucidate the involvement of non-cholinergic mechanisms in CPF-induced developmental neurotoxicity. These studies revealed that cell signaling pathways, transcription factors and neurotrophic growth factors that are known to regulate neuronal cell replication and differentiation can be disturbed by CPF, shedding light on mechanistic understanding of CPF-induced developmental neurotoxicity.

The adenylyl cyclase-cAMP transduction pathway has been suggested as the main pathway that is affected by CPF in newborn rats (Song et al., 1997; Adigun et al., 2010a). Besides the CPF-induced alterations in the expression and activity of adenylyl cyclase itself, the function of G-proteins, which link cyclase activity to neurotransmitter receptors, and the expression and function of neurotransmitter receptors acting through this cascade are also affected by CPF. CPF also interferes with the expression and function of specific nuclear transcription factors such as AP-1, Sp1 (Crumpton et al., 2000; Slotkin and Seidler, 2007c; Adigun et al., 2010a; 2010b) and Ca^{+2}/cAMP response element binding protein (CREB) (Schuh et al., 2002; Slotkin and Seidler, 2007c). These transcription factors are downstream effectors of this pathway and are involved in the regulation of genes involved in brain cell proliferation. CPF has also been shown to interfere with calcium homeostasis suggesting that this is the prior cellular effect that can lead later to disturbance of CREB pathway, whereas CPO failed to do so (Meijer et al., 2014). Protein kinase C is another signaling cascade with important role in neuronal development through neurotrophic factor regulation that is disturbed both in vitro and in vivo by CPF (Slotkin and Seidler, 2009b; Yanai et al., 2010). Genes belonging to mitogen-activated protein kinase (MAPK) pathway have also been found to be altered after CPF acute exposure in rats PND 7 (Ray et al., 2010).

Neurotrophic factors are known to play pivotal role in neuronal cell development and for this reason the effects of CPF on their protein and the mRNA levels have been studied. The levels of the nerve growth factor (NGF) in the forebrain are decreased after CPF treatment (1.5 or 3 mg/kg) of rats PND 1 for 6 days whereas, the levels of the brain derived neurotrophic factor (BDNF) are not affected (Betancourt and Carr, 2004; Betancourt et al., 2006). Slotkin et al. (2007b) have found significant changes after CPF exposure on the levels of mRNAs coding for a number of different neurotrophic factors belonging to the fibroblast growth factor (FGF) superfamily that is considered important in neuronal development. Another microarray study from the same research group revealed that neonatal rat and PC12 cell exposure to CPF affects FGF, neurotrophins, Wingless-type (WNT) and Frizzled homolog (FZD) gene families and to lesser extend affects BDNF and NGF groups (Slotkin et al., 2008).

It is evident that developmental exposure to CPF leads to interference with various cell signaling pathways, transcription and neurotrophic factors that are known to be important for brain cell proliferation. All these molecular findings are associated but not necessarily correlated with CPF antiproliferative effects.

Effects of Developmental Exposure of CPF on Apoptosis through Disruption of Intrinsic and Extrinsic Pathways

Although in the cells of most organs and tissues programmed cell death occurs throughout life, in nervous system, the vast majority of neurons undergo their last round of cell division early in development. Neurons start to differentiate right after exiting the cell cycle. At this phase about half of the neurons are highly susceptible and undergo programmed cell death (apoptosis). After this phase and during maturation the tendency of neurons to undergo apoptosis progressively diminishes. Special mechanisms that involve both intrinsic (mitochondrial related pathways) and extrinsic pathways (death receptor related pathways) are involved in developmental regulated apoptotic machinery (reviewed in Creeley and Olney, 2013).

CPF induces apoptosis in the neuroepithelium of cultured rat embryos at embryonic day 9.5 (Roy et al., 1998). This ability of CPF to induce apoptosis has also been demonstrated by using primary cortical neurons from embryonic day 17/18 or from newborn rats. Embryonic cortical neuronal cultures show higher sensitivity to CPF-induced apoptosis compared to the cells derived from newborn rats that show resistance to apoptotic ability of CPF (Caughlan et al., 2004). Although CPO causes greater apoptotic effect than CPF, CPF-induced apoptosis is not attributed to inhibition of AChE (Caughlan et al., 2004). CPO also induces at least a doubling in the apoptotic markers caspase 3 and the tumor protein 53 (p53) in mouse neuroprogenitor cell line mCNS but has no effect in the correspondent human cell line ReN CX (Culbreth et al., 2012).

Postnatal administration of CPF to newborn rats (PND 1-4) affects the expression of genes related to apoptosis (Slotkin and Seidler, 2007c; Slotkin and Seidler, 2012). More specific, CPF, at doses that cause little or no AChE inhibition, increases the expression of genes encoding for caspases 1, 4, 9 and 12, p53 and certain bcl 2-associated proteins (Slotkin and Seidler, 2007c). The undifferentiated and differentiated PC12 cells have been used to compare the effects of CPF on gene transcription known to be involved in apoptosis with the associated findings in neonatal rats, revealing 60-70% overlap in the pattern of affected genes between the two models (Slotkin and Seidler, 2012). Finally, in vivo data revealed that the expression of a group of genes associated with mitochondrial dysfunction is altered after CPF acute exposure in rats PND 7, suggesting the possible involvement of intrinsic pathways in apoptotic effect induced by CPF (Ray et al., 2010).

Effects of Developmental Exposure of CPF on Axonogenesis (Neurite Outgrowth)

CPF has been found to impair neurite outgrowth in in vivo studies both after pre- and postnatal exposure. Administration of CPF to rat pups PND 1 for 4 days at a daily dose of 1 mg/kg decreases neurite outgrowth as it has been determined by measuring the ratio of membrane protein to total protein (Slotkin et al., 2006). Prenatal exposure to CPF at doses of 1 or 5 mg/kg daily on gestational days 17-20 also causes contraction of neurites (Qiao et al., 2003).

In vitro, CPF has been shown to impair neurite outgrowth in both neurotypic cell lines and primary neuronal cultures. PC12 cells induced to differentiate by NGF in the presence of 50 μg/ml CPF for 2 days (Song et al., 1998) or to 10 μg/ml CPF for 7 days (Das and Barone, 1999) show reduced number of neurite extensions. CPF also affects neurite outgrowth in cultured mouse N2a neuroblastoma cells induced to differentiate by serum withdrawal and the addition of dibutyryl cAMP. In these cells, CPF, at a concentration of 3 μM, is capable of causing significant inhibition of neurite outgrowth after only 4 hours (Sachana et al., 2001; 2005). The same cell line after longer exposure (24 h) to CPF indicated that the IC_{50} value for neurite outgrowth inhibition is 13 μM (Axelrad et al., 2003). In another mouse neuroblastoma cell line named N-18, CPF also inhibits neurite outgrowth with an IC_{50} value of 44 μM (Henschler et al., 1992). In primary cultures of embryonic rat sympathetic neurons derived from superior cervical ganglia and in sensory neurons derived from embryonic rat dorsal root ganglia, CPF after 24 h exposure, at a concentration as low as 0.001 μM, interferes with the outgrowth of axonal processes by decreasing axonal length, without affecting the number of axonal processes (Howard et al., 2005; Yang et al., 2008). CPO, at a concentration 10 ng/ml, impairs the extension of neurites in PC12 cells line after 7 days of treatment (Das and Barone, 1999). CPO, at a low concentration as 0.001 nM, also decreases within 24 hours axonal length in primary neuronal cultures derived from embryonic rat superior cervical ganglia (Howard et al., 2005) whereas, at a concentration of 0.01 nM, inhibits axonal outgrowth in primary cultures of neurons derived from embryonic rat dorsal root ganglia (Yang et al., 2008).

Effects of Developmental Exposure of CPF on Synaptogenesis

Synaptogenesis is a multi-step process that is crucial for brain development and involves the formation of synapses. It follows axonal migration, at which stage presynaptic and postsynaptic differentiation occurs (Garner et al., 2002). "Synaptic assembly" that refers to the gathering of the appropriate components and "synaptic formation" that is defined by the mechanisms involved in recruitment of molecules required for differentiation, stabilization and maturation of synapse, are the main phases that characterise synaptogenesis (Colón-Ramos, 2009). Elimination is also another step involved in synaptogenesis and is relevant to synapses that fail to get stabilised and mature. The effects of CPF on synaptic development have involved markers of both the cholinergic and the monoaminergic (catecholaminergic and serotonergic) pathways.

Administration of CPF to rat pups PND 1 for 4 days at a daily dose of 1 mg/kg causes within 24 hours a decrease in choline acetyltransferase (ChAT) activity, without affecting hemicholinium-3 binding to the choline transporter and expression levels of m_2-muscarinic receptors (Dam et al., 1999). This rapid onset of alterations in cholinergic pathway following CPF exposure persists into adolescence and adulthood and ultimately negatively affects cholinergic synaptic activity, as indicated by decrease in hemicholinium-3 binding (Slotkin et al., 2001) and neurobehavioral alterations (Levin et al., 2001). Exposure of pregnant rats on gestational days 17-20 at a daily dose of CPF (1 mg/kg) that does not affect fetal brain AChE activity, disturbs cholinergic system development in forebrain, which is a brain region involved in cognition (Qiao et al., 2003). These deficits extend into adolescence and adulthood (PND 60) and correspond to the observed long-term impairments of memory (Levin et al., 2002). On the other hand, CPF does not cause any alterations in the development of cholinergic nerve terminals, as ChAT activity has not been affected. Additionally, these deficits are no longer present by PND 60 (Qiao et al., 2003). The lasting deterioration of synaptic activity is not compensated by up-regulation of cholinergic receptors, as m_2-muscarinic receptor binding does not change.

Richardson and Chambers (2003; 2004; 2005) evaluated the effects of CPF (7-1 mg/kg) after gestational and post natal exposure to presynaptic cholinergic endpoints such as ChAT, mAChR, high affinity presynaptic choline transporter (HAChT) and vesicular Ach transporter (VAChT). During gestational CPF treatment, the changes at these presynaptic indicators have been found to last even after a month from the treatment and at conditions where inhibited AChE returned to physiological levels. Following postnatal CPF administration, the same presynaptic parameters have been affected at doses that significantly inhibited the activity of AChE (Richardson and Chambers, 2005).

The effects of CPF on the developing cholinergic system have also been studied at the transcriptional level. The use of gene microarray technology has revealed a number of effects on the transcription of genes related to ACh synthesis, storage, degradation and receptors following OP administration to neonatal rats at doses causing no or less than 20 percent AChE inhibition (Slotkin and Seidler, 2007c). The most important effect of CPF is a reduction in the expression of the gene coding for ChAT, an effect consistent with the reductions in the activity of this enzyme noted in the previous studies. The reduction in the expression of the gene encoding ChAT is accompanied by changes in the expression of genes coding for the choline transporter indicating that CPF interferes with the development of the ACh phenotype. The latter effect has been suggested to cause miswiring of brain circuits during development and behavioral abnormalities (Slotkin and Seidler, 2007c). CPF also induces large decreases in the expression of the genes encoding the m_2-muscarinic and α7-nicotinic receptors.

Besides cholinergic synapses, a number of studies suggest that serotonergic as well as dopaminergic and noradenergic synapses might be affected by CPF (Dam et al., 1999; Raines et al., 2001; Slotkin et al., 2002; Aldridge et al., 2003; 2004; 2005b, 2005c; Slotkin and Seidler, 2005; 2007a; 2007b; Eells and Brown, 2009; Chen et al., 2011; Slotkin et al., 2014). Gestational or postnatal exposure to CPF can alter 5HT1 transporters and 5HT transporter binding in different brain regions. More specifically, early postnatal exposure of rat pups to CPF causes increase in the expression of 5HT synaptic proteins as an adaptive response to deficits in 5HT turnover (index of presynaptic impulse activity) from PND 30 to 150 (Slotkin et al., 2014). In PC12 cells, CPF (30 μM) increases the expression of most 5HT receptor subtypes, reduces the expression of 5HT transporter genes and induces tryptophan hydroxylase that all contribute to the augment of extracellular 5HT (Slotkin and Seidler,

2008). Early (7.5-11.5 GD) and late (13-17 GD) gestational administration of CPF elicits a decrease in levels of dopamine in hippocampus and cortex at 14 and 60 PND (Chen et al., 2011). Repeated exposure to CPF during PND 1-21 alters dopamine metabolism that can last up to one month (Eells and Brown, 2009).

Effects of Developmental Exposure of CPF on Gliogenesis

Glial cells play vital role in axonal guidance, synaptogenesis and architectural modeling of the developing brain (LoPachin and Aschner, 1999; Ullian et al., 2004). Disturbance of gliogenesis by chemical substances can consequently alter these processes that continue well into childhood, inducing developmental neurotoxicity. Initial indications of glial involvement in CPF-induced developmental neurotoxicity have been drawn from in vivo studies showing that its adverse effects in the intact brain occur at a time of vigorous glial development. CPF has been found to inhibit DNA synthesis (Whitney et al., 1995; Dam et al., 1998) causing brain cell loss (Campbell et al., 1997) after completion of neurogenesis but during the gliogenesis process. CPF effects on cell signaling (Song et al., 1997) and synaptic development and activity (Dam et al., 1999) occur during the period of gliogenesis. Further support is provided by the findings that neurobehavioral effects are induced by CPF when exposure occurs at times of active gliogenesis (Moser and Padilla., 1998; Dam et al., 2000; Levin et al., 2001).

At the peak of glial development (PNDs 11-14), the administration of CPF to rats suppresses the levels of the specific astrocyte marker glial fibrillary acidic protein (GFAP) in all brain areas, suggesting specific and targeted reduction of normal glial development (Garcia et al., 2002). In contrast, administration of CPF during gestation when limited gliogenesis occurs has no effect on fetal brain GFAP. In aggregating cell cultures of fetal rat telenchephalon, activity levels of glutamine synthase (a marker of astrocytes) and cyclic nucleotide phosphohydrolase (a marker of oligodendrocytes) as well as GFAP have been altered after treatment with CPO (Monnet-Tschudi et al., 2000; Zurich et al., 2000).

CPF causes immediate inhibition of DNA synthesis in C6 cells (Qiao et al., 2001; Garcia et al., 2001), a marked effect that is independent of cholinergic stimulation. DNA synthesis in the same cell culture model is also inhibited by CPO (Qiao et al., 2001). CPF and its in vivo oxon metabolite also inhibit DNA synthesis in 1321N1 cells (Guizzetti et al., 2005).

CPF affects the adenylyl cyclase transduction cascade, a signaling pathway that plays pivotal role in glial cell proliferation and differentiation. CPF causes considerable impairment of G-protein signaling in undifferentiated C6 cells, leading to persistent changes in the catalytic properties of adenylyl cyclase (Garcia et al., 2001). CPF disturbs G-protein functioning in differentiating C6 cells, indicating that its adverse effects extend beyond the phase of glial cell replication and last up to the glial differentiation stage. Additionally, CPF interferes with the expression of the transcription factor Sp1, a downstream target of the adenylyl cyclase pathway which is necessary for C6 cell differentiation (Garcia et al., 2001). On the other hand, CPF does not affect in cultured astrocytes the activation of the transcription factor CREB, another target of the adenylyl cyclase pathway that is important in cell differentiation (Schuh et al., 2002).

Administration of CPF to rats on PNDs 1-6, at doses causing considerable AChE inhibition, decreases the expression of the gene encoding the oligodendrocyte marker myelin-associated glycoprotein, but elicits an increase in the expression of the gene encoding GFAP, an effect indicative of increased astrocyte reactivity due to the high dosages employed (Betancourt et al., 2006). Treatment of cultures of human fetal astrocytes with CPF for 7 days has a major impact on the expression of genes coding for various developmentally important molecules, with a total of 35 genes being affected (Mense et al., 2006). Administration of CPF to rats on PNDs 1-4, at doses causing little or no AChE inhibition, also affects the expression of a number of genes related to glial development. CPF elicits a marked decrease in the expression of the gene encoding GFAP, indicative of a direct, specific effect on glia in the absence of neuronal cell injury (Slotkin and Seidler, 2007c).

The ability of CPF to interfere with the differentiation of glial cells has also been shown at the morphological level, as these compounds can impair the development of extensions from cultured C6 cells under differentiation-promoting conditions. CPF inhibits the extension outgrowth from differentiating C6 cells exposed to the OPs for a total of 20 days (Henschler et al., 1992). CPF and CPO at biological relevant concentrations suppresses within 24 hours extension outgrowth in differentiating C6 cells (Sachana et al., 2008).

Comparison of the effects of CPF on glial development with those exerted on the development of neuronal cells reveals that glial cells are more sensitive to CPF. CPF is more effective in inhibiting DNA synthesis in gliotypic C6 cells than in neuronotypic PC12 cells (Qiao et al., 2001; Garcia et al., 2001). In vivo, the oral administration of CPF to rats during 1-6 PND affects the expression of genes related to glial-specific markers and not to β-III tubulin, which is a neuronal marker throughout development (Betancourt et al., 2006). However, there are reports suggesting that the effects of CPF on developing glial cells are not as significant as those detected on developing neuronal cells. For example, CPF alter the expression of transcription factors AP-1 and Sp1 in undifferentiated PC12 cells (Crumpton et al., 2000), but causes no change on replicating C6 cells (Garcia et al., 2001). In addition, CPF elevates the levels of transcription factor CREB in cortical and hippocampal neurons, without inducing alterations in cultured astrocytes derived from developing rats (Schuh et al., 2002). Finally, CPF and CPO affect to lesser extend the expression of glial-specific markers than the expression of neuronal-specific markers in aggregating brain cell cultures (Monnet-Tschudi et al., 2000).

CONSIDERATIONS FOR THE EMBRYOTOXICITY RISK ASSESSMENT OF CPF

The information available in the scientific literature regarding CPF-induced embryotoxicity is considered contradictory. Most of the studies performed, which followed the OECD Guidelines for developmental or reproductive toxicity in vivo, have reported embryotoxicity only at doses with pronounced maternal toxicity (FAO/WHO, 1999; Eaton et al., 2008). Furthermore, epidemiological studies have not proved conclusive outcome since Mink et al. (2012) reviewed four independent epidemiological cohort reports and found no causal association between CPF exposure during pregnancy and measures of fetal growth. In vitro approaches, including expression of marker genes during early differentiation at concentrations that do not affect viability might address the neurodevelopmental risk of CPF (Estevan et al., 2013). The above-stated discrepancies might be explained on the basis of the route of exposure. Indeed, the OECD Guidelines almost exclusively consider oral exposure (Estevan et al., 2011), where the first passage of administered CPF through the liver of the mother would massively bioactivate CPF to CPO, which is a more powerful AChE inhibitor and would cause clinical cholinergic signs before reaching embryotoxic concentrations. However, other animal studies have considered different routes, such as subcutaneous exposure, where less CPF reaches the liver and CPO appears more slowly than in oral exposure, allowinghigher and potentially toxic CPF concentrations to reach the embryo with no significant maternal toxicity signs. Similar situations may appear during human exposure to CPF, which can essentially be through inhalation or skin, therefore entailing greater risk. In order to assess the potential risks associated to the exposure of CPF, the classical approach has been followed (van Leeuwen and Vermeire, 2007), including the characterization of the hazard profile of the substance and the determination of the realistic exposure via the existing routes. The hazard characterization of the substance includes a description of the relevant effects, in this case focusing on the developmental hazard assessment as well as the classification triggered by it. The exposure in different use scenarios and sources, as well as the combined exposure, and an attempt to extrapolate from in vitro concentrations to in vivo doses are described in the sections below.

Developmental Hazard Assessment of CPF

In January 2014, the European Food Safety Agency (EFSA, 2014) published a human health risk assessment on CPF and its conclusions were reached on the basis of the evaluation of the authorized representative uses of CPF as an insecticide. Based on this report that relied on newly available toxicological data, a decrease of the reference values that had been previously agreed has been recommended. The new reference values such as Acceptable Daily Intake, Acceptable Operator Exposure Level and Acute Reference Dose were based on red blood cells (RBC) ChE inhibition data derived from rats. The EFSA report considered that CPF does not show genotoxic and carcinogenic potential and it is not a reproductive and developmental toxicant. CPF has also been evaluated earlier by the U.S. Environmental Protection Agency (U.S. Environmental Protection Agency, 2000), addressing its developmental toxicity in rats, mice and rabbits. CPF's developmental properties are unclear as contradicted findings have been reported. The studies included in the U.S. Environmental Protection Agency dossier showed increased post-implantation loss after exposure of rats to CPF (15 mg/kg/day) associated with maternal toxicity while another study fails to observe similar developmental effects at the same dose. Developmental effects at high CPF-metabolite trichloropyridinol (TClP) doses (up to 250 mg/kg/day) are not observed in mice (Hanley et al., 2000). Similarly, rabbits receiving 140 mg/kg/day of CPF showed decrease fetal weights and crown rump lengths, and unossified xiphisternum and/or 5th sternebra (U.S. Environmental Protection Agency, 2000). However, in both mice and rabbits, the developmental effects occured at maternally toxic doses as indicated by reduced weight gain and food consumption in both species, including increased mortality in case of mouse dams. In an in vivo developmental neurotoxicity study, CPF was associated with alterations in brain development in rat offspring that involved gestational exposure (Icenogle et al., 2004). Specifically, female pups of the 1 mg/kg/day group exhibited significant treatment-related decreases in measurements of the parietal cortex at PND 66. The only maternal effect at that dose was plasma and RBC ChE inhibition. At higher doses, pups of the 5 mg/kg/day group exhibited decreased body weight/body weight gain and food consumption in both sexes, reductions in pup viability, delays in development, decreased brain weight and morphometric alterations in the brain. However, these effects were observed in the presence of maternal toxicity as evidenced by fasciculations, hyperpnea and hyperactivity, in addition to reduced body weight gain. Several studies in the peer reviewed literature mentioned above are supportive of the possibility that CPF exposure may affect brain development (e.g., altered synaptic development, alterations in DNA, RNA, and protein synthesis, inhibition of mitosis and mitotic figures, and disruption of the structural architecture of the brain). Most importantly, there are suggestive data that these effects may arise independently of ChE inhibition.

Classification of CPF

CPF is classified according to the European Chemicals Agency in the Hazard Class Acute Tox. Cat 3 since it is toxic if swallowed. Based on its environmental toxicity, CPF is classified as Aquatic Acute Cat 1 and Aquatic Chronic Cat 1, because it is very toxic to aquatic life causing long lasting effects.

In what regards the classification for reproductive toxicity or developmental effects, CPF is not classified for effects on or via lactation or for reproductive toxicity since the evidences of the substance producing those effects have not been considered sufficient for classification.

CPF Exposure

The potential sources of exposure to CPF rely on its use that is mainly restricted nowdays in agriculture and consequently may contaminate food and drinking water. The residential use of CPF has been banned since 2001 in US. The Environmental Protection Agency concluded in its report (U.S. Environmental Protection Agency, 2000) that acute, short-term, aggregate CPF exposure through food and water does not exceed the level of concern for the general public. Furthermore, the chronic dietary risk estimate associated with CPF exposure was below EPA's level of concern. These aggregate assessments performed by EPA contained mitigation plans that would reduce potential CPF exposures through food and in the residential/recreational environment.

The correlation found between in vivo exposures and in vitro doses is always hard to estimate. Nonetheless, several attempts have been done in order to correlate the in vitro doses with an equivalent in vivo exposure. Albers et al. (2007) suggested that a person occupationally exposed to the Threshold Limit Value (TLV) (200 μg/m^3) at a respiratory rate of 10 m^3 air/working day via the inhalatory route, by assuming 100% of pulmonary absorption, would be exposed to a systemic dose of 2000 μg CPF/day. Assuming that CPF is distributed according to a monocompartimental model, which seems logical since OPs do not bioaccumulate (WHO, 1986), by contemplating a corporal volume of 38 l for a person weighing 70 kg (Lehman-McKeeman, 2009), the expected systemic concentration of CPF would be around 0.2 μM. On the other hand, it was estimated that adult exposures to CPF could be around 11.4 μg/kg/day (Cochran, 2002). If the considerations as above are assumed, then the systemic CPF concentrations could be around 0.1 μM. To conclude, the estimated systemic concentration in both cases is still far from the lowest CPF concentrations causing significant alterations to gene expressions, of either 10 μM for 12 hours exposure or 20 μM for 3 days exposure (Estevan et al., 2013). Other studies have considered that the acute reference dose for CPF is 0.1 mg/kg, estimated based on inhibition of blood ChEs and by assuming a safety factor of 10 (FAO/WHO, 1999). If we use the same toxicokinetic parameters as those above to estimate the systemic concentration needed to reach the acute reference dose, we can conclude that it may be around 0.5 μM.

It is remarkable that the above-stated reference values (TLV and acute reference dose) were obtained on the basis of statistically significant alterations of blood ChE and by applying a safety factor of 10. However, statistically significant alterations of blood ChE do not necessarily imply the observation of any clinical signs since they do not appear below the 70% inhibition of brain ChE (Clegg and van Gemert, 1999). Therefore, exposure scenarios where occupational hygiene is not strictly observed may cause CPF exposures which, without clinical maternal symptoms because AChE inhibition is not high enough, reach similar systemic concentrations to those causing significantly altered gene expressions. The highest urinary excretion of TClP reported for workers involved in manufacturing CPF was 325 μg/day (Chen et al., 2011). By assuming 1 l of urine/day and the above-stated toxicokinetic and physiological parameters (70 kg, 38 l of corporal volume and no bioaccumulation), the estimated systemic TClP concentration could be around 43 nM. This figure is several orders of magnitude lower than that which caused significantly altered gene expressions. Thus, the risk of embryotoxicity due to CPF exposure is not expected through its metabolite. Regarding CPO, it is also remarkable to note that gene expressions alter only at concentrations that cause severe AChE inhibition and, therefore, severe maternal toxicity.

AGE-RELATED IMPLICATIONS FOR CPF RISK ASSESSMENT

The weight of evidence derived from experimental studies provides appreciable support for the increased sensitivity that young rodents demonstrate towards the neurotoxic effects of CPF compared to adult and for the susceptibility of the developing brain to CPF. Various in vivo studies clearly indicate that at a given oral dose the response of young animals to both cholinergic- and non-cholinergic-related effects of CPF (as defined biochemically and behaviorally) is more pronounced than the ones recorded in adult animals.

At high acute doses, CPF is fatal to rat pups, but produces no lethality and little or no behavioral changes in adult rats (e.g. the LD10 and MTD doses are 15 mg/kg and 100-136 mg/kg for neonates and adults, respectively). At LD10 or MTD doses, neonates are up to ~5-fold more sensitive than adult rats to ChE inhibition (brain and blood) and clinical/behavioral effects-induced by CPF. Furthermore, at a single treatment of CPF (15 mg/kg), the down-regulation of the cholinergic (muscarinic) receptors was more extensive in pups than in adults treated even with higher dose (80 mg/kg) (Qiao et al., 2003). The magnitude of change, the effective time points, and the brain regions involved were different in pups compared to adult rats, suggesting that the cholinergic receptors are more readily altered in pups following CPF treatment. Although the consequence of this is unknown, cholinergic system is well established to play an important role in normal brain development and might explain partially the increased sensitivity of developing organisms towards CPF neurotoxicity.

The increase in sensitivity between young and adult animals appears to occur even at acute doses of CPF below 15 mg/kg. The study by Zheng et al. (2000) using low dose levels ranging from 0.15 mg/kg to 15 mg/kg provides AChE inhibition data from 7-day old animals and adult male rats. Young animals show ChE inhibition up to ~3-fold in RBC and plasma and at least 5-fold in brain than adults. At the high dose of CPF (15 mg/kg) adult brain AChE activity remained unaltered, meaning that a difference in response greater than 5-fold cannot be ruled out between young and adult animals regarding this biochemical parameter (Zheng et al., 2000). Due to lack of data, the extent of differences in brain AChE activity between pups and pregnant female rats remains uncertain. Although young animals appear to recover at least two times faster than adult animals from AChE inhibition induced by acute CPF treatment, other neurotoxicological manifestations (e.g. delays in brain development, neubehavioral effects) may persist or appear at a later time.

Repeated dosing with CPF does not appear to result in higher brain or blood ChE inhibition in neonates relative to adults (Lassiter et al., 1998) with one exception. Based on ED50s, there is a 1.5-fold difference in brain AChE inhibition reported from PND 7 pups compared to adult males (Zheng et al., 2000). In contrast to the rapid recovery from AChE inhibition observed after acute CPF treatment of neonates (Pope and Liu, 1997), repeated dosing with CPF (every other day, 11 treatments during PND 1 to PND 21) indicated that AChE inhibition persists from approximately 9 to more than 19 days depending on the administered dose (Tang et al., 1999). According to the studies submitted in the U.S. Environmental Protection Agency report in 2000, changes in body weight changes and behavior might occur at doses up to 3 times lower in neonates than in adult rats after repeated treatment with CPF at doses from 3 mg/kg/day.

It is apparent that AChE activity is inhibited in the fetus if the dam is treated with CPF doses which can be absorbed by the fetus. The magnitude of brain, plasma, and RBC ChE inhibition in the fetus is less or equal to that observed in dams that received acute or repeated treatment with CPF. The lack of an apparent differential response of the fetus or neonates that received repeated dosing of CPF versus the maternal system may be attributed to the increased new synthesis or more rapid turnover of inhibited molecules of AChE in the fetal brain than in the adult (Lassiter et al., 1998; Mortensen et al., 1998).

Differences in detoxification between young and adult animals may explain the increased sensitivity of exposed pups to CPF toxicity. CPF and its oxon metabolite CPO are detoxified by binding to carboxylesterases and by hydrolysis from A-esterases. Young animals possess minimal activity of these detoxification enzymes compared to adult animals. The precise influence of these enzymes on sensitivity to CPF treatment has not been established yet. Because detoxification enzyme activities increase with age, the enzymatic profile of newborn rats raises concern related to the sensitivity of newborns even in comparison with older neonates that receive acute CPF treatment. There is some evidence (albeit at high doses) suggesting that the magnitude of the differential sensitivity between young and adult animals depends on the age of the animal. Based on the LD10 values derived from Zheng et al. (2000) and the ChE inhibition data from Zheng et al. (2000) and Moser and Padilla (1998), the order of sensitivity would be PND 7 > PND 17 > PND 27 > adult female > adult male. Therefore, given that 7-day old rats are the youngest animals that have been evaluated until now in relation to these parameters, it is uncertain whether the magnitude of differential sensitivity would be greater in pups exposed earlier than PND 7 to CPF.

A variety of in vitro and in vivo studies published in the peer reviewed literature (reviewed in Slotkin, 1999) showed the possibility of CPF altering synthesis of macromolecules, neuronal activity, neurotransmitter levels, neurite outgrowth and branching, and cell signaling in the developing rat brain. These studies did not include measures of direct adverse effects such as functional effects but rather used biomarkers. Nevertheless they raised concern that CPF can potentially affect processes influencing cell replication and differentiation needed for normal function that occur in early and late developmental periods of brain growth. Furthermore, several of the key responses observed are highly significant and robust (e.g., effects on norepinephrine turnover, DNA synthesis, adenylyl cyclase transduction). The available data suggest a selective action of CPF on the developing brain, given the regional and temporal pattern of responses (U.S. Environmental Protection Agency, 2000). Thus, it seems unlikely that the observed effects are due to nonspecific toxicity.

The mechanisms of action for the CPF-induced changes (e.g., macromolecular synthesis, cell signaling) are unclear. However, since these effects can be found after intracisternal injection ofCPF, with *in vitro* TCIP exposure, and in vitro PC12 cell cultures with limited capability to activate CPF to its ChE-inhibiting oxon (CPO), raises the issue of whether these effects can occur independently of ChE inhibition. Although it is not possible to link each effect reported with another effect or with a functional outcome, the data show a consistent pattern of the potential for CPF to produce qualitatively different effects in the CNS of young versus adult animals (U.S. Environmental Protection Agency, 2000). Potential implications of the effects include alteration of synaptic responses that are programmed by neural input, disruption of cell replication and differentiation, and temporary or persistent delays in the development of CNS structures.

Overall, the weight of the evidence raises concern for an increase in both the sensitivity and susceptibility of the fetus or young animals to adverse biochemical, morphological, or behavioral alterations from exposure to CPF during brain development. With respect to ChEinhibition, an increase in sensitivity of the young compared to adults was seen all along the dose response curve, even at relatively low doses. There is a clear differential response (2- to approximately 5-fold) in the young compared to the adult animal after an acute exposure to a relatively low dose of CPF. There was also increased sensitivity found after repeated dosing (up to 9-fold), but at the LD10 and MTD. It is important to point out that uncertainty remains concerning the magnitude of the differential response, given that newborn animals (less than PND 7) have not been characterized for sensitivity. Effects on the developing CNS indicative of the unique susceptibility to the young animal included changes in macromolecular synthesis, altered cell signaling and muscarinic receptor down-regulation, as well as morphological alterations in brain development. The effects observed raise a high degree of concern that the fetus or young animal is particularly susceptible to adverse outcome if exposed to CPF.

CONCLUSION

Accumulating evidence suggests that CPF induces AChE-related and -unrelated adverse effects leadingto cognition, motor activity and disturbance of emotional responses in developing organisms and a mapping of the different molecular, cellular and organ events involved is presentedin Figure 1. Disturbance of neurotransmission has been suggested to be associated to this adverse effect that is not limited to cholinergic system. At organ level the reduced neuronal network formation due to severe effect of CPF on synaptogenesis can also explain the impairment of learning and memory induced by CPF. At cellular level, neural and/or glial cell loss is one of the most common findings that underlie CPF-developmental neurotoxicity in both vitro and in vivo models. Antimitotic and pro-apoptotic mechanisms that operate within a critical window of neurodevelopment seem to be involved covering specific cell signaling pathways, transcription and neurotrophic factors. However, these mechanistic findings may be the direct effect of CPF on brain development or indirect consequence associated with alterations at hormonal level.

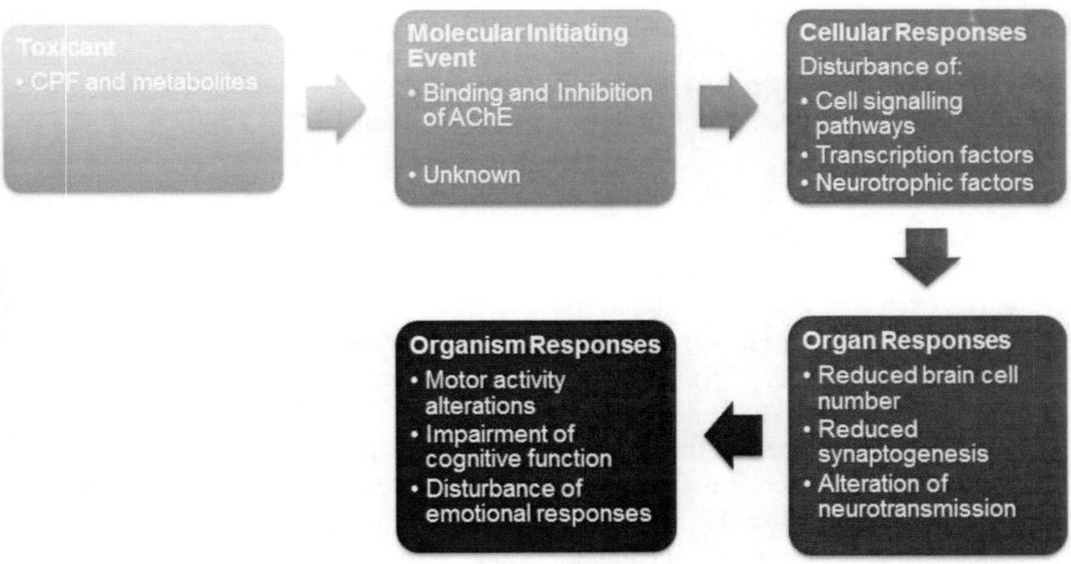

Figure 1. Mapping of the different molecular, cellular and organ events involved in AChE-related and -unrelated adverse effects induced by CPF during neurodevelopment.

The assessment of the risk of exposure to CPF by general population during development remains an issue to be further investigated. Our understanding about CPF's ability to produce developmental neurotoxicity and other toxicities related to developing organisms is not complete and consequently not sufficient for classification during the different evaluation processes of this pesticide. However, as the mechanistic understanding related to CPF-induced developmental neurotoxicity is increasing, IATAs will emerge and regulators will start to gain experience in using mechanistic data in decision making protecting more efficiently developing organisms not only from CPF but also by the hazardous substances with similar properties.

References

Abou-Donia, M. B., Khan, W. A., Dechkovskaia, A. M., Goldstein, L. B., Bullman, S. L., & Abdel-Rahman, A. (2006) In utero exposure to nicotine and chlorpyrifos alone, and in combination produces persistent sensorimotor deficits and Purkinje neuron loss in the cerebellum of adult offspring rats. *Arch. Toxicol.* 80: 620-631.

Adigun, A. A., Seidler, F. J., & Slotkin, T. A. (2010a) Disparate developmental neurotoxicants converge on the cyclic AMP signaling cascade, revealed by transcriptional profiles in vitro and in vivo. *Brain Res.* 1316: 1-16.

Adigun, A. A., Ryde, I. T., Seidler, F. J., & Slotkin, T. A. (2010b). Organophospahte exposure during a critical developmental stage reprograms adenylyl cyclase signaling in PC12 cells. *Brain Res.* 1329: 36-44.

Albers, J. W., Garabrant, D. H., Mattsson, J. L., Burns, C. J., Cohen, S. S., Sima, C., Garrison, R. P., Richardson, R. J., & Berent, S. (2007) Dose-effect analyses of occupational chlorpyrifos exposure and peripheral nerve electrophysiology. *Toxicol. Sci.* 97: 196-204.

Aldridge, J. E., Seidler, F. J., Meyer, A., Thillai, I., & Slotkin, T. A. (2003) Serotonergic systems targeted by developmental exposure to chlorpyrifos: Effects during different critical periods. *Environ. Health Perspect.* 111: 1736-1743.

Aldridge, J. E., Seidler, F. J., & Slotkin, T. A. (2004) Developmental exposure to chlorpyrifos elicits sex-selective alterations of serotonergic synaptic function in adulthood: critical periods and regional selectivity for effects on the serotonin transporter, receptor subtypes, and cell signaling. *Environ. Health Perspect.* 112: 148-155.

Aldridge, J. E., Levin, E. D., Seidler, F. J., & Slotkin, T. A. (2005a) Developmental exposure of rats to chlorpyrifos leads to behavioral alterations in adulthood, involving serotonergic mechanisms and resembling animal models of depression. *Environ. Health Perspect.* 113: 527-531.

Aldridge, J. E., Meyer, A., Seidler, F. J., & Slotkin, T. A. (20005b) Alterations in central nervous system serotonergic and dopaminergic synaptic activity in adulthood after prenatal or neonatal chlorpyrifos exposure. *Environ. Health Perspect.* 113: 1027-1031.

Aldridge, J. E., Meyer, A., Seidler, F. J., & Slotkin, T. A. (2005c) Developmental exposure to terbutaline and chlorpyrifos: Pharmacotherapy of preterm labor and an environmental neurotoxicant converge on serotonergic systems in neonatal rat brain regions. *Toxicol. Appl. Pharmacol.* 203: 132-144.

Axelrad, J. C., Howard, C. V., & McLean, W. G. (2003) The effects of acute pesticide exposure on neuroblastoma cells chronically exposed to diazinon. *Toxicology* 185: 67-78.

Betancourt A. M., & Carr R. L. (2004) The effect of chlorpyrifos and chlorpyrifos-oxon on brain cholinesterase, muscarinic receptor binding, and neurotrophin levels in rats following early postnatal exposure. *Toxicol. Sci.* 77: 63-71.

Betancourt, A. M., Burgess, S. C., & Carr, R. L. (2006) Effect of developmental exposure to chlorpyrifos on the expression of neurotrophin growth factors and cell-specific markers in neonatal rat brain. *Toxicol. Sci.* 92: 500-506.

Billauer-Haimovitch, H., Slotkin, T. A., Dotan, S., Langford, R., Pinkas, A., & Yanai, J. (2009) Reversal of chlorpyrifos neurobehavioral teratogenicity in mice by nicotine administration and neural stem cell transplantation. *Behav. Brain Res.* 205: 499-504.

Braquenier, J. B., Quertemont, E., Tirelli, E., Plumier, J. C. (2010) Anxiety in adult female mice following perinatal exposure to chlorpyrifos. *Neurotoxicol. Teratol.* 32: 234-239.

Campbell, C. G., Seidler, F. J., & Slotkin, T. A. (1997) Chlorpyrifos interferes with cell development in rat brain regions. *Brain Res. Bull.* 43: 179-189.

Carr, R. L., Chambers, H. W., Guarisco, J. A., Richardson, J. R., Yang, J., & Chambers, J. E. (2001). Effects of repeated oral postnatal exposure to chlorpyrifos on open-field behavior in juvenile rats. *Toxicol. Sci.* 59: 260-267.

Carr, R. L., Borazjani, A., & Ross, M. K. (2011) Effect of developmental chlorpyrifos exposure on endocannabinoid metabolizing enzymes. *Toxicol. Sci.* 122: 112-120.

Carr, R. L., Adams, A. L., Kepler, D. R., Ward, A. B., & Ross, M. K. (2013) Induction of endocannabinoid levels in juvenile rat brain following developmental chlorpyrifos exposure. *Toxicol. Sci.* 135: 193-201.

Carr, R. L., Graves, C. A., Mangum, L. C., Nail, C. A., & Ross, M. K. (2014) Low level chlorpyrifos exposure increases anandamide accumulationin juvenile rat brain in the absence of brain cholinesterase inhibition. *Neurotoxicology* 43: 82-89.

Caughlan, A., Newhouse, K., Namgung, U., & Xia, Z. (2004) Chlorpyrifos induces apoptosis in rat cortical neurons that is regulated by a balance between p38 and ERK/JNK MAP Kinases. *Toxicol. Sci.* 78: 125-134.

Chakraborti, T. K., Farrar, J. D., & Pope, C. N. (1993) Comparative neurochemical and neurobehavioral effects of repeated chlorpyrifos exposures in young and adult rats. *Pharmacol. Biochem. Behav.* 46: 219-224.

Chanda S. M., & Pope C. N. (1996) Neurochemical and neurobehavioral effects of repeated gestational exposure to chlorpyrifos in maternal and developing rats. *Pharmacol. Biochem. Behav.* 53: 771-776.

Chen, X. P., Wang, X., & Dong, J. Y. (2011) Different reaction patterns of dopamine content to prenatal exposure to chlorpyrifos in different periods. *J. Appl. Toxicol.* 31: 355-359.

Clegg D. J., & van Gemert M. (1999) Determination of the reference dose for chlorpyrifos: Proceedings of an expert panel. *J. Toxicol. Environ. Health B* 2: 211-255.

Cochran, R. C. (2002) Appraisal of risks from nonoccupational exposure to chlorpyrifos. *Regul. Toxicol. Pharmacol.* 35: 105-121.

Creeley, C. E., & Olney, J. W. (2013) Drug-Induced Apoptosis: Mechanism by which Alcohol and Many Other Drugs Can Disrupt Brain Development. *Brain Sci.* 3: 1153-1181.

Cole, T. B., Fischer, J. C., Burbacher, T. M., Costa, L. G., & Furlong, C. E. (2012) Neurobehavioral assessment of mice following repeated postnatal exposure to chorpyrifos-oxon. *Neurotoxicol. Teratol.* 34: 311-322.

Colón-Ramos, D. A. (2009) Synapse formation in developing neural circuits. *Curr. Top. Devel. Biol.* 87: 53-79.

Council of the European Communities (1991) Council Directive 91/414/EEC of 15 July 1991 concerning the placing of plant protection products on the market. *Official Journal* 230: 1-32.

Crumpton, T. L., Seidler, F. J., & Slotkin, T. A. (2000). Developmental neurotoxicity of chlorpyrifos in vivo and in vitro: effects on nuclear transcription factors involved in cell replication and differentiation. *Brain Res.* 857: 87-98.

Culbreth, M. E., Harrill, J. A., Freudenrich, T. M., Mundy, W. R., & Shafer, T. J. (2012) Comparison of chemical-induced changes in proliferation and apoptosis in human and mouse progenitor cells. *Neurotoxicology* 33: 1499-1510.

Dam, K., Seidler, F. J., & Slotkin, T. A. (1998) Developmental neurotoxicity of chlorpyrifos: delayed targeting of DNA synthesis after repeated administration. *Brain Res. Dev. Brain Res.* 108: 39-45.

Dam, K., Garcia, S. J., Seidler, F. J., & Slotkin, T. A. (1999) Neonatal chlorpyrifos exposure alters synaptic development and neuronal activity in cholinergic and catecholaminergic pathways. *Brain Res. Dev. Brain Res.* 116: 9-20.

Dam, K., Seidler, F. J., & Slotkin, T. A. (2000) Chlorpyrifos exposure during a critical neonatal period elicits gender-selective deficits in the development of coordination skills and locomotor activity. *Brain Res. Dev. Brain Res.* 121: 179-187.

Das, K. P., & Barone, S. (1999) Neuronal differentiation in PC 12 cells is inhibited by chlorpyrifos and its metabolites: Is acetylcholinesterase inhibition the site of action? *Toxicol. Appl. Pharmacol.* 160: 217-230.

De Angelis, S., Tassinari, R., Maranghi, F., Eusepi, A., Di Virgilio, A., Chiarotti, F., Ricceri, L., Venerosi Pesciolini, A., Gilardi, E., Moracci, G., Calamandrei, G., Olivieri, A., & Mantovani, A. (2009) Developmental exposure to chlorpyrifos induces alterations in thyroid and thyroid hormone levels without other toxicity signs in CD-mice. *Toxicol. Sci.* 108: 311-319.

De Felice, A., Venerosi, A., Ricceri, L., Sabbioni, M., Scattoni, M. L., Chiarotti, F., & Calamandrei, G. (2014) Sex-dimorphic effects of gestetional exposure to the organophosphate insecticide chlorpyrifos on social investigation in mice. *Neurotoxicol. Teratol.* 28: 32-39.

Eaton, D. L., Daroff, R. B., Autrup, H., Bridges, J., Buffler, P., Costa, L. G., Coyle, J., McKhann, G., Mobley, W. C., Nadel, L., Neubert, D., Schulte-Hermann, R., & Spencer, P. S. (2008) Review of the toxicology of chlorpyrifos with an emphasis on human exposure and neurodevelopment. *Crit. Rev. Toxicol.* 38: 1-125.

Eells, J. B., & Brown, T. (2009) Repeated developmental exposure to chlorpyrifos and methyl parathion causes persistent alterations in nicotinic acetylcholine subunit mRNA expression with chlorpyrifos altering dopamine metabolite levels. *Neurotoxicol. Teratol.* 31: 98-103.

EFSA (2014) Conclusion on the peer review of the pesticide human health risk assessment of the active substance chlorpyrifos. *EFSA Journal* 2014; 12(4): 3640.

Estevan, C., Pamies, D., Sogorb, M. A., & Vilanova, E. (2011) OECD guidelines and validated methods for in vivo testing of reproductive toxicity. In: Gupta, R.G. (Ed.), Reproductive and Developmental Toxicology. Elsevier, Burlington, MA. USA, pp. 123–133.

Estevan, C., Vilanova, E., & Sogorb, M. A. (2013) Chlorpyrifos and its metabolites alter gene expression at non-cytotoxic concentrations in D3 mouse embryonic stem cells under in vitro differentiation: Considerations for embryotoxic risk assessment. *Toxicol. Lett.* 217: 14-22.

European Commission, 2005. Commission Directive 2005/72/EC of 21 October 2005 amending Council Directive 91/414/EEC to include chlorpyrifos, chlorpyrifosmethyl, mancozeb, maneb, and metiram as active substances. *Official Journal* 279, 0063–0069.

European Commission, 2009. Regulation (EC) No 1107/2009 of the European Parliament and of the Council of 21 October 2009 concerning the placing of plant protection products on the market and repealing Council Directives79/117/EEC and 91/414/EEC.

Food and Agriculture Organization of the United Nations and World Health Organization (1999) Chlorpyrifos. In: Food and Agriculture Organization of the United Nations and World Health Organization (Eds.), Pesticides Residues in Food.

Garcia, S. J., Seidler, F. J., Crumpton, T. L., & Slotkin, T. A. (2001). Does the developmental neurotoxicity of chlorpyrifos involve glial targets? Macromolecule synthesis, adenylyl cyclase signaling, nuclear transcription factors, and formation of reactive oxygen in C6 glioma cells. *Brain Res.* 891: 54-68.

Garcia, S. J., Seidler, F. J., Qiao, D., & Slotkin, T. A. (2002). Chlorpyrifos targets developing glia: effects on glial fibrillary acidic protein. *Dev. Brain Res.* 133: 151-161.

Garner, C. C., Zhai, R. C., Gundelfinger, E. D., & Ziv, N. E. (2002) Molecular mechanisms of CNS synaptogenesis. *Cell Press* 25: 243-250.

Grandjean, P., Landrigan, P. J. (2014) Neurobehavioural effects of developmental toxicity. *Lancet Neurol.* 13: 330-338.

Guizzetti, M., Pathak, S., Giordano, G., & Costa, L. G. (2005). Effect of organophosphorus insecticides and their metabolites on astroglial cell proliferation. *Toxicology* 215: 182-190.

Hanley, T. R. Jr, Carney, E. W., & Johnson, E. M. (2000) Developmental toxicity studies in rats and rabbits with 3,5,6-trichloro-2-pyridinol, the major metabolite of chlorpyrifos. *Toxicol. Sci.* 53:100-118.

Haviland, J. A., Butz, D. E., & Porter, W. P. (2010) Long-term sex selective hormonal and behavior alterations in mice exposed to low doses of chlorpyrifos in utero. *Reprod. Toxicol.* 29: 74-79.

Henschler, D., Schmuck, G., Van Aerssen, M., & Schiffmann, D. (1992) The inhibitory effect of neuropathic organophosphate esters on neurite outgrowth in cell cultures: a basis for screening for delayed neurotoxicity. *Toxicol. In Vitro* 6: 327-335.

Howard, A. S., Bucelli, R., Jett, D. A. Bruun, D., Yang, D., & Lein, P. J. (2005) Chlorpyrifos exerts opposing effects on axonal and dendritic growth in primary neuronal cultures. *Toxicol. Appl. Pharmacol.* 207: 112-124.

Icenogle, L. M., Christopher, N. C., Blackwelder, W. P., Caldwell, D. P., Qiao, D., Seidler, F. J., Slotkin, T. A., & Levin, E. D. (2004) Behavioral alterations in adolescent and adult rats caused by a brief subtoxic exposure to chlorpyrifos during neurulation. *Neurotoxicol. Teratol.* 26: 95-101.

Jett, D. A., Navoa, R. V., Beckles, R. A., & McLemore, G. L. (2001) Cognitive function and cholinergic neurochemistry in weanling rats exposed to chlorpyrifos. *Toxicol. Appl. Pharmacol.* 174: 89-98.

Johnson, D. E., Seidler F. J., & Slotkin, T. A. (1998) Early biochemical detection of delayed neurotoxicity resulting from developmental exposure to chlorpyrifos. *Brain Res. Bull.* 45: 143-147.

Johnson F. O., Chambers J. E., Nail C. A., Givaruangsawat S., & Carr R. L. (2009) Developmental chlorpyrifos and methyl parathion exposure alters radial-arm maze performance in juvenile and adult rats. *Toxicol. Sci.* 109: 132-142.

Lassiter, T. L., Padilla, S., Mortensen, S. R., Chanda, S. M., Moser, V. C & Barone, S. Jr. (1998) Gestational exposure to chlorpyrifos: apparent protection of the fetus? *Toxicol. Appl. Pharmacol.* 152: 56-65.

Lauder, J. M., & Schambra, U. B. (1999) Morphogenetic roles of acetylcholine. *Environ. Health Perspect.* 107: 65-69.

Lehman-McKeeman, L. D. (2009) Absorption, distribution and excretion of toxicants. K. Klaassen (Ed.), Casarett & Doull's Toxicology the Basic Science of Poisons, McGraw Hill, Kansas. USA, pp. 31-159.

Levin E. D., Addy N., Nakajima A., Christopher N. C., Seidler F. J., & Slotkin T. A. (2001) Persistent behavioral consequences of neonatal chlorpyrifos exposure in rats. *Brain Res. Dev. Brain Res.* 130: 83-89.

Levin E. D., Addy N., Baruah A., Elias A., Christopher N. C., Seidler F. J., & Slotkin T. A. (2002) Prenatal chlorpyrifos exposure in rats causes persistent behavioral alterations. *Neurotoxicol. Teratol.* 24: 733-741.

Levin, E. D., Cauley, M., Johnson, J. E., Cooper, E. M., Stapleton, H. M., Ferguson, P. L., Seidler, F. J., & Slotkin, T. A. (2014) Prenatal dexamethasone augments the neurobehavioral teratology of chlorpyrifos: significance for maternal stress and preterm labor. *Neurotoxicol. Teratol.* 41: 35-42.

Li, A. A., Kimberly, A. L., McIntosh, L. J., & Mink, P. J. (2012) Evaluation of epidemiology and animal data for risk assessment: chlorpyrifos developmental neurobehavioral outcomes. *J. Toxicol. Environ. Health Part B* 15: 109-184.

LoPachin, R. M., & Aschner, M. (1999). Neuronal-glial interactions as potential targets of neurotoxicant effect. In Neurotoxicology. (Tilson, H.,A., and Harry, G.,J., Eds.), Taylor & Francis., New York. pp 53-80.

Maurissen, J. P., Hoberman A. M., Garman R. H., & Hanley T. R., Jr. (2000) Lack of selective developmental neurotoxicity in rat pups from dams treated by gavage with chlorpyrifos. *Toxicol. Sci.* 57: 250-263.

Meijer, M., Hamers, T., & Westerink, R. H. (2014) Acute disturbance of calcium homeostasis in PC12 cells as a novel mechanism of action for (sub)micromolar concentrations of organophosphate insecticides. *Neurotoxicology* 43: 110-116.

Mense, S. M., Sengupta, A., Lan, C., Zhou, M., Bentsman, G., Volsky, D. J., Whyatt, R. M., Perera, F. P., & Zhang, L. (2006) The common insecticides cyfluthrin and chlorpyrifos alter the expression of a subset of genes with diverse functions in primary human astrocytes. *Toxicol. Sci.* 93: 125-135.

Mink, P. J., Kimmel, C. A., & Li, A. A. (2012) Potential effects of chlorpyrifos on fetal growth outcomes: implications for risk assessment. *J. Toxicol. Environ. Health. Part B: Critical Reviews* 15: 281-316.

Monnet-Tschudi F., Zurich, M. G., Schilter, B., Costa, L. G., & Honegger, P. (2000) Maturation-dependent effects of chlorpyrifos and parathion and their oxygen analogs on acetylcholinesterase and neuronal and glial markers in aggregating brain cell cultures. *Toxicol. Appl. Pharmacol.* 165: 175-183.

Mortensen, S. R., Hooper, M. J., & Padilla, S. (1998) Rat brain acetylcholinesterase activity: developmental profile and maturational sensitivity to carbamate and organophosphorus inhibitors. *Toxicology* 125: 13-19.

Moser, V. C., & Padilla S. (1998) Age- and gender-related differences in the time course of behavioral and biochemical effects produced by oral chlorpyrifos in rats. *Toxicol. Appl. Pharmacol.* 149: 107-119.

Ohishi, T., Wang, L., Akane, H., Itahashi, M., Nakamura, D., Yafune, A., Mitsumori, K., Shibutani, M. (2013) Reversible effects of maternal exposure to chlorpyrifos on the intermediate granule cell progenitors in the hippocampal dentate gyrus of rat offspring. *Reprod. Toxicol.* 35: 125-136.

Ostrea, E. M., Morales, V., Ngoumgna, E., Prescilla, R., Tan, E., Hernandez, E., Ramirez, G.B., Cifra, H. L., Manlapaz, M. L. (2002) Prevalence of fetal exposure to environmental toxin as determined by meconium analysis. *Neurotoxicology* 23: 329-339.

Phang, W. (2002). Studies of developmental neurotoxicity and their use in establishing acute reference doses and acceptable daily intakes. Pesticide residues in food - 2002- Joint FAO/WHO meeting on pesticide residues. Available at: http://www.inchem.org/documents/jmpr/jmpmono/ 2002pr15.htm. Accessed 26-11-2014.

Pope, C. N., & Liu, J. (1997) Age-related differences in sensitivity to organophosphorus pesticides. *Environ. Toxicol. Pharmacol.* 4: 309-314.

Qiao, D., Seidler, F. J., & Slotkin, T. A. (2001) Developmental neurotoxicity of chlorpyrifos modelled in vitro: comparative effects of metabolites and other cholinesterase inhibitors on DNA synthesis in PC12 and C6 cells. *Environ. Health Perspect.* 109: 909-913.

Qiao, D., Seidler, F. J., Padilla, S., & Slotkin, T. A. (2002) Developmental neurotoxicity of chlorpyrifos: What is the vulnerable period? *Environ. Health Perspect.* 110: 1097-1103.

Qiao, D., Seidler, F. J., Tate, C. A., Cousins, M. M., & Slotkin, T. A. (2003) Fetal chlorpyrifos exposure: Adverse effects on brain cell development and cholinergic biomarkers emerge postnatally and continue into adolescence and adulthood. *Environ. Health Perspect.* 111: 536-544.

Raines K. W., Seidler F. J., & Slotkin T. A. (2001) Alterations in serotonin transporter expression in brain regions of rats exposed neonatally to chlorpyrifos. *Brain Res. Dev. Brain Res.* 130: 65-72.

Ray, A., Liu, J., Ayoubi, P., Pope, C. (2010) Dose-related gene expression changes in forebrain following acute, low-level chlorpyrifos exposure in neonatal rats. *Toxicol. Appl. Pharmacol.* 248: 144-155.

Ricceri, L., Markina, N., Valanzano, A., Fortuna, S., Cometa, M. F., Meneguz A., & Calamandrei G. (2003) Developmental exposure to chlorpyrifos alters reactivity to environmental and social cues in adolescent mice. *Toxicol. Appl. Pharmacol.* 191: 189-201.

Ricceri L., Venerosi Capone A. F., Cometa M. F., Lorenzini P., Fortuna S., & Calamandrei G. (2006) Developmental neurotoxicity of organophosphorous pesticides: Fetal and neonatal exposure to chlorpyrifos alters sex-specific behaviors at adulthood in mice. *Toxicol. Sci.* 93: 105-113.

Richardson J., & Chambers J. (2003) Effects of gestational exposure to chlorpyrifos on postnatal central and peripheral cholinergic neurochemistry. *J. Toxicol. Environ. Health A* 66: 275-289.

Richardson J. R., & Chambers J. E. (2004) Neurochemical effects of repeated gestational exposure to chlorpyrifos in developing rats. *Toxicol. Sci.* 77: 83-90.

Richardson J. R., & Chambers J. E. (2005) Effects of repeated oral postnatal exposure to chlorpyrifos on cholinergic neurochemistry in developing rats. *Toxicol. Sci.* 84: 352-359.

Roegge, C. S., Timofeeva, O. A., Seidler, F. J., Slotkin, T. A., & Levin, E. D. (2008) Developmental diazinon neurotoxicity in rats: Later effects on emotional response. *Brain Res. Bull.* 75: 166-172.

Roy, T. S., Andrews, J. E., Seidler, F. J., & Slotkin, T. A. (1998) Chlorpyrifos elicits mitotic abnormalities and apoptosis in neuroepithelium of cultured rat embryos. *Teratology* 58: 62-68.

Sachana, M., Flaskos, J., Alexaki, E., Glynn, P., & Hargreaves, A. J. (2001) The toxicity of chlorpyrifos towards differentiating mouse N2a neuroblastoma cells. *Toxicol. In Vitro* 15: 369-372.

Sachana, M., Flaskos, J., & Hargreaves, A. J. (2005) Effects of chlorpyrifos and chlorpyrifos-methyl on the outgrowth of axon-like processes, tubulin and GAP-43 in N2a cells. *Toxicol. Mech. Methods* 15: 405-410.

Sachana, M., Flaskos, J., Sidiropoulou, E., Yavari, C. A., & Hargreaves, A. J. (2008) Inhibition of extension outgrowth in differentiating rat C6 glioma cells by chlorpyrifos and chlorpyrifos oxon: effects on microtubule proteins. *Toxicol. In Vitro* 22: 1387-1391.

Schuh, R. A., Lein, P. J., Beckles, R. A., & Jett, D. A. (2002). Noncholinesterase mechanism of chlorpyrifos neurotoxicity: altered phosphorylation of Ca^{+2}/cAMP response element binding protein in cultured neurons. *Toxicol. Appl. Pharmacol.* 182: 176-185.

Slotkin, T. A. (1999) Developmental cholinotoxicants: nicotine and chlorpyrifos. *Environ. Health Perspect.* 107: 71-80.

Slotkin, T. A., Cousins, M. M., Tate, C. A., & Seidler, F. J. (2001) Persistent cholinergic presynaptic deficits after neonatal chlorpyrifos exposure. *Brain Res.* 902: 229-243.

Slotkin T. A., Tate C. A., Cousins M. M., & Seidler F. J. (2002) Functional alterations in CNS catecholamine systems in adolescence and adulthood after neonatal chlorpyrifos exposure. *Brain Res. Dev. Brain Res.* 133: 163-173.

Slotkin, T. A., Levin, E. D., & Seidler, F. J. (2006) Comparative developmental neurotoxicity of organophosphate insecticides: effects on brain development are separable from systemic toxicity. *Environ. Health Persp.* 114: 746-751.

Slotkin T. A., & Seidler F. J. (2005) The alterations in CNS serotonergic mechanisms caused by neonatal chlorpyrifos exposure are permanent. *Brain Res. Dev. Brain Res.* 158: 115-119.

Slotkin T. A., & Seidler F. J. (2007a) Prenatal chlorpyrifos exposure elicits presynaptic serotonergic and dopaminergic hyperactivity at adolescence: Critical periods for regional and sex-selective effects. *Reprod. Toxicol.* 23: 421-427.

Slotkin T. A., & Seidler F. J. (2007b) Developmental exposure to terbutaline and chlorpyrifos, separately or sequentially, elicits presynaptic serotonergic hyperactivity in juvenile and adolescent rats. *Brain Res. Bull.* 73: 301-309.

Slotkin, T. A., & Seidler, F. J. (2007c). Comparative developmental neurotoxicity of organophosphates in vivo: transcriptional responses of pathways for brain cell development, cell signaling, cytotoxicity and neurotransmitter systems. *Brain Res. Bull.* 72: 232-274.

Slotkin, T. A., McKillop, E. A., Ryde, I. T., Tate, C. A., & Seidler, F. J. (2007a) Screening for developmental neurotoxicity using PC12 cells: comparisons of organophosphates with a carbamate, an organochlorine, and divalent nickel. *Environ. Health Perspect.* 115: 93-101.

Slotkin, T. A., Seidler, F. J., & Fumagalli, F. (2007b). Exposure to organophosphates reduces the expression of neurotrophic factors in neonatal rat brain regions: similarities and differences in the effects of chlorpyrifos and diazinon on the fibroblast growth factor superfamily. *Environ. Health Perspect.* 115: 909-916.

Slotkin, T. A., & Seidler, F. J. (2008) Developmental neurotoxicants target neurodifferentiation into the serotonin phenotype: Chlorpyrifos, diazinon, dieldrin and divalent nickel. *Toxicol. Appl. Pharmacol.* 233: 211-219.

Slotkin, T. A., Seidler, F. J., & Fumagalli, F. (2008) Targeting of neurotrophic factors, their receptors, and signaling pathways in the developmental neurotoxicity of organosphosphates in vivo and in vitro. *Brain Res. Bull.* 76: 424-438.

Slotkin, T. A., & Seidler, F. J. (2009) Oxidative and excitatory mechanisms of developmental neurotoxicity: transcriptional profiles for chlorpyrifos, diazinon, dieldrin, and divalent nickel in PC12 cells. *Environ. Health Perspect.* 117: 587-596.

Slotkin, T. A., Lobner, D., & Seidler, F. J. (2010) Transcriptional profiles for glutamate transporters reveal differences between organophosphates but similarities with unrelated neurotoxicants. *Brain Res. Bull.* 83: 79-83.

Slotkin, T. A., & Seidler, F. J. (2012) Developmental neurotoxicity of organophosphates targets cell cycle and apoptosis, revealed by transcriptional profiles in vivo and in vitro. *Neurotoxicol. Teratol.* 34: 232-241.

Slotkin, T. A., Cooper, E. M., Stapleton, H. M., & Seidler, F. J. (2013) Does thyroid disruption contribute to the developmental neurotoxicity of chlorpyrifos? *Environ. Toxicol. Pharmacol.* 36: 284-287.

Slotkin, T.A., Card, J., & and Seidler, F. J. (2014) Prenatal dexamethasone, as used in preterm labor, worsens the impact of postnatal chlorpyrifos exposure on serotonergic pathways. *Brain Res. Bull.* 100: 44-54.

Song, X., Seidler, F. J., Saleh, J. L., Zhang, J. Padilla, S., & Slotkin T. A. (1997) Cellular mechanisms for developmental toxicity of chlorpyrifos: targeting the adenylyl cyclase signaling cascade. *Toxicol. Appl. Pharmacol.* 145: 158-174.

Song, X., Violin, J. D., Seidler, F. J., & Slotkin, T. A. (1998) Modeling the developmental neurotoxicity of chlorpyrifos in vitro: macromolecule synthesis in PC12 cells. *Toxicol. Appl. Pharmacol.* 151: 182-191.

Stanton, M. E., Mundy, W. R., Ward, T., Dulchinos, V., & Barry, C. C. (1994). Time-dependent effects of acute chlorpyrifos administration on spatial delayed alternation and cholinergic neurochemistry in weanling rats. *Neurotoxicology* 15: 201-208.

Tait, S., Ricceri, L., Venerosi, A., Maranghi, F., Mantovani, A., & Calamandrei G. (2009) Long-term effects on hypothalamic neuropeptides after developmental exposure to chlorpyrifos in mice. *Environ. Health Perspect.* 117: 112-116.

Tang J., Carr R. L., & Chambers J. E. (1999) Changes in rat brain cholinesterase activity and muscarinic receptor density during and after repeated oral exposure to chlorpyrifos in early postnatal development. *Toxicol. Sci.* 51: 265-272.

Ullian, E. M., Christopherson, K. S., & Barres, B. A. (2004) Role for glia in synaptogenesis. *Glia* 47: 209-216.

U.S. Environmental Protection Agency. (2000) Office of Pesticide Programs Health Effects Division.

van Leeuwen C. J. & Vermeire T. G. (2007) The risk management process. In: Risk Assessment of Chemicals: An Introduction. Springer Science & Business Media. USA. pp 2-10.

Venerosi, A., Calamandrei, G., & Ricceri, L. (2006) A social recognition test for female mice reveals behavioral effects of developmental chlorpyrifos exposure. *Neurotoxicol. Teratol.* 28: 466-471.

Venerosi, A., Cutuli, D., Colonnello, V., Cardona D., Ricceri, L., & Calamandrei, G. (2008) Neonatal exposure to chlorpyrifos affects maternal responses and maternal aggression of female mice in adulthood. *Neurotoxicol. Teratol.* 30: 468-474.

Venerosi, A., Ricceri, L., Scattoni, M. L., & Calamandrei, G. (2009) Prenatal chlorpyrifos exposure alters motor behavior and ultrasonic vocalization in CD-1 mouse pups. *Environ. Health.* 8: 12.

Venerosi, A., Ricceri, L., Rungi, A., Sanghez, V., & Calamandrei, G. (2010) Gestational exposure to the organophosphate chlorpyrifos alters social-emotional behaviour and impairs responsiveness to the serotonin transporter inhibitor fluvoxamine in mice. *Psychopharmacology (Berlin)* 208: 99-107.

Whitney, K. D., Seidler, F. J., & Slotkin, T. A. (1995) Developmental neurotoxicity of chlorpyrifos cellular mechanism. *Toxicol. Appl. Pharmacol.* 134: 53-62.

WHO (1986) Organophosphorous insecticides a general introduction. In: World Health Organization (Eds.), *Environmental Health Criteria*, vol. 63.

Yanai, J., Brick-Turin, Y., Dotan, S., Langford, R., Pinkas, A., & Slotkin, T. A. (2010) A mechanism-based complementary screening approach for the amelioration and reversal of neurobehavioral teratogenicity. *Neurotoxicol. Teratol.* 32: 109-113.

Yang, D., Howard, A., Bruun, D., Ajua-Alemanj, M., Pickart, C., & Lein, P. J. (2008) Chlorpyrifos and chlorpyrifos-oxon inhibit axonal growth by interfering with the morphogenic activity of acetylcholinesterase. *Toxicol. Appl. Pharmacol.* 228: 32-41.

Zheng Q., Olivier K., Won Y. K., & Pope C. N. (2000) Comparative cholinergic neurotoxicity of oral chlorpyrifos exposures in preweanling and adult rats. *Toxicol. Sci.* 55: 124-132.

Zurich, M. G., Honegger, P., Schilter, B., Costa, L. G., & Monnet-Tschudi, F. (2000) Use of aggregating brain cell cultures to study developmental effects of organophosphorus insecticides. *Neurotoxicology* 21: 599-606.

In: Chlorpyrifos
Editor: Caitlin Mayes

ISBN: 978-1-63482-111-7

Chapter 7

FATE AND DISPOSITION OF CHLORPYRIFOS IN HUMAN

Khaled Abass[1,2,3*], Manal Abd-El-Razik[1] and Gamal Abouelghar[1]
[1]Department of Pesticides, Menoufia University, Egypt
[2]Pharmacology and Toxicology Department, University of Oulu, Finland
[3]Medical Research Center Oulu, Oulu University Hospital and University of Oulu, Finland

This work is dedicated to the late Prof. Hosni Radwan and to Prof. Hany Kamal who died in a tragic accident on March 14, 2014

ABSTRACT

Chlorpyrifos is a widely used broad-spectrum phosphorothioate insecticide. Chlorpyrifos fate in human is the process by which it is biotransformed through enzymatic catalysis to metabolites that are eliminated directly or after conjugation with endogenous cofactors. Biotransformation plays a major role in chlorpyrifos mediated toxicity and risk assessment. Chlorpyrifos is bioactivated by human cytochrome P450 (CYP) enzymes to the potent neurotoxic inhibitor chlorpyrifos-oxon and detoxified via dearylation to 3,5,6-trichloro-2-pyridinol. Human cytochrome P450 enzymes that mediate these reactions and the response mechanisms are under genetic and epigenetic control and influenced by other environmental stressors. Potential interactions of chlorpyrifos and its metabolites may occur in human during occupational and dietary exposure. This chapter provides a comprehensive knowledge on the fate, disposition and interactions of chlorpyrifos in human.

* Corresponding author: khaled.megahed@oulu.fi, khaled.m.abass@gmail.com.

Introduction

Xenobiotics are substances foreign to an entire biological system and cover wide range of chemicals and pollutants such as pharmaceutical, industrial chemicals and agrochemicals. Xenobiotics are usually lipid soluble and not ideally suited for excretion. After absorption in the body, they may undergo biotransformation to products which are rapidly excreted and eliminated from the body. The metabolic reactions involved in biotransformation are usually divided into two reactions, Phase I and Phase II reactions. Although biotransformation reactions are often *referred to as* detoxification process, these reactions, especially Phase I, sometimes enhance the toxicity of compounds. Cytochrome P450 enzymes (CYPs) are the main enzymes involved in Phase I reactions. CYPs are found in high concentration in the liver and are also presented in other tissues, i.e. gastrointestinal tract, lung, brain, kidney and skin (Adams et al., 1991; Bergh and Strobel, 1992; Hjelle et al., 1986; Lawton et al., 1990). The CYP superfamily has been divided on the basis of structural homology into families and subfamilies. There are 18 CYP families and 44 subfamilies in humans covering 57 functional CYP genes. More details about CYP classifications are found in the website, http://drnelson.utmem.edu/CytochromeP450.html. Human CYPs have broad substrate specificity. Metabolism and biotransformation were the main routes of elimination of the top 200 prescribed drug in US and of that 75 % are mediated by the group of human CYPs, Figure 1, (Wienkers and Heath, 2005). Enzymes in families 1-3 mostly are involved in the mediated metabolism of a wide variety of drugs and environmental agents (Abass et al., 2012a; Pelkonen et al., 2008; Zanger et al., 2008).

Chlorpyrifos and Human Cytochrome P450 Enzymes System

Chlorpyrifos (CP) (*O,O*-diethyl-*O*-[3,5,6-trichloro-2-pyridyl], phosphorothioate) is a widely used broad-spectrum insecticide. In mammals, chlorpyrifos has moderate acute oral toxicity (LD_{50} for rats: 96–475 mg/kg body weight (IPCS, 1999).Chlorpyrifos mechanism of toxicity is related to the inhibition of acetylcholinesterase enzyme.

In case of acetylcholinesterase, cholinergic toxicity happens either because of extremely high levels of inhibition or inhibition plus aging. A substantial level of acetylcholinesterase inhibition on its own is sufficient to produce cholinergic toxicity and death. When aging of inhibited acetylcholinesterase also occurs (net loss of ligand from the phosphorus of the OP-enzyme conjugate, leaving a negatively charged phosphoryl moiety attached to the active site), the qualitative nature of the toxicity does no changed (Figure 2). Instead, the inhibited acetylcholinesterase becomes intractable to reactivation, rendering therapy with oximes, such as 2-pralidoxime methiodide (2-PAM) ineffective (Moser et al., 2008; Thompson, 2004).

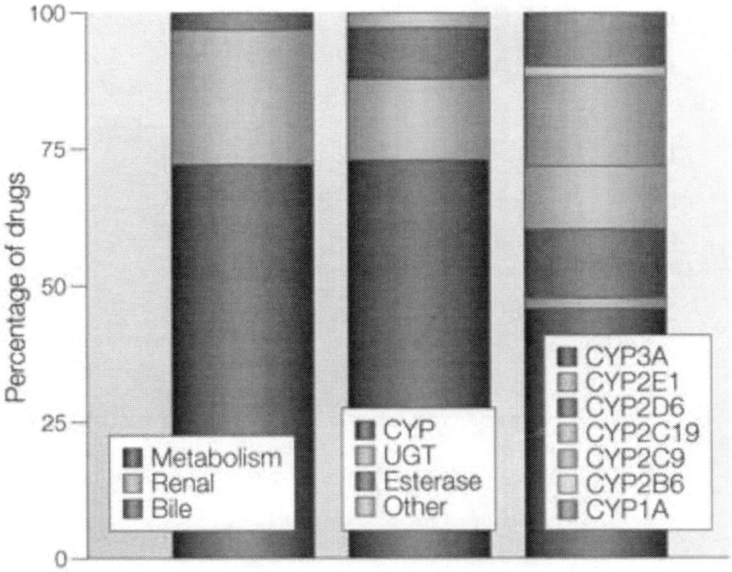

Reprinted with permission from Nature Publishing Group.

Figure 1. Metabolism represents the listed clearance mechanism for ~ 73% of the top 200 drugs. Of the drugs cleared via metabolism, about three-quarters are metabolized by members of the cytochrome P450 (CYP) superfamily. For the CYP-mediated clearance mechanisms, the majority of drug oxidations (46%) were carried out by members of the CYP3A family; followed by 16% by CYP2C9; 12% for both CYP2C19 and CYP2D6; 9% for members of the CYP1A family; and 2% for both CYP2B6 and CYP2E1. UGT, uridine diphosphate glucuronyl transferase. (Wienkers and Heath, 2005) Nature Reviews Drug Discovery 4, 825-833 (2005)).

There is sufficient evidence that neurotoxicity is also the primary endpoint for repeated organophosphate exposures (Farahat et al., 2010; Farahat et al., 2011; Farahat et al., 2003; Jiang et al., 2010; Middlemore-Risher et al., 2010; Rohlman et al., 2011; Speed et al., 2012).

During the last few years several articles on the role of CYPs in the metabolism of a wide variety of pesticides have appeared (Abass et al., 2012a; Abass et al., 2007; Abass et al., 2014; Buratti et al., 2006; Buratti and Testai, 2007; Leoni et al., 2008; Sams et al., 2000). Many CYPs are involved in the metabolism of pesticides and the metabolism may result in activation and/or detoxification reactions. A database covering human Phase I enzymes and their isoforms that metabolize pesticides and related compounds has been published by Abass et al. (Abass et al., 2012a).

Chlorpyrifos and other organophosphate compounds are lipophilic and ready to enter the nerve system. If the parent compound and/or metabolites are reactive enough, they can phosphorylate neural target proteins, such as various serine hydrolases (Quistad et al., 2005).

Chlorpyrifos is used worldwide with the highest exposures reported in an Egyptian agricultural population (Farahat et al., 2010; Farahat et al., 2011; Farahat et al., 2003). This population was found to have neurobehavioral deficits when compared to a reference population (Abdel Rasoul Gaafar et al., 2008; Farahat et al., 2003). Exposure to chlorpyrifos, based on epidemiological and *in vivo* studies, is connected to possible developmental delays (Eaton et al., 2008; Levin et al., 2002; Whyatt et al., 2005).

Chlorpyrifos inhibits neural development at levels lower than those associated with acetylcholinesterase inhibition in *in vitro* studies (Howard et al., 2005; Sachana et al., 2008).

Moreover chlorpyrifos is a potent inhibitor of estradiol and testosterone metabolism in human liver microsomes *in vitro* (Tang et al., 2004; Usmani et al., 2004).

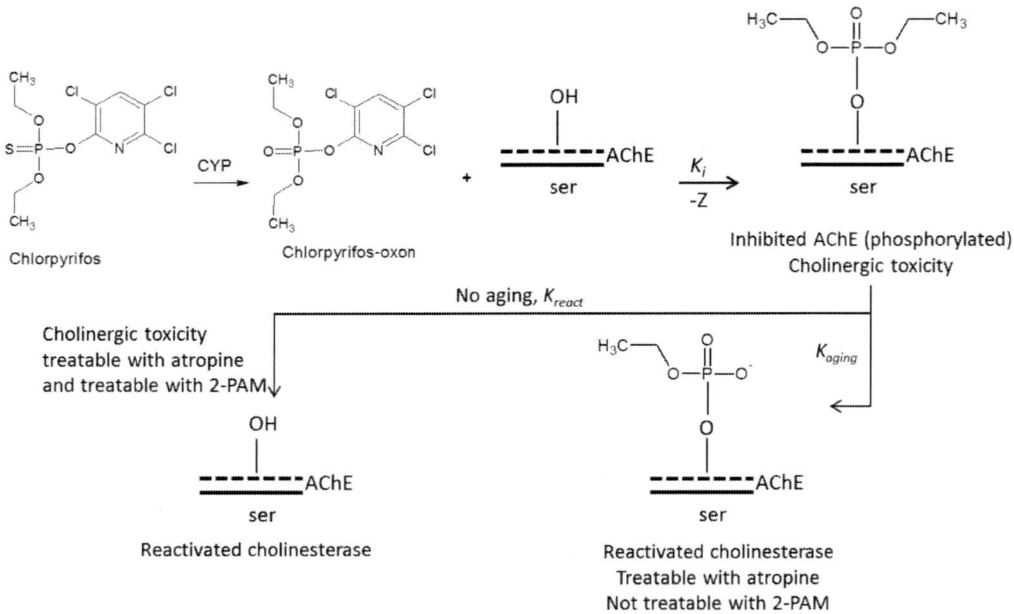

Abbreviations: CYP, cytochrome P450 enzymes; k_{aging}, aging (nonreactivation) rate constant; k_i, inhibition rate constant; k_{react}, reactivation rate constant.
Modified from Moser et al., 2008; Thompson et al., 2010.

Figure 2. Formation of chlorpyrifos-oxon and inhibition of acetylcholinesterase. Aging does not alter the type of toxicity but it obviates 2-PAM therapy. Non aging is treatable with atropine and 2-PAM.

Biotransformation by cytochrome P450 enzymes plays a major role in chlorpyrifos mediated toxicity and risk assessment. It is well characterized that chlorpyrifos is bioactivated in human by cytochrome P450s to the potent neurotoxic inhibitor chlorpyrifos oxon. CYPs mediated oxidative desulfuration of chlorpyrifos and thus generate the highly electrophilic intermediate, chlorpyrifos-oxon. The activated intermediate chlorpyrifos-oxon has higher affinity towards the active site of serine-dependent ester hydrolases than the parent compound, chlorpyrifos. The inhibition of acetylcholinesterase results in an excessive accumulation of acetylcholine both in synapses and neuromuscular junctions leading to overstimulation and potentially paralysis of muscarinic and nicotinic receptors.

Thus, the primary manifestations of acute cholinesterase inhibitor toxicity are those of cholinergic (neurotransmitter) hyperactivity (Thompson, 2004). Chlorpyrifos detoxification process is also catalyzed by CYPs via the cleavage of the aromatic pyridinol moiety (oxidative dearylation) for both parent compounds and the active metabolites. CYP-dependent metabolism of chlorpyrifos is illustrated in Figure 3.

BIOTRANSFORMATION OF CHLORPYRIFOS IN *IN VITRO* HUMAN HEPATIC MODELS

Buratti and coworkers (Buratti et al., 2002) studied the role of different cytochrome P450 isoforms in the bioactivation of chlorpyrifos via desulfuration to chlorpyrifos-oxon at levels representative of actual human exposure.

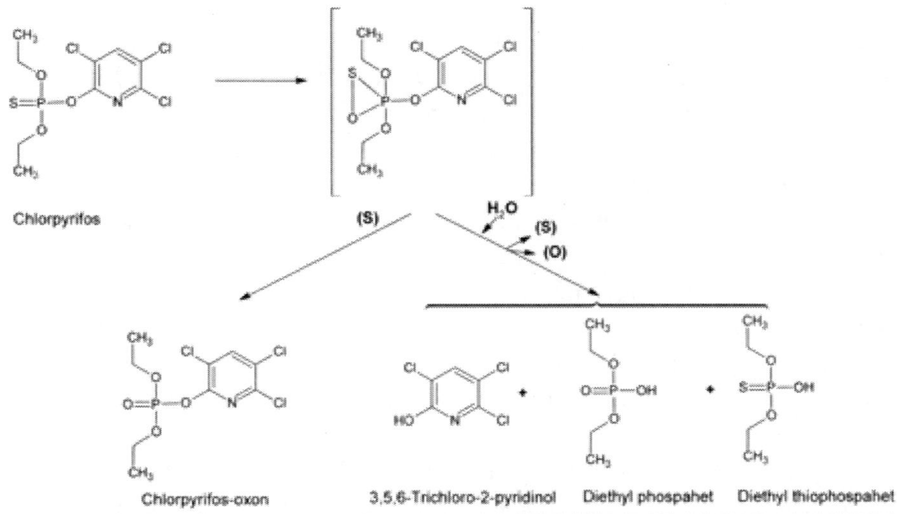

Figure 3. Cytochrome P450 enzymes-dependent metabolism of chlorpyrifos.

Several approaches have been employed to identify CYP enzymes responsible in biotransformation reactions. The detection of oxon formation was measured by acetylcholinesterase inhibition assay which able to detect nM levels of oxon formed. Chlorpyrifos was bioactivated at low concentration mainly by CYP2B6, CYP1A1, CYP1A2, CYP2C19 and CYP3A4.

Taking into account the average human hepatic microsomal protein amount and the measured kinetic values, CYP1A2 found to be the major enzyme responsible for oxon formation followed by CYP2C19. Based on kinetics, CYP3A4 contribution to oxon formation could be relevant in poisoning scenario when individuals are exposed at high doses of chlorpyrifos. Correlation of chlorpyrifos oxon formation with some CYP-catalyzed activities measured in a panel of 20 human hepatic microsomes at 5 μM chlorpyrifos showed significant positive correlation with Caffeine N3-Demethylation (CYP1A2), Dextromethorphan N- Demethylation (CYP 3A4), and S-Mephenytoin N-Demethylation (CYP 2B6). Results from immune-inhibition studies indicated that anti-CYP2B6, CYP3A4, and CYP1A2 caused a significant inhibition of chlorpyrifos-oxon formation. The combined addition indicated that anti-CYP2B6 and anti-CYP1A2 were together able to cause a dramatic reduction of chlorpyrifos activation (95–98%). Human inter-individual studies showed that CYP-mediated oxon formation from chlorpyrifos is efficiently catalyzed by human liver microsomes and displayed a high variability, 30-fold, among individuals (Buratti et al., 2002; Buratti et al., 2003).

Meanwhile, Mutch and Williams, 2006, described the activation and detoxification of chlorpyrifos and other organophosphate insecticides, diazinon and parathion, by recombinant P450 isozymes and by human liver microsomes. No Significant correlations between chlorpyrifos oxon formation in human hepatic microsomes with any of the P450 marker activities were observed. Chlorpyrifos dearylation correlated with one of the CYP3A4/5, CYP2C8, CYP2C19 and CYP1A2 mediated model reactions. It is probable that the chlorpyrifos-oxon produced by human liver microsomes was further degraded to trichloro-2-pyridinol by paraoxonase 1 and carboxylesterases and this confounded the relationship with the P450-mediated marker reactions. Karanth et al. (2001) highlighted the importance of both paraoxonase 1 and carboxylesterases in the effective detoxification of chlorpyrifos-oxon.

Recombinant CYP2D6, CYP3A5, CYP2B6 and CYP3A4 were best at chlorpyrifos activation and CYP2C19, CYP2D6, CYP3A5 and CYP3A4 at chlorpyrifos detoxification. Therefore, inter-individual variation in expression of CYPs, due to genetic and environmental influences, may result in sub-populations or individuals who produce higher oxon levels with increased susceptibility to organophosphates toxicity.

The rate of chlorpyrifos metabolism in human hepatic microsomes and recombinant human P450s were investigated by Foxenborg et al. (Foxenberg et al., 2007) to identify specific isoforms responsible for metabolism and to create CYP-specific kinetic data to predict biotransformation *in vivo*. CYP1A2, CYP2B6, CYP2C19, CYP3A4, CYP3A5, and CYP3A7 were found to be active to a widely varying degree in chlorpyrifos biotransformation. CYP2B6 also shows a higher affinity and activity for chlorpyrifos metabolism to chlorpyrifos-oxon, with a K_m value of 0.81 μM and V_{max} of 12.544 pmol/ min/nmol P450.

In contrast, CYP2C19 has a high affinity and activity for chlorpyrifos detoxification, with a K_m of 1.63 μM and a V_{max} of 13.128 pmol/min/nmol P450. CYP2B6 has greater activity in the formation of the oxon (CL_{int} = 15.560) and, thus, may be more important in assessing risk because it preferentially forms the toxic compound over the direct detoxification of the chlorpyrifos at low level exposures. On the other hand, CYP2C19 has the highest V_{max}/K_m (CL_{int} = 8.075) in detoxification, and thus plays a prominent role in the chlorpyrifos detoxification reaction.

Human hepatocytes and human S9 fractions were employed to investigate the metabolism of chlorpyrifos and chlorpyrifos-oxon (Choi et al., 2006).

From LC/MS/MS spectra, 15 chlorpyrifos related and 6 chlorpyrifos-oxon related metabolites were detected. Overall, four different metabolic pathways might be considered for chlorpyrifos and chlorpyrifos-oxon as following: dearylation, deethylation, glucuronidation and glutathione conjugation. Both parent compound and its main metabolites, chlorpyrifos and chlorpyrifos-oxon, had qualitatively similar biotransformation pathways. Results suggested that human liver plays an important role in detoxification rather than activation of chlorpyrifos. Both compounds undergoes not only cytochrome dependent metabolism (Phase I reactions) but also Phase II metabolism (conjugations) in human *in vitro* hepatic models, human hepatocytes and human liver S9 fractions.

Biotransformation of Chlorpyrifos by Human Foetal CYP3A7 Isoforms

Expression of CYP3A7 protein is mainly confined to fetal and newborn livers, although in rare cases CYP3A7 mRNA has been detected in adults (Hakkola et al. 2001; Kitada and Kamataki 1994; Schuetz et al. 1994). CYP3A7 is responsible for the metabolism of endogenous substrates such as testosterone and DHEA-S in fetuses (Kitada et al., 1987a; Kitada et al., 1987b) as well as a few pesticides (Abass et al., 2012a; Casabar et al., 2006; Foxenberg et al., 2007). While CYP3A4 is the major CYP isoform expressed in adult liver and intestine, CYP3A7 is mainly expressed in the foetal liver (Hakkola et al., 2001). CYP3A7 expression is switched off gradually after birth in favor of CYP3A4 (Stevens et al., 2003). CYP3A7 has been also detected in very small amounts in some adult livers and intestine (Burk et al., 2002; Schuetz et al., 1994). The role of CYP3A7 in metabolism and biotransformation in adult is thought to be negligible, but it can assume a pivotal role during prenatal developmental stages. Therefore, Buratti and coworkers (Buratti et al., 2006) investigated the foetal and adult human CYP3A isoforms in the bioactivation of chlorpyrifos by using recombinant enzymes. CYP3A7 was able to produce significant levels of chlorpyrifos-oxon and the kinetics of the reaction was linear up to the highest concentration tested with chlorpyrifos. The efficiency of the CYP3A forms in catalyzing the desulfuration reaction was, based on CL_{int} values, 3A4 > 3A5 > 3A7 for CPF. The results clearly indicated that CYP3A7, the major foetal isoform, responsible for most of the P450-mediated prenatal hepatic activity toward xenobiotics (Stevens et al., 2003), is able to produce significant levels of chlorpyrifos oxon when tested at concentrations (0.05–2.5 μM), representative of actual exposure conditions (Buratti et al., 2006).

Biotransformation of Chlorpyrifos in Human Small Intestine *In Vitro*

In human, xenobiotics are absorbed mainly in the small intestine following oral exposure. Several cytochrome P450 isoforms, phase II enzymes (conjugation enzymes), as well as transporters (i.e. the multidrug efflux pump P-glycoprotein) (Paine et al., 2006; von Richter et al., 2004; Zhang and Benet, 2001) are expressed in the small intestine. Intestinal pre-systemic first-pass metabolism affects the oral bioavailability of several drugs (e.g., cyclosporine, midazolam, and verapamil) (Galetin and Houston, 2006; Kivistö et al., 2004; Kivistö et al., 2004; Obach et al., 2001; von Richter et al., 2004; Zhang and Benet, 2001).

However, limited data are available on chlorpyrifos metabolism in human intestine. In human CYP3A subfamily members, 3A4 and 3A5, are the most abundant P450 subfamily in the small intestine, representing about 80% of the total CYP intestinal content while it is represent about 40 % of total hepatic CYP. CYP enzymes in human small intestine display a large inter-individual variation. Not only CYP3A are expressed in human intestine but also CYP2C9 (15%), CYP2C19 (2.9%), CYP2J2 (1.4%) and CYP2D6 (1%), as well as the other metabolizing enzymes such as paraoxonase 1 and carboxylesterases. These enzymes play an important role in chlorpyrifos metabolism at the actual level of oral exposure (Paine et al., 2006; Shamir et al., 2005; Taketani et al., 2007). Leoni et al. (Leoni et al., 2012) investigated the human small intestine role in the biotransformation of chlorpyrifos in microsomes from duodenum and ileum/jejunum of 11 individual donors. CYP3A4 and its related activity, measured by testosterone hydroxylation and quantified by Western blotting, were detected in 10/11 samples.

Only an individual showed also the presence of CYP2C9. Chlorpyrifos was activated to chlorpyrifos-oxon in 10/11 human duodenum microsomes with CL_{int} highest values of 0.75 pmol/ (mg protein min), while detoxification reaction was negligible. Most chlorpyrifos activation were detected in human duodenum microsomes and declined toward the distal ileum.

Even though chlorpyrifos bioactivation in the small intestine is much lower than in the hepatic biotransformation, human small intestine role in the pre-systemic chlorpyrifos clearance is important (Leoni et al., 2012).

CHLORPYRIFOS BIOTRANSFORMATION AND HUMAN CYTOCHROME P450 ENZYMES POLYMORPHISMS

All cytochrome P450 enzymes are able to mediate both activation and detoxification of chlorpyrifos at some levels. Based on literature, results differ as to the capacity of CYP1A2, CYP2A6, CYP2C9 and CYP2D6 to metabolize chlorpyrifos. In contrast, CYP2B6, CYP2C19 and CYP3A4 have consistently been identified as major metabolizing isoforms. In short, CYP2B6 is the predominant enzyme in chlorpyrifos bioactivation while CYP2C19 is the predominant enzyme in detoxification process (Buratti et al., 2003; Croom et al., 2010; Foxenberg et al., 2007; Sams et al., 2000; Tang et al., 2001). CYP2B6 and CYP2C19 have also been detected in human brain tissue, the target organ of toxicity for chlorpyrifos (Miksys, 2003; Miksys and Tyndale, 2002). The other main CYP involved in chlorpyrifos metabolism is CYP3A4. CYP3A4 is the highest cytochrome P450 enzymes expressed in human liver and it is the predominant enzyme of the low-affinity component of chlorpyrifos metabolism (Croom et al., 2010; Foxenberg et al., 2007). However, CYP2B6 is a highly polymorphic enzyme which mediated clinically important drugs, i.e. bupropion, cyclophosphamide, efavirenz, propofol, selegiline as well as many pesticides (Abass et al., 2012a; Hodgson and Rose, 2007; Pelkonen et al., 2008; Turpeinen et al., 2006b; Zanger et al., 2008). CYP2B6 enzyme variants have the potential to modulate chlorpyrifos *in vivo*. The most prevalent allelic, An *allele* is an alternative form of a gene (one member of a pair) that is located at a specific position on a specific chromosome, isoforms of CYP2B6 is CYP2B6*6 (15631G>T and 18053A>G) at a frequency of 20–31% in Caucasian populations and up to 60% in African and other populations (Zanger et al., 2008). The 15631G>T SNP in this variant codes for amino acid change Q172H in exon 4 but the important functional consequence is thought to result from erroneous splicing leading to decreased hepatic expression and activity (Hofmann et al., 2008). However, the 18053A>G SNP [K262R] also found in CYP2B6*6 and other variants appears to confer increased substrate turnover and some substrate dependent alterations in metabolism (Bumpus et al., 2005).

CYP2C19 is also a polymorphic enzyme with the most common variant being the null variant CYP2C19*2 at a frequency of 12–13% in Caucasians and up to 30% in Asians (Goldstein et al., 1997; Mizutani, 2003).

To study in more details the role of CYP variants in the biotransformation of chlorpyrifos and other relevant environmental contaminants, Dai et al. (Dai et al., 2001) has sequenced the coding regions and intron-exon junctions of CYP3A4 in DNAs from 72 individuals from three diverse racial groups. A total of 28 SNPs (single nucleotide polymorphism) in CYP3A4 were detected. Five SNPs produced amino acid substitutions including R162Q in exon 6; F189S in exon 7; L293P in exon 10; M445T in exon 12; and P467S in exon 12. P189S and M445T were detected only in Caucasians with frequencies of 2% and 4%, respectively. It has been reported that M445T is detected in Asians (Sata et al., 2000). The L293P and P467S alleles were found only in Asians, both at frequencies of 2% (L293 occurred in Chinese with a frequency of 10% while P467S occurred in Indo-Pakistani with a frequency of 12.5%). The cDNAs for the F189S, L293P, M445T, and P467S mutant alleles were constructed by site-directed mutagenesis and expressed in an *Escherichia coli* expression system. CYP3A4 F189S exhibited lower activities toward bioactivation and detoxification of chlorpyrifos, while CYP3A4 L293P had higher activities. The yields of the CYP3A4 M445T and P467S alleles in metabolizing chlorpyrifos were not significantly different from those of wild-type CYP3A4. The L293P allele could possibly increase the toxicity of chlorpyrifos and other similar organophosphate pesticides to individuals carrying this allele. Individuals might potentially have alterations in their ability to metabolize chlorpyrifos and polymorphisms of CYP enzymes could potentially influence the risk of different populations from various environmental compounds (Dai et al., 2001).

CYP2B6, the predominant enzyme in bioactivation, and CYP2C19, the predominant enzyme in detoxification process, are polymorphic. CYP2B6 and CYP2C19 levels display a large inter-individual variability and consequently may modulate the risk of chlorpyrifos intoxication. It is an established fact that cytochrome P450 enzymes display a large inter-individual variability and several polymorphisms genes have been found to impact drug metabolism.

For instance CYP2B6*6 is a clinical relevant polymorphism with reduced function. CYP2C19*2 is common polymorphism with loss of function (Hofmann et al., 2008; Itoh et al., 1999).

Croom and coworkers (Croom et al., 2010) investigated human inter-individual variability in chlorpyrifos bioactivation to chlorpyrifos-oxon by human hepatic microsomes from 17 donors.

The incubation with 20 μM chlorpyrifos illustrated 14-fold variability between individuals, while incubation with 100 μM chlorpyrifos resulted in 57-fold variability. Similarly, detoxification pathway varied 9-fold and 19- fold at 20 μM and 100 μM chlorpyrifos, respectively.

Several approaches were employed to investigate the role of CYP isoforms in the chlorpyrifos biotransformation. The correlation studies between chlorpyrifos-oxon formation and CYP relative activities resulted in positive correlated with CYP2B6, CYP2C19 and CYP3A4. The highest correlation was obtained with CYP2B6 at 20 μM chlorpyrifos concentrations while CYP3A4 correlated well with chlorpyrifos-oxon formation at 100 μM. The formation of the less toxic metabolite, 3,5,6-trichloro-2-pyridinol was best correlated with CYP3A4 activity at all chlorpyrifos concentrations. Chemical inhibition studies with ketoconazole, the specific inhibitor of CYP3A4, and ticlopidine, the soecicif inhibitor of CYP2C19, significantly inhibited the metabolism of chlorpyrifos. CYP2B6 genotype did not has an effect on chlorpyrifos-oxon formation. Overall, individuals with high CYP2B6 levels may be at greatest risk from chlorpyrifos exposure.

Crane et al. (Crane et al., 2012) studied the effects of CYP2B6 and CYP2C19 genotype on chlorpyrifos metabolism in 22 individual human liver microsomes and the effect of CYP2B6 genotype on chlorpyrifos metabolism in recombinant enzymes. Chlorpyrifos biactivation to chlorpyrifos-oxon was decreased in CYP2B6*6/*6 specimens as compared to CYP2B6*1/*1 specimens and this decrease in metabolism corresponds to a decrease in protein expression in CYP2B6*6 human liver microsomes. No change in the detoxification of chlorpyrifos was seen in CYP2C19*2/*2 specimens. The finding illustrated those individuals with CYP2B6*6 genotype may be less susceptible to chlorpyrifos toxicity.

Chlorpyrifos Interactions with Carbaryl and Carbofuran *In Vitro*

Carbaryl biotransformation to carbaryl methylol is primarily mediated by CYP2B6 (Tang et al., 2002). Chlorpyrifos is activated to chlorpyrifos-oxon through a CYP2B6-catalyzed desulfuration reaction, although other isoforms are involved. The sulfur atom released from chlorpyrifos during the activation is highly reactive and binds immediately to the heme iron of CYP and inhibit its activity. Co-incubations of carbaryl and chlorpyrifos in HLM greatly inhibited carbaryl metabolism (Tang et al., 2002).

The ability of human hepatic microsomes to metabolize carbaryl was reduced by pre-incubation with chlorpyrifos due to the completion between carbaryl and chlorpyrifos for the active site of CYP2B6. Therefore, environmental contaminates biotransformation in humans is reflected by the concentration of CYP isoforms and its interactions with other contaminants.

Carbofuran oxidation is catalyzed in human mainly by CYP3A4 and to lesser extend by CYP1A2 and 2C19. Other human CYP isoforms have minimal or no activity toward carbofuran. It is well known that chlorpyrifos acts as an irreversible non-competitive inhibitor of CYP3A4. The co-incubation of carbofuran with chlorpyrifos resulted in inhibition of the metabolism of carbofuran to 3-hydroxycarbofuran in human hepatic microsomes with 50% inhibition occurring at a concentration of IC_{50} of 39 μM. Carbofuran inhibition occurred rapidly at lower concentrations until 60% inhibition was achieved, after which it was difficult to achieve further inhibition. This is due to the fact that these chemicals inhibited CYP3A4 but had lesser effects on the other isoforms involved in carbofuran metabolism (Usmani et al., 2004).

Chlorpyrifos Metabolism in Human *In Vivo*

Humans are exposed to toxic substances through different routes like ingestion, inhalation and dermal absorption, and exposure is dependent on the concentration of the contaminant and frequency and duration of the contact. People are exposed daily to multiple pesticides, at low levels, mainly via the diet. Intake of pesticides is regulated by using limits such as maximum residue levels and acceptable daily intakes. However, these limit values are necessarily set by consideration of the toxicology of each individual compound. This section will summarized in vivo metabolism studies of chlorpyrifos in human.

In vivo human study that dosed six men once with 0.5 mg/kg orally and four weeks later dosed five of these same men with 5 mg/kg dermally, and one man with 0.5 mg/kg dermally was reviewed by US-EPA, Health Effects Division (US-EPA, 2000). Blood chlorpyrifos concentrations were extremely low (less than 30 ng/ml), and no unchanged chlorpyrifos was found in the urine following either route of administration. Mean blood Trichloro-2-pyridinol concentrations peaked at 0.93 μg/ml 6 hr after ingestion of the oral dose and at 0.063 μg/ml 24 hr after the 5.0 mg/kg dermal dose.

Trichloro-2-pyridinol was cleared from the blood and eliminated in the urine with a half-life of 27 hours following both the oral and dermal doses. An average of 70% of the oral dose but less than 3% of the dermal dose was excreted in the urine as Trichloro-2-pyridinol; thus only a small fraction of the dermally applied chlorpyrifos was absorbed. Chlorpyrifos and its principal metabolite were rapidly eliminated and therefore have a low potential to accumulate in man on repeated exposure (Nolan et al., 1984).

The elimination kinetics of an oral dose (1 mg = 2852 nmol) and a dermal dose (28.59 mg = 81567 nmol) of chlorpyrifos were investigated in five human volunteers (Griffin et al., 1999). The concentrations of two diethylphosphate and diethylthiophosphate were measured in urine samples. The elimination half-life of the two metabolites was 15.5 and 30 hrs after the oral dose and the dermal, consequently. The average amount of urinary metabolites recovered from volunteers after oral dose of chlorpyrifos was 93% while only 53% was recovered after dermal application.

Excretion of chlorpyrifos urine metabolites was slower after skin application compared with oral dose. In another study aimed to develop a Physiologically Based Pharmacokinetic and Pharmacodynamic (PBPK/PD) model for chlorpyrifos in rats and humans, single oral-dose pharmacokinetic studies were conducted in rats and human (Timchalk et al., 2002). After treated humans with 0.5–2 mg/kg of chlorpyrifos, trichloro-2-pyridinol and chlorpyrifos were detected by ionization–tandem mass spectrometry (GC–NCI-MS) in blood, while chlorpyrifos-oxon was not detected.

However, in a case study of intoxicated 59 years- old female, fifteen metabolites of chlorpyrifos and its bioactivated intermediate, chlorpyrifos-oxon were detected by liquid chromatography–electrospray ionisation-tandem mass spectrometry in the intoxicated 59 years -old female urine.

Based on the findings, chlorpyrifos was biotransformed via three possible pathways: 1) cleavage reactions at the aromatic phosphoester bond, 2) cleavage reactions at the alkyl phosphoester bonds, and 3) glutathione (GSH) dependent nucleophilic substitution of the 6-chlorine at the aromatic moiety (Bicker et al., 2005) (Figure 4).

In order to investigate potential interactions that may occur during dietary exposure, human volunteers were orally administrated by deltamethrin and pirimicarb together with chlorpyrifos, at the acceptable daily intake (Sams and Jones, 2011). Deltamethrin (0.01 mg/kg/day) and chlorpyrifos (0.01 mg/kg/ day) were administrated as a mixed dose for five volunteers. Pirimicarb (0.02 mg/kg/day) and chlorpyrifos-methyl (0.01 mg/kg/day) were administrated as a mixed dose for four volunteers.

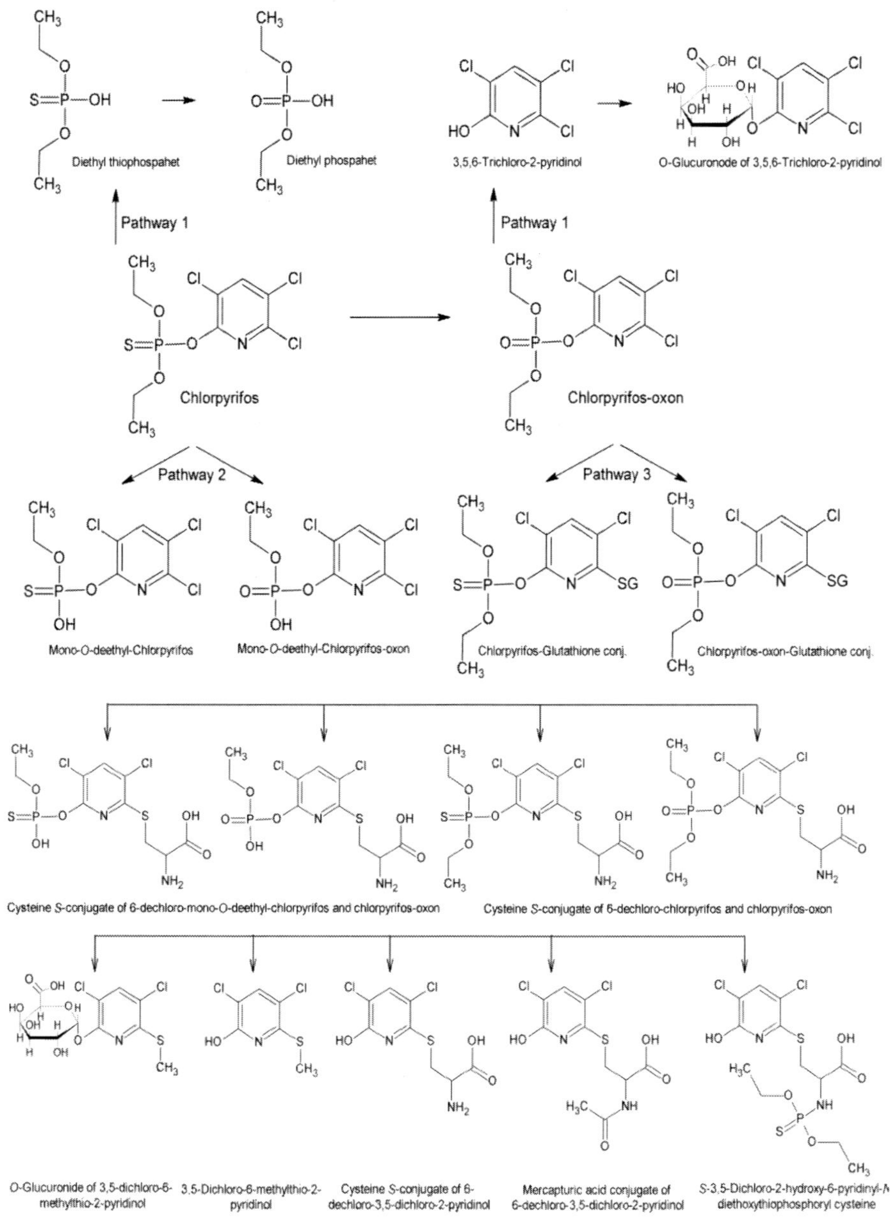

Modified from Bicker et al., 2005.

Figure 4. Tentative scheme of chlorpyrifos biotransformation in human. Most of metabolites were identified by liquid chromatography–electrospray ionisation-tandem mass spectrometry (LC–ESI-MS/MS) in urine of an intoxicated 59 years old female.

The urinary excretion profile of chlorpyrifos detoxified metabolite (3,5,6-Trichloro-2-pyridinol) was similar to that found for a single dose of chlorpyrifos, a mixed exposure to chlorpyrifos-methyl and deltamethrin, and a mixed exposure to chlorpyrifos and pirimicarb.

In vitro data obtained using human liver have shown that chlorpyrifos interact and potently inhibit the metabolism of carbofuran and carbaryl (Tang et al., 2002; Usmani et al., 2004).

Therefore, Sams and Jones (2011) stated that the conflicting results between the *in vitro* and *in vivo* studies may be explained by a dose-dependent interaction. It is difficult to scale up from the concentration of compounds used for *in vitro* incubations to the available dose following oral administration to human volunteers.

However, while *in vitro* studies have clearly demonstrated the potential for metabolic interactions to occur, the human *in vivo* data indicated that the administered doses of pesticides were below the threshold, where any significant effects could be observed.

SUMMARY OF HUMAN CYTOCHROME P450 ENZYMES INVOLVED IN *IN VITRO* CHLORPYRIFOS BIOTRANSFORMATION

The roles of human CYPs involved in the biotransformation of chlorpyrifos have been studied in several articles as discussed above. Table 1 and Table 2 listed human CYPs and their isoforms catalysing chlorpyrifos biotransformation in *in vitro* human settings.

CHLORPYRIFOS INHIBITORY INTERACTIONS WITH HUMAN CYTOCHROME P450 ENZYMES *IN VITRO*

CYP-xenobiotic interactions involve either induction or inhibition of metabolizing enzymes. Inhibitory interactions could cause destruction of pre-existing enzymes or inhibition of enzyme synthesis and thus inactivating the metabolizing enzyme.

Elucidation of the inhibitory interactions is one of the approaches used in determining whether P450s are involved in the hepatic and extrahepatic metabolism of xenobiotics.

Table 1. Desulfuration (activation) of chlorpyrifos to chlorpyrifos-oxon by human cytochrome P450 isoforms *in vitro*

Human CYP enzymes	Detection method	References
CYP1A2, CYP2B6, CYP2C19, CYP3A4	AChE inhibition and HPLC	(Buratti et al., 2002)
CYP1A2, CYP2B6, CYP2C9, CYP2C19, CYP2C19*1B, CYP3A4	LC-UV	(Tang et al., 2001)
CYP3A4, CYP3A4-F189S, CYP3A4-L293P, CYP3A4-M445T, CYP3A4-P467S	LC-UV	(Dai et al., 2001)
CYP1A2, CYP2B6, CYP2C19, CYP3A4, CYP3A5, CYP3A7	LC-UV	(Foxenberg et al., 2007)
CYP3A4, CYP3A5, CYP2B6, CYP1A2, CYP2C8, CYP2C19, CYP2D6*1	LC-UV	(Mutch and Williams, 2006)
CYP3A4,CYP3A5, CYP3A7	AChE inhibition and HPLC	(Buratti et al., 2006)
CYP2B6, CYP2C9, CYP2C19, CYP3A4, CYP3A5, CYP3A7	LC-UV	(Croom et al., 2010)

Table 2. Dearylation (detoxification) of chlorpyrifos by human cytochrome P450 isoforms *in vitro*

Human CYP enzymes	Detection method	References
CYP1A2, CYP2B6, CYP2C9, CYP2C19, CYP2C19*1B, CYP2C19*8, CYP2C19*6, CYP2C19*5, CYP3A4.	LC-UV	(Tang et al., 2001)
CYP3A4, CYP3A4-F189S, CYP3A4-L293P, CYP3A4-M445T, CYP3A4-P467S	LC-UV	(Dai et al., 2001)
CYP1A2, CYP2B6, CYP2C19, CYP3A4, CYP3A5, CYP3A7	LC-UV	(Foxenberg et al., 2007)
CYP3A4, CYP3A5, CYP2B6, CYP1A2 CYP2C8, CYP2C19, CYP2D6*1	LC-UV	(Mutch and Williams, 2006)
CYP2B6, CYP2C9, CYP2C19, CYP3A4, CYP3A5, CYP3A7		(Croom et al., 2010)

It is also contributed to the characterization of metabolic factors, e.g. metabolic stability, metabolizing enzymes, inter-individual variability, and interspecies differences, factors which provide important quantitative data for chemical risk assessment (Abass, 2013).

Moreover, determination of the potential of new chemical entity to inhibit CYPs and cause drug-drug or xenobiotic-drug interactions is very important during the development of a new chemical entity. By comparing the potential interactions effects on CYPs of a chemical to the respective effects of specific inhibitors, a prediction of the *in vivo* situation can be made (Pelkonen and Raunio, 2005; Pelkonen et al., 2005).

CYPs interaction studies have been usually conducted in animals or tissue samples from pesticide-treated animals (Paolini et al., 1996; Paolini et al., 1999; Pozzetti et al., 1999; Sun et al., 2005). However, species variability and the specific compound in question must be considered regarding pesticide-CYP interactions since species differences are significant in terms of activities and selectivities of specific CYPs. Furthermore, human interactions data are needed for human risk assessment.

We must rely upon an *in vitro* human CYPs system to study interactions since humans cannot be used as study subjects except in occasional circumstances (Abass, 2013). Inhibition interactions can be determined either by single substrate assay or cocktail assay. In single substrate assay, traditionally CYP interaction studies are performed using specific assays for each CYP isoforms. Decrease in probe metabolite formation produced by inhibition is usually analysed by UV-LC, LC-MS or fluorimeter. Detailed information about compounds frequently employed as inhibitory probes for specific human isoforms and the experimental conditions were published by Abass et al. (Abass et al. 2007). N-in-one assay is employed due to the dramatically increased number of new chemicals and the demand for efficient and rapid screening techniques to analyse inhibition interactions (Turpeinen et al., 2005). Modern mass spectrometers have a capability to analyse multiple compounds simultaneously. This is applied to inhibition studies and several cocktail-type N-in-one-assays have been developed for screening of CYP inhibitions (Abass et al., 2010; Abass et al., 2012b; Abass and Pelkonen, 2013; Abass et al., 2014; Turpeinen et al., 2005; Turpeinen et al., 2006a). The most extensive of these is the LC-MS-MS assay developed by Tolonen et al. (Tolonen et al., 2005) and it contains all the nine major CYP isoforms with 13 metabolic probe reactions. In this assay, ten probe substrates are added to single incubation with tested substance and metabolites were analysed from one run using triple quadrupole instrument in MRM mode.

Even though this method has no reference metabolites for quantitation it nevertheless provides reliable information about inhibition potencies towards various CYP-selective activities of a studied compound by comparing peak areas in cocktail samples with different inhibitor concentrations. It has to be stressed that, if detailed and more accurate information is wanted, the traditional single substrate technique would still be the recommended assay. N-in-one assay does not enable discrimination on the basis of mechanism of inhibition (Abass and Pelkonen, 2013).

Human hepatic microsomes were employed to study potential inhibitory interactions of chlorpyrifos with major human hepatic cytochrome P450 enzymes (Abass et al., 2009; Abass and Pelkonen, 2013). In these studies, inhibitory interactions were determined by both single substrate assay and cocktail assay. The enzyme activities in the presence of chlorpyrifos were compared with the control incubations into which only solvent were added.

The IC_{50} values for chlorpyrifos (concentration causing 50% reduction of control activity) were determined from triplicate incubations by linear regression analysis from the plot of the logarithm of inhibitor concentration versus percentage of the activity remaining after inhibition. The inhibition (IC_{50}) of major human hepatic cytochrome P450 enzymes by chlorpyrifos by both assays are summarized in Table 3.

Table 3. The inhibition of major human hepatic cytochrome P450 enzymes by chlorpyrifos by single[1] and N-in-one assay[2]

Human CYP isoforms	N-in-one assay	Single substrate assay
CYP1A2	3.0	2.9
CYP 2A6	284	28.0
CYP 2B6	2.8	2.5
CYP 2C8	212	22.2
CYP 2C9	113	89.9
CYP2C19	83.4	89.9
CYP 2D6	113	3.3
CYP2E1	3.6	977
CYP3A4[3]	82.3	4.0
CYP3A4[4]	37.9	32.2

Data obtained from Abass et al. (2009), Abass and Pelkonen (2013).

[1]Chlorpyrifos was added at final concentrations of 1, 5, 25, 50 and 100 μM to the incubation mixture for the IC_{50} values determination. The IC_{50} values were determined from three technical replicates from the same human microsomal pool by linear regression analysis from the plot of the logarithm of inhibitor concentration versus percentage of the activity remaining after inhibition using MicroCal Origin 6.0 (MicroCal Software, Inc., Northampton, MA). The enzyme activities in the presence of inhibitors were compared with the control incubations (incubations with the solvent only). IC_{50} values higher than 100 μM were extrapolated based on the linear regression analysis obtained from tested concentrations.

[2]IC_{50} values (μM) measured by single substrate assays were obtained from our previous study (Abass et al., 2009) for comparison between the assays.

[3]1-OH-MDZ.

[4]SO2–OME.

CONCLUSION

In summary, cytochrome P450 enzymes are playing a key role in the biotransformation of chlorpyrifos. Chlorpyrifos is biotransformed in human to the more active and potent inhibitor chlorpyrifos-oxon and to the less toxic metabolite, 3,5,6-trichloro-2-pyridinol.

CYP1A2 and CYP2B6 are mainly responsible for chlorpyrifos activation at low concentrations while CYP3A4 and CYP2C19 are the main isoforms responsible at higher chlorpyrifos concentration. CYP2C19 was the main enzymes, beside CYP3A4, in chlorpyrifos detoxification.

Several metabolites were detected in human *in vivo* and chlorpyrifos was biotransformed via three possible pathways: 1) cleavage reactions at the aromatic phosphoester bond, 2) cleavage reactions at the alkyl phosphoester bonds, and 3) glutathione (GSH) dependent nucleophilic substitution of the 6-chlorine at the aromatic moiety. Chlorpyrifos inhibited relatively potently at CYP1A4, CYP2B6, CYP3A4-selective activity, which may have relevance for potential interactions in occupational exposures.

REFERENCES

Abass, K., 2013. From in vitro hepatic metabolic studies towards human health risk assessment: Two case studies of diuron and carbosulfan. *Pesti. Biochem. Physiol.* 107, 258-265.

Abass, K., Turpeinen, M., Rautio, A., Hakkola, J., Pelkonen, O., 2012a. Metabolism of Pesticides by Human Cytochrome P450 Enzymes In Vitro – A Survey. In: Perveen, F. (Ed.), *Insecticides - Advances in Integrated Pest Managemen* ISBN: 978-953-307-780-2. InTech, Austria.

Abass, K., Lämsä, V., Reponen, P., Küblbeck, J., Honkakoski, P., Mattila, S., Pelkonen, O., Hakkola, J., 2012b. Characterization of human cytochrome P450 induction by pesticides, *Toxicology* 294, 17-26.

Abass, K., Pelkonen, O., 2013. The inhibition of major human hepatic cytochrome P450 enzymes by 18 pesticides: Comparison of the N-in-one and single substrate approaches, *Toxicol. in Vitro* 27, 1584-1588.

Abass, K., Reponen, P., Mattila, S., Pelkonen, O., 2010. Metabolism of carbosulfan II. Human interindividual variability in its in vitro hepatic biotransformation and the identification of the cytochrome P450 isoforms involved, *Chem. Biol. Interact.* 185, 163-173.

Abass, K., Reponen, P., Mattila, S., Rautio, A., Pelkonen, O., 2014. Human variation and CYP enzyme contribution in benfuracarb metabolism in human in vitro hepatic models, *Toxicol. Lett.* 224, 300-309.

Abass, K., Reponen, P., Turpeinen, M., Jalonen, J., Pelkonen, O., 2007. Characterization of diuron N-demethylation by mammalian hepatic microsomes and cDNA-expressed human cytochrome P450 enzymes, *Drug Metab. Disposition* 35, 1634-1641.

Abass, K., Turpeinen, M., Pelkonen, O., 2009. An evaluation of the cytochrome P450 inhibition potential of selected pesticides in human hepatic microsomes, *J. Environ. Sci. Health, Part B: Pestic. Food Contam. Agric. Wastes* 44, 553.

Abdel Rasoul Gaafar, M., Abou Salem Mahmoud, E., Mechael, A. A., Hendy, O. M., Rohlman, D. S., Ismail, A. A., 2008. Effects of occupational pesticide exposure on children applying pesticides. *Neurotoxicology* 29, 833-838.

Adams, D. R., Jones, A. M., Plopper, C. G., Serabjit-Singh, C. J., Philpot, R. M., 1991. Distribution of cytochrome P-450 monoxygenase enzymes in the nasal mucosa of hamster and rat, *Am. J. Anat.* 190, 291-298.

Bergh, A. F., Strobel, H. W., 1992. Reconstitution of the brain mixed function oxidase system: Purification of NADPH-cytochrome P450 reductase and partial purification of cytochrome P450 from whole rat brain, *J. Neurochem.* 59, 575-581.

Bicker, W., L\ammerhofer, M., Genser, D., Kiss, H., Lindner, W., 2005. A case study of acute human chlorpyrifos poisoning: novel aspects on metabolism and toxicokinetics derived from liquid chromatography-tandem mass spectrometry analysis of urine samples. *Toxicol. Lett.* 159, 235-251.

Bumpus, N. N., Sridar, C., Kent, U. M., Hollenberg, P. F., 2005. The naturally occurring cytochrome P450 (P450) 2B6 K262R mutant of P450 2B6 exhibits alterations in substrate metabolism and inactivation. *Drug metab. dispos.: the biological fate of chemicals* 33, 795-802.

Buratti, F. M., Leoni, C., Testai, E., 2006. Foetal and adult human CYP3A isoforms in the bioactivation of organophosphorothionate insecticides, *Toxicol. Lett.* 167, 245-255.

Buratti, F. M., Testai, E., 2007. Evidences for CYP3A4 autoactivation in the desulfuration of dimethoate by the human liver, *Toxicology* 241, 33-46.

Buratti, F. M., Volpe, M. T., Fabrizi, L., Meneguz, A., Vittozzi, L., Testai, E., 2002. Kinetic parameters of OPT pesticide desulfuration by c-DNA expressed human CYPs, Environ. *Toxicol. Pharmacol.* 11, 181-190.

Buratti, F. M., Volpe, M. T., Meneguz, A., Vittozzi, L., Testai, E., 2003. CYP-specific bioactivation of four organophosphorothioate pesticides by human liver microsomes, *Toxicol. Appl. Pharmacol.* 186, 143-154.

Burk, O., Tegude, H., Koch, I., Hustert, E., Wolbold, R., Glaeser, H., Klein, K., Fromm, M. F., Nuessler, A. K., Neuhaus, P., Zanger, U. M., Eichelbaum, M., Wojnowski, L., 2002. Molecular mechanisms of polymorphic CYP3A7 expression in adult human liver and intestine. *J. bio. chem.* 277, 24280-24288.

Casabar, R. C. T., Wallace, A. D., Hodgson, E., Rose, R. L., 2006. Metabolism of endosulfan-{alpha} by human liver microsomes and its utility as a simultaneous in vitro probe for CYP2B6 and CYP3A4, *Drug Metab. Dispos.* 34, 1779-1785.

Choi, K., Joo, H., Rose, R. L., Hodgso, E., 2006. Metabolism of chlorpyrifos and chlorpyrifos oxon by human hepatocytes*, *J. Biochem. Mol. Toxicol.* 20, 279-291.

Crane, A. L., Klein, K., Zanger, U. M., Olson, J. R., 2012. Effect of CYP2B6*6 and CYP2C19*2 genotype on chlorpyrifos metabolism, *Toxicology* 293, 115-122.

Croom, E. L., Wallace, A. D., Hodgson, E., 2010. Human variation in CYP-specific chlorpyrifos metabolism, *Toxicology* 276, 184-191.

Dai, D., Tang, J., Rose, R., Hodgson, E., Bienstock, R. J., Mohrenweiser, H. W., Goldstein, J. A., 2001. Identification of variants of CYP3A4 and characterization of their abilities to metabolize testosterone and chlorpyrifos, *J. Pharmacol. Exp. Ther.* 299, 825-831.

Eaton, D. L., Daroff, R. B., Autrup, H., Bridges, J., Buffler, P., Costa, L. G., Coyle, J., McKhann, G., Mobley, W. C., Nadel, L., Neubert, D., Schulte-Hermann, R., Spencer, P.

S., 2008. *Review of the toxicology of chlorpyrifos with an emphasis on human exposure and neurodevelopment.*

Farahat, F. M., Ellison, C. A., Bonner, M. R., McGarrigle, B. P., Crane, A. L., Fenske, R. A., Lasarev, M. R., Rohlman, D. S., Anger, W. K., Lein, P. J., Olson, J. R., 2011. Biomarkers of chlorpyrifos exposure and effect in Egyptian cotton field workers. *Environ. Health Perspect.* 119, 801-806.

Farahat, F. M., Fenske, R. A., Olson, J. R., Galvin, K., Bonner, M. R., Rohlman, D. S., Farahat, T. M., Lein, P. J., Anger, W. K., 2010. Chlorpyrifos exposures in Egyptian cotton field workers. *Neurotoxicology* 31, 297-304.

Farahat, T. M., Abdelrasoul, G. M., Amr, M. M., Shebl, M. M., Farahat, F. M., Anger, W. K., 2003. Neurobehavioural effects among workers occupationally exposed to organophosphorous pesticides. *Occup. Environ. Med.* 60, 279-286.

Foxenberg, R. J., McGarrigle, B. P., Knaak, J. B., Kostyniak, P. J., Olson, J. R., 2007. Human hepatic cytochrome P450-specific metabolism of parathion and chlorpyrifos, *Drug Metab. Dispos.* 35, 189-193.

Galetin, A., Houston, J. B., 2006. Intestinal and hepatic metabolic activity of five cytochrome P450 enzymes: impact on prediction of first-pass metabolism. *J. Pharmacol. Exp. Ther.* 318, 1220-1229.

Goldstein, J. A., Ishizaki, T., Chiba, K., de Morais, S. M., Bell, D., Krahn, P. M., Evans, D. A., 1997. Frequencies of the defective CYP2C19 alleles responsible for the mephenytoin poor metabolizer phenotype in various Oriental, Caucasian, Saudi Arabian and American black populations. *Pharmacogenetics* 7, 59-64.

Griffin, P., Mason, H., Heywood, K., Cocker, J., 1999. Oral and dermal absorption of chlorpyrifos: a human volunteer study, *Occup. Environ. Med.* 56, 10-13.

Hakkola, J., Raunio, H., Purkunen, R., Saarikoski, S., Vähäkangas, K., Pelkonen, O., Edwards, R., Boobis, A., Pasanen, M., 2001. Cytochrome P450 3A expression in the human fetal liver: evidence that CYP3A5 is expressed in only a limited number of fetal livers. *Biol. Neonate*. 3, 193-201.

Hjelle, J., Hazelton, G., Klaassen, C., Hjelle, J., 1986. Glucuronidation and sulfation in rabbit kidney, *J. Pharmacol. Exp. Ther.* 236, 150-156.

Hodgson, E., Rose, R. L., 2007. The importance of cytochrome P450 2B6 in the human metabolism of environmental chemicals, *Pharm. Thera.* 113, 420-428.

Hofmann, M. H., Blievernicht, J. K., Klein, K., Saussele, T., Schaeffeler, E., Schwab, M., Zanger, U. M., 2008. Aberrant splicing caused by single nucleotide polymorphism c.516G>T [Q172H], a marker of CYP2B6*6, is responsible for decreased expression and activity of CYP2B6 in liver. *J. Pharmacol. Exp. Ther.* 325, 284-292.

Howard, A. S., Bucelli, R., Jett, D. A., Bruun, D., Yang, D., Lein, P. J., 2005. Chlorpyrifos exerts opposing effects on axonal and dendritic growth in primary neuronal cultures. *Toxicol. Appl. Pharmacol.* 207, 112-124.

IPCS, 1999. *IPCS InChem JMPR Monographs and Evaluation, Chlorpyrifos.* (http://www.inchem.org/documents/jmpr/jmpmono/v99pr03.htm) accessed Septermber 25th 2014.

Itoh, K., Inoue, K., Yanagiwara, S., Kyoya, H., Suzuki, T., 1999. A rapid and simple detection of genetic defects responsible for the phenotypic polymorphism of cytochrome P450 2C19. *Biological and pharmaceutical bulletin* 22, 77-79.

Jiang, W., Duysen, E. G., Hansen, H., Shlyakhtenko, L., Schopfer, L. M., Lockridge, O., 2010. Mice treated with chlorpyrifos or chlorpyrifos oxon have organophosphorylated tubulin in the brain and disrupted microtubule structures, suggesting a role for tubulin in neurotoxicity associated with exposure to organophosphorus agents. *Toxicological sciences*, 115, 183-193.

Karanth, S., Olivier, K., Liu, J., Pope, C., 2001. In vivo interaction between chlorpyrifos and parathion in adult rats: sequence of administration can markedly influence toxic outcome. *Toxicol. Appl. Pharmacol.* 177, 247-255.

Kitada, M., Kamataki, T., Itahashi, K., 1987a. P-450 HFLa, a form of cytochrome P-450 purified from human fetal livers, is the 16α-hydroxylase of dehydroepiandrosterone 3-sulfate, *J. Biol. Chem.* 262, 13534-13537.

Kitada, M., Kamataki, T., Itahashi, K., Rikihisa, T., Kanakubo, Y., 1987b. Significance of cytochrome P-450 (P-450 HFLa) of human fetal livers in the steroid and drug oxidations, *Biochem. Pharmacol.* 36, 453-456.

Kivistö, K., Niemi, M., Fromm, F., 2004. Functional interaction of intestinal CYP3A4 and P-glycoprotein. *Fundamental and clinical pharmacology* 18, 621-626.

Lawton, M., Gasser, R., Tynes, R., Hodgson, E., Philpot, R., 1990. The flavin-containing monooxygenase enzymes expressed in rabbit liver and lung are products of related but distinctly different genes, *J. Biol. Chem.* 265, 5855-5861.

Leoni, C., Balduzzi, M., Buratti, F. M., Testai, E., 2012. The contribution of human small intestine to chlorpyrifos biotransformation, *Toxicol. Lett.* 215, 42-48.

Leoni, C., Buratti, F. M., Testai, E., 2008. The participation of human hepatic P450 isoforms, flavin-containing monooxygenases and aldehyde oxidase in the biotransformation of the insecticide fenthion, *Toxicol. Appl. Pharmacol.* 233, 343-352.

Levin, E. D., Addy, N., Baruah, A., Elias, A., Christopher, N. C., Seidler, F. J., Slotkin, T. A., 2002. Prenatal chlorpyrifos exposure in rats causes persistent behavioral alterations, *Neurotoxicol. Teratol.* 24, 733-741.

Middlemore-Risher, M. L., Buccafusco, J. J., Terry, A. V., 2010. Repeated exposures to low-level chlorpyrifos results in impairments in sustained attention and increased impulsivity in rats. *Neurotoxicol. Teratol.* 32, 415-424.

Miksys, S., 2003. Smoking, alcoholism and genetic polymorphisms alter CYP2B6 levels in human brain, *Neuropharmacology* 45, 122-132.

Miksys, S. L., Tyndale, R. F., 2002. Drug-metabolizing cytochrome P450s in the brain. *Journal of psychiatry and neuroscience: JPN* 27, 406-415.

Mizutani, T., 2003. PM Frequencies of Major CYPs in Asians and Caucasians, *Drug Metab. Rev.* 35, 99-106.

Mizutani, T., PM frequencies of major CYPs in Asians and Caucasians. *Drug Metab. Rev.* 35, 99-106.

Moser, V., Aschner, M., Richardson, R., Philber M., 2008. Toxic responses of the nervous system. In: Klaassen, C. D. (Ed.), *Casareet and Doull's Toxicology: The Basic Science of Poisons*, 7th ed.

Mutch, E., Williams, F. M., 2006. Diazinon, chlorpyrifos and parathion are metabolised by multiple cytochromes P450 in human liver, *Toxicology* 224, 22-32.

Nolan, R. J., Rick, D. L., Freshour, N. L., Saunders, J. H., 1984. Chlorpyrifos: pharmacokinetics in human volunteers. *Toxicol. Appl. Pharmacol.* 73, 8-15.

Obach, R. S., Zhang, Q. Y., Dunbar, D., Kaminsky, L. S., 2001. Metabolic characterization of the major human small intestinal cytochrome P450S, *Drug Metab. Disposition* 29, 347-352.

Paine, M. F., Hart, H. L., Ludington, S. S., Haining, R. L., Rettie, A. E., Zeldin, D. C., 2006. The human intestinal cytochrome P450 pie. *Drug metabolism and disposition: the biological fate of chemicals* 34, 880-886.

Paolini, M., Mesirca, R., Pozzetti, L., Sapone, A., Cantelli-Forti, G., 1996. Biomarkers of effect in evaluating metalaxyl cocarcinogenesis. Selective induction of murine CYP 3A isoform, *Mutat. Res.* 361, 157-164.

Paolini, M., Pozzetti, L., Perocco, P., Mazzullo, M., Cantelli-Forti, G., 1999. Molecular non-genetic biomarkers of effect related to methyl thiophanate cocarcinogenesis: organ- and sex-specific cytochrome P450 induction in the rat, *Cancer lett.* 135, 203-213.

Pelkonen, O., Turpeinen, M., Hakkola, J., Honkakoski, P., Hukkanen, J., Raunio, H., 2008. Inhibition and induction of human cytochrome P450 enzymes: current status, *Arch. Toxicol.* 82, 667-715.

Pelkonen, O., Raunio, H., 2005. In vitro screening of drug metabolism during drug development: can we trust the predictions? *Expert Opin. Drug Metab. Toxicol.* 1, 49-59.

Pelkonen, O., Turpeinen, M., Uusitalo, J., Rautio, A., Raunio, H., 2005. Prediction of drug metabolism and interactions on the basis of in vitro investigations, *Basic Clin. Pharmacol. Toxicol.* 96, 167-175.

Pozzetti, L., Paolini, M., Barillari, J., Cantelli-Forti, G., 1999. Induction and suppression of murine CYP-mediated biotransformation by dithianon: organ- and sex-related differences, *Cancer Lett.* 141, 47-56.

Quistad, G. B., Fisher, K. J., Owen, S. C., Klintenberg, R., Casida, J. E., 2005. Platelet-activating factor acetylhydrolase: selective inhibition by potent n-alkyl methylphosphonofluoridates, *Toxicol. Appl. Pharmacol.* 205, 149-156.

Rohlman, D. S., Anger, W. K., Lein, P. J., 2011. Correlating neurobehavioral performance with biomarkers of organophosphorous pesticide exposure. *Neurotoxicology* 32, 268-276. doi: 10.1016/j.neuro.2010.12.008.

Sachana, M., Flaskos, J., Sidiropoulou, E., Yavari, C. A., Hargreaves, A. J., 2008. Inhibition of extension outgrowth in differentiating rat C6 glioma cells by chlorpyrifos and chlorpyrifos oxon: effects on microtubule proteins. *Toxicology in vitro: an international journal published in association with BIBRA* 22, 1387-1391.

Sams, C., Mason, H. J., Rawbone, R., 2000. Evidence for the activation of organophosphate pesticides by cytochromes P450 3A4 and 2D6 in human liver microsomes, *Toxicol. Lett.* 116, 217-221.

Sams, C., Jones, K., 2011. Human volunteer studies investigating the potential for toxicokinetic interactions between the pesticides deltamethrin; Pirimicarb and chlorpyrifos-methyl following oral exposure at the acceptable daily intake, *Toxicol. Lett.* 200, 41-45.

Sata, F., Sapone, A., Elizondo, G., Stocker, P., Miller, V. P., Zheng, W., Raunio, H., Crespi, C. L., Gonzalez, F. J., 2000. CYP3A4 allelic variants with amino acid substitutions in exons 7 and 12: evidence for an allelic variant with altered catalytic activity. *Clin. Pharmacol. Ther.* 67, 48-56.

Schuetz, J. D., Beach, D. L., Guzelian, P. S., 1994. Selective expression of cytochrome P450 CYP3A mRNAs in embryonic and adult human liver, *Pharmacogenetics* 4, 11-20.

Shamir, R., Hartman, C., Karry, R., Pavlotzky, E., Eliakim, R., Lachter, J., Suissa, A., Aviram, M., 2005. Paraoxonases (PONs) 1, 2, and 3 are expressed in human and mouse gastrointestinal tract and in Caco-2 cell line: selective secretion of PON1 and PON2. *Free radical biology and medicine* 39, 336-344.

Speed, H. E., Blaiss, C. A., Kim, A., Haws, M. E., Melvin, N. R., Jennings, M., Eisch, A. J., Powell, C. M., 2012. Delayed reduction of hippocampal synaptic transmission and spines following exposure to repeated subclinical doses of organophosphorus pesticide in adult mice. *Toxicological sciences: an official journal of the Society of Toxicology* 125, 196-208.

Stevens, J. C., Hines, R. N., Gu, C., Koukouritaki, S. B., Manro, J. R., Tandler, P. J., Zaya, M. J., 2003. Developmental expression of the major human hepatic CYP3A enzymes. *J. Pharmacol. Exp. Ther.* 307, 573-582.

Sun, G., Thai, S., Tully, D. B., Lambert, G. R., Goetz, A. K., Wolf, D. C., Dix, D. J., Nesnow, S., 2005. Propiconazole-induced cytochrome P450 gene expression and enzymatic activities in rat and mouse liver, *Toxicol. Lett.* 155, 277-287.

Taketani, M., Shii, M., Ohura, K., Ninomiya, S., Imai, T., 2007. Carboxylesterase in the liver and small intestine of experimental animals and human. *Life Sci.* 81, 924-932.

Tang, J., Cao, Y., Rose, R. L., Brimfield, A. A., Dai, D., Goldstein, J. A., Hodgson, E., 2001. Metabolism of chlorpyrifos by human cytochrome P450 isoforms and human, mouse, and rat liver microsomes, *Drug Metab. Dispos.* 29, 1201-1204.

Tang, J., Cao, Y., Rose, R. L., Hodgson, E., 2002. In vitro metabolism of carbaryl by human cytochrome P450 and its inhibition by chlorpyrifos, *Chem. Biol. interact.* 141, 229-241.

Tang, J., Tang, J., Amin Usmani, K., Hodgson, E., Rose, R. L., 2004. In vitro metabolism of fipronil by human and rat cytochrome P450 and its interactions with testosterone and diazepam, *Chem. Biol. Interact.* 147, 319-329.

Thompson, C. M., Richardson, R. J., 2004. Anticholinesterase insecticides. In: Marrs, T. C., Ballantyne, B. (Ed.), *Pesticide Toxicology and International Regulation.* John Wiley and Sons, Inc, Chichester.

Thompson, C. M., Prins, J. M., George, K. M., 2010. Mass spectrometric analyses of organophosphate insecticide oxon protein adducts. *Environ. Health Perspect.* 118, 11-19.

Timchalk, C., Nolan, R. J., Mendrala, A. L., Dittenber, D. A., Brzak, K. A., Mattsson, J. L., 2002. A Physiologically Based Pharmacokinetic and Pharmacodynamic (PBPK/PD) Model for the Organophosphate Insecticide Chlorpyrifos in Rats and Humans, *Toxicological Sciences* 66, 34-53.

Tolonen, A., Turpeinen, M., Uusitalo, J., Pelkonen, O., 2005. A simple method for differentiation of monoisotopic drug metabolites with hydrogen–deuterium exchange liquid chromatography/electrospray mass spectrometry, *Eur. J. Pharm. Sci.* 25, 155-162.

Turpeinen, M., Korhonen, L. E., Tolonen, A., Uusitalo, J., Juvonen, R., Raunio, H., Pelkonen, O., 2006a. Cytochrome P450 (CYP) inhibition screening: Comparison of three tests, *Eur. J. Pharm. Sci.* 29, 130-138.

Turpeinen, M., Raunio, H., Pelkonen, O., 2006b. The functional role of CYP2B6 in human drug metabolism: substrates and inhibitors in vitro, in vivo and in silico. *Curr. Drug Metab.* 7, 705-714.

Turpeinen, M., Uusitalo, J., Jalonen, J., Pelkonen, O., 2005. Multiple P450 substrates in a single run: rapid and comprehensive in vitro interaction assay, *Eur. J. Pharm. Sci.* 24, 123-132.

US-EPA, H., 2000. US-EPA- Health Effects Division - Revised Human Health Risk Assessment Chlorpyrifos.

Usmani, K. A., Hodgson, E., Rose, R. L., 2004. In vitro metabolism of carbofuran by human, mouse, and rat cytochrome P450 and interactions with chlorpyrifos, testosterone, and estradiol, *Chem. Biol. Interact.* 150, 221-332.

Von Richter, O., Burk, O., Fromm, M. F., Thon, K. P., Eichelbaum, M., Kivist\o Kari T, 2004. Cytochrome P450 3A4 and P-glycoprotein expression in human small intestinal enterocytes and hepatocytes: a comparative analysis in paired tissue specimens. *Clin. Pharmacol. Ther.* 75, 172-183.

Whyatt, R. M., Camann, D., Perera, F. P., Rauh, V. A., Tang, D., Kinney, P. L., Garfinkel, R., Andrews, H., Hoepner, L., Barr, D. B., 2005. Biomarkers in assessing residential insecticide exposures during pregnancy and effects on fetal growth. *Toxicol. Appl. Pharmacol.* 206, 246-254.

Wienkers, L. C., Heath, T. G., 2005. Predicting in vivo drug interactions from in vitro drug discovery data, *Nature Reviews Drug Discovery* 4, 825-833.

Zanger, U., Turpeinen, M., Klein, K., Schwab, M., 2008. Functional pharmacogenetics/ genomics of human cytochromes P450 involved in drug biotransformation, *Anal. Bioanal. Chem.* 392, 1093-1108.

Zhang, Y., Benet, L. Z., 2001. The gut as a barrier to drug absorption: combined role of cytochrome P450 3A and P-glycoprotein. *Clin. Pharmacokinet.* 40, 159-168.

In: Chlorpyrifos
Editor: Caitlin Mayes

ISBN: 978-1-63482-111-7

Chapter 8

CHLORPYRIFOS INTERACTIONS WITH SEDIMENTS FROM LAKE VICTORIA: SORPTION ON CONTAINER WALLS AND CONCENTRATION, SORPTION AND DEGRADATION IN SEDIMENTS

J. Wasswa[1], P. Nkedi-Kizza[2,*], G. N. Kasozi[1] and B. T. Kiremire[1]

[1]Department of Chemistry, Makerere University, Kampala Uganda
[2]Soil and Water Science Department,
University of Florida, Gainesville FL, US

ABSTRACT

In this chapter, the role of biogeochemical processes in influencing the fate of chlorpyrifos in a lake environment is described using a case study. Lake Victoria being the second largest freshwater body in the world was selected for this purpose. Since chlorpyrifos is not readily soluble in water, sediments which serve as sink to strongly hydrophobic chemicals (SHOCs) were selected as a matrix of study. An understanding of sorption processes is key to describing the fate of chlorpyrifos in aquatic environments and was therefore investigated both on the walls of the containers used during laboratory analysis as well as on the lake sediments. In addition, biotic and abiotic degradation processes for chlorpyrifos in the lake sediments were studied. Surface sediment samples were collected using a core sampler, extracted for chlorpyrifos residues using solid dispersion methods and extracts cleaned up using gel permeation chromatograph (GPC). Analysis for Chlorpyrifos residues was done using a temperature programmed Gas Chromatograph equipped with an Electron Capture Detector. Validation of samples was done using a Gas Chromatograph equipped with a Mass Selective Detector. The average concen tration of chlorpyrifos in sediments was 17μg/kg. Sorption experiments were conducted in mixed solvents (methanol-water) and sorption coefficients in aqueous systems (using HPLC/UV) were calculated based on the Solvophobic model. Degradation of chlorpyrifos was defined as the disappearance of extractable

* P. Nkedi-Kizza, corresponding author. Soil and Water Science Department, University of Florida, Gainesville, FL 32611. Tel: (352) 294-3106. E-mail: (Kizza@ufl.edu)

concentrations with time. Experiments were conducted using both sterilized unsterilized samples. Although working in mixed solvents has been believed to eliminate sorption on container walls, results from this chapter indicate that even at volume fraction of 0.4 (40% methanol- 60% water) there was considerable sorption of chlorpyrifos on Teflon lined centrifuge tubes. Thus, sorption on containers can cause grave errors in the estimation of sorption coefficients, especially for SHOCs. To account for these effects for chlorpyrifos, sorption isotherms on sediments, were determined in methanol-water at volume fractions (f_c) greater than 0.5. Chlorpyrifos has been found to have a relatively high affinity to the lake sediment, with the sorption coefficients (K_m) decreasing with increasing f_c, while increasing with organic matter content at $f_c = 0$. Log K_{OC} values for chlorpyrifos were in the range of 3.29 – 3.81. Degradation studies indicate that chlorpyrifos degrades faster in unsterilized than in sterilized sediment samples of Lake Victoria, suggesting that microorganisms contribute to the degradation processes within the lake sediments. Thus, while many parameters are known to affect the degradation of organic pesticides in the environment, this chapter demonstrates that microbial presence and organic matter content play key roles in the degradation of chlorpyrifos in Lake Victoria sediments. The pesticide half-life ranged from 22-39 days.

Keywords: Solvophobic theory; strongly hydrophobic organic chemicals (SHOCs); half-life (t1/2); abiotic and biotic processes; sorption coefficient

BACKGROUND

Chlorpyrifos [O,O-diethyl O-(3,5,6-trichloro-2-pyrdinyl)-phosphorothioate] is one of the most widely-used active ingredients in agricultural insect control products in the world and is registered in about 100 countries worldwide. More than 50 crops, many of which are dietary staples for entire nations, are protected from insect infestation with chlorpyrifos products (Anankware et al., 2012). However, despite of its usefulness in pest control, chlorpyrifos is also known to be moderately toxic following acute oral, dermal and inhalation exposures (toxicity category II). It affects the nervous system by reversibly inhibiting the activity of cholinesterase (ChE), an enzyme necessary for the proper functioning of the nervous system (Salyha, 2010). Because of possible health risks to humans, the pesticide is being phased out for virtually all nonagricultural uses in most countries.

In the developing world, the use of chlorpyrifos especially its use in ensuring and sustaining a high quality and quantity of agricultural produce, has been steadily increasing over the years. Lake Victoria basin has been in particular ear-marked as the region with the greatest number of activities involving the use of chlorpyrifos (Wasswa, 2009; Stoytcheva, 2011). Unfortunately, there has been a history of abuse of the pesticide within the basin (NEMA, 2007), which causes justified worry. Inability to monitor, or irregular monitoring of chlorpyrifos residues in food crops and the environment because of the expenses involved as well as absence of the required technological expertise has also been a big challenge. In addition, the complexity of the fate of chlorpyrifos in the environment and the lack of data on the processes regulating its bioavailability in the Lake's environment has also been a big concern.

The Lake which is the second largest freshwater body in the world with an area of 68,800 km^2, has a mean depth of 40m (maximum depth being 84m) and an irregular shoreline of about 3,440 km in length lying in catchments of about 184,000 km^2 in area (Payne, 1986). It is bordered by three riparian states (Kenya 6%, Tanzania 51%, and Uganda 43%, by area) (Johnson and Odada, 1996; Ogutu-Ohwayo, 2001). The Lake basin is blessed with a favorable climate, which attract extensive practice of both commercial and subsistence farming along its catchment area. Believed to contain 350 fish species, the Lake has also provided tons of harvestable fishes since the 1940s (Hecky, 1993; LVBC, 2006), making it one of the major sources of income to the riparian countries and to one of the most dense populations in the world (Cohen *et al.*, 1996), in addition to being a source of domestic and industrial water. The geographical setting of the lake is thus blown up with complex topography and land surface heterogeneities. Consequently, the Lake Basin constitutes a unique environmental setting in which a myriad of studies have and can be carried out. In this chapter we concentrated on analysis of sediments because they act as sink as well as a source of a number of pollutants more so for those which are sparingly soluble. Sorption studies are crucial in describing the fate of such compounds and were therefore investigated both on the walls of the containers used during laboratory analysis as well as on the lake sediments. In addition, degradation processes for chlorpyrifos in the lake sediments were studied. Because degradation can occur by both biotic and abiotic processes (Calamari and Barg, 1993), these two courses of action were investigated.

MATERIALS AND METHODS

Study Area

Lake Victoria stretches across the Equator at 0° 30' N - 3° 00' S and 31°39' W - 34° 53' E. The area of study was the Uganda side of the Lake (Figure 1). The ecology of this area has seen rapid changes resulting from agricultural practices which have resulted into massive deforestation. The Lake shores which were historically fringed by extensive papyrus (*Cyperus papyrus*) dominated wetlands, have seen much of the wetland degraded through excessive resource use by a rapidly rising human population. As a result of the increased population growth, this area has seen increasing pressure on the resources of Lake Victoria and agricultural activities, leading to catchment degradation. Four bays of the Lake: Murchison, Napoleon, Thurston and Waiya (Figure 1) were chosen as suitable sites for the study because of the large human settlements, industrial and agricultural activities within their immediate vicinities.

Sample Collection and Preservation

A total of 117 sediment samples were collected at quarterly intervals for a period of four years. Sediments were collected in triplicate at 0-20 cm depths per sampling site, as such sediments were expected to be the most contaminated, and have the greatest potential for exchange with the water column (Nowell et al., 1999). The samples were taken using a sediment corer, wrapped in aluminium foil and placed in labelled, air-tight bags. They were kept in ice-coolers and transported to the laboratory where they were kept at -20°C before extraction.

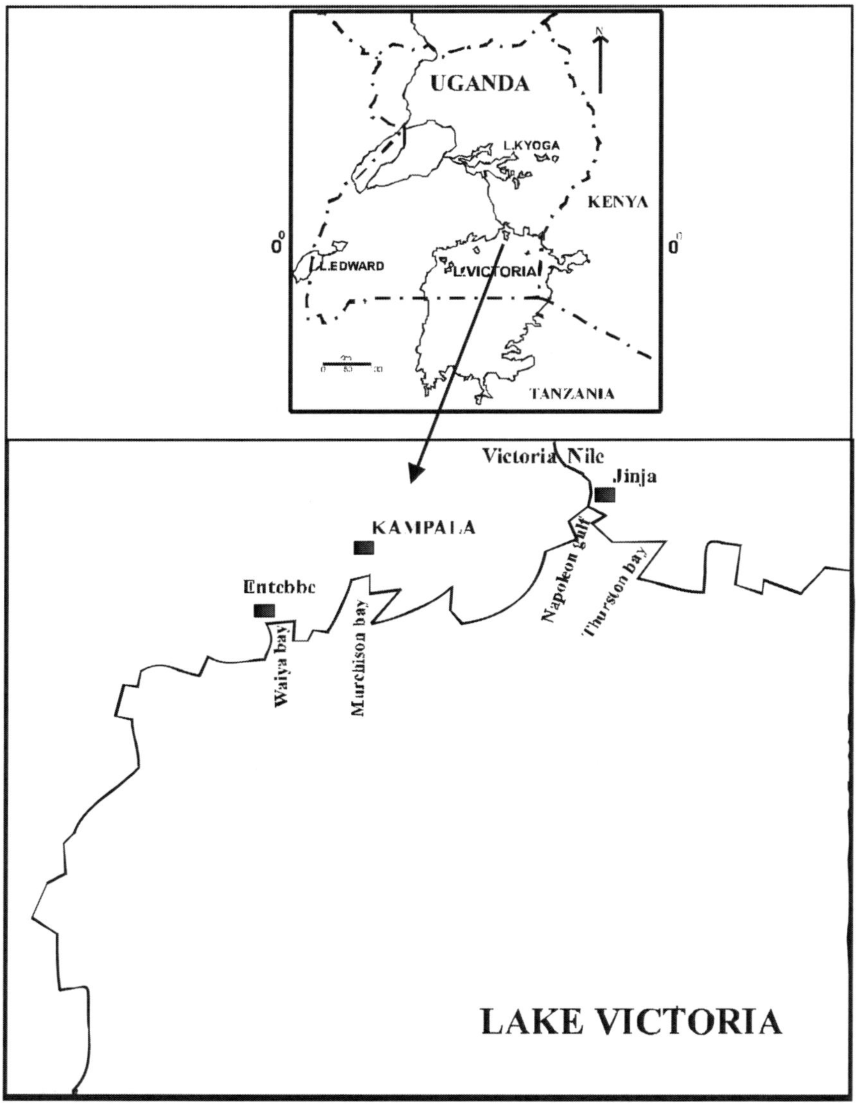

Figure 1. Study area.

Sediment Extraction Procedure

Sediment samples were extracted for pesticides residues using a dispersion method (Åkerblom, 1995). Gravel was removed from each sample prior to extraction. After the sample had been thoroughly homogenised, two portions (10 g each) were separately weighed. The first portion was placed in a pre-weighed petri-dish and left overnight to dry in an oven kept at 105°C. After cooling, the portion was re-weighed to obtain the dry weight. The second portion was ground with sodium sulphate (40 g), and the powder extracted by successive shaking with 1:1 (*v/v*) acetone/cyclohexane (50, 30, 20, 20 mL). The combined extract was filtered into a separation funnel and the filtrate shaken with saturated sodium chloride solution (200 mL). The mixture was then left to separate and the aqueous layer further extracted with ethyl acetate/cyclohexane (50 mL; 15:18 *v/v*). The organic layer obtained from the second extraction was combined with that obtained from the first in an Erlenmayer flask containing sodium sulphate (20 g). The flask was shaken and the sodium sulphate left to settle. The organic phase was then decanted through glass-wool into an evaporating flask and the sodium sulphate washed with ethyl acetate/cyclohexane (20 mL, 15:18 *v/v*) system which was then passed through the same glass-wool. The organic solvents were then evaporated off at 40°C to dryness using a vacuum rotary-evaporator. The extract was dissolved in cyclohexane/ethyl acetate (2 mL; 1:1 *v/v*), for clean-up.

Clean-up of Extracts

The sediment extract clean-up was performed using a gel permeation chromatograph (GPC). The GPC consisted of a 50 cm x 1 cm *i.d.* chromatographic tube, two adapters, teflon tubing and a six-way valve. Calibration of the GPC was done using ethion and *p,p'*-DDT as indicator substances. These substances were detected in volumes collected between the 14th and the 25th mL. The extract (1 mL) was loaded on the GPC and eluted at a rate of 1 mL min^{1} using ethyl acetate:cyclohexane (1:1 *v/v*). The final fraction (14th-26th mL) collected was cautiously concentrated to dryness with nitrogen gas and dissolved in cyclohexane (1 mL) for GC analysis.

Gas Chromatographic Analysis

Chlorpyrifos residues were analysed using a Varian (CP-3800, Palo Alto, CA, USA) gas chromatograph, equipped with an electron capture detector (ECD) and fitted with a DB-1 non-polar capillary column 50 m x 0.53 mm *i.d.* The column was later interchanged with a CP-Sil 19 CB semi-polar one of 30 m x 0.53 mm *i.d.* The GC temperature was initially set at 90°C for 1 min, raised to 180°C at a rate of 30 deg C min^{-1}, then to 260°C at 4 deg C min^{-1} and finally maintained at that temperature for 10 min. The temperatures of the injector and ECD detector were 230°C and 300°C, respectively. The injection volume was 1 μL. Reference standard for chlorpyrifos pesticides was obtained from Protocol Analytical Company (USA). Confirmation of Chlorpyrifos residues was done using a GC/MS (Agilent model 6890N) equipped with a quadruple Agilent 5975 inert XL mass selective detector and a fused silica-capillary column (HP-5MS). Inter-laboratory analysis of the samples was done at the Food and Environmental Toxicology Laboratory of the University of Florida, USA, using a Perkin Elmer Auto System with helium and nitrogen as the carrier as the make-up gases, respectively. Further confirmatory tests for selected samples were done on GC/MS, at the same laboratory, using an ion trap detector (ITD) in electron ionization mode (EI). A full-scan mode was selected for identification purposes.

Quality Control Procedures

Recovery tests were conducted by spiking sediment samples (200 g) with a standard solution equivalent of 10 μg of standard Chlorpyrifos. The sediments used for this purpose had been previously analyzed and found not to contain Chlorpyrifos. Detection limit - the lowest concentration of the analyte that can be detected above the noise from blank chromatograms - was estimated by considering the lower level of detection (LLD) defined as the concentration that produces a signal sufficiently large that 99% of the trials with that amount will produce detectable signals. LLD for Chlorpyrifos was determined according to Eaton et al. (1995)

Sorption Experiments

The key properties of chlorpyrifos used for sorption experiments are provided in Table 1. Chlorpyrifos is sparingly soluble in water; therefore direct determination of its sorption properties which is based on the concentration of the chemical in water is problematic due to sorption on container walls in aqueous systems. For this reason, various mixtures of methanol and water were used in this chapter as the binary solvents. Methanol, bought from British Drug House (BDH), was chosen to represent an organic cosolvent that is completely miscible with water. Analytical grade chlorpyrifos was obtained from Protocol Analytical Company. The calculated solubility of chlorpyrifos in various methanol-water mixtures is given in Table 2.

Because of the low solubility of chlorpyrifos in water and its potential to adsorb on container walls, recovery tests were made to account for the amount of the pesticide that adsorbed on the walls of the containers. The recovery experiments were conducted using 50 ml Teflon lined centrifuge tubes as follows. A volume of 1 ml of chlorpyrifos standard concentration (7 μg/ml in methanol) was pipetted into the various 50 ml centrifuge tubes. Various volumes of methanol and water were added to the centrifuge tubes to make a total volume of 50 ml and different volume fractions as shown in Table 2. This gave a solution of initial concentrations (C_o) of 0.14 μg/ml.

The systems were shaken for 3 hours using a Glass-Col at 30 % revolution and then left to equilibrate (up to the 24th hour) after which the suspensions were analyzed with a High Performance Liquid Chromatography (HPLC), equipped with a UV Diode Array Detector and C-18 column at a wavelength of 214 nm with Acetonitrile: Water (80: 20 v/v) as the mobile phase. HPLC conditions were based on methods developed by Abu-qare and Tariq (Abu-qare 2001; Tariq, *et al.*, 2004). The difference between the initial concentration and the concentration in solution at equilibrium was used to calculate the amount of the pesticide that had been adsorbed on the walls of the container at that fraction of cosolvent (f_c).

Batch experiment technique (Rao *et al.*, 1985; Nkedi-Kizza *et al.,* 1985) were used to determine sorption of chlorpyrifos by the lake sediment. The equilibrium concentrations [C_e (μg/ml)] were determined using a High Performance Liquid Chromatography (HPLC) at conditions described above, and the adsorbed concentrations [S_e (μg/g)] were determined by difference from the original concentration. The mass of sediment [m (g)] that would give equilibrium concentration (C_e) was estimated based on the mass balance equation (1).

$$VC_o = mS_e + VC_e \quad (1)$$

where C_o (μg/ml) = Initial pesticide concentration in solution and V (ml) = Total volume of solution.

Table 1. Important properties of chlorpyrifos

Property	Value
Aqueous solubility at 25°C ($\mu g\ ml^{-}$)	2[a]
Solubility in methanol at 25°C ($\mu g\ ml^{-}$)	116,498[b]
Log K_{OC}	3.57- 4.49[c]
k (ergs K^{-1})	$1.38*10^{-16}$
T (K)	298
Chemical structure 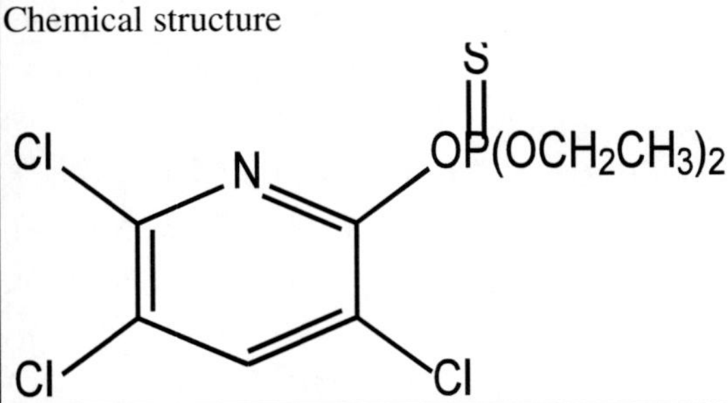	

a- (Verschueren,1983); b-(Kulkarni et al., 2000); c-(Racke *et al*., 1996).

Table 2. Volume fractions for sorption studies on walls of Teflon lined centrifuge tubes and calculated solubility (S^M) in mixtures of methanol + water

f_c	Volume of standard *7μg/ml Chlorpyrifos* added (ml)	Volume of methanol added (ml)	Volume of water added (ml)	Total volume (ml)	Calculated Solubility of *Chlorpyrifos* (S^M, $\mu g\ ml^{-1}$)
0.0					2
0.1	1	4	45	50	6
0.2	1	9	40	50	19
0.3	1	14	35	50	57
0.4	1	19	30	50	174
0.5	1	24	25	50	524
0.6	1	29	20	50	1523
0.7	1	34	15	50	4682
0.8	1	39	10	50	13974
0.9	1	44	5	50	41695

Assuming Freundlich linear isotherm;

$$S_e = K_d C_e \tag{2}$$

where K_d (ml/g) is the adsorption coefficient. Substituting equation (2) in equation (1) and solving for C_o gives equation (3).

$$C_o = C_e\left(K_d \frac{m}{V} + 1\right) \tag{3}$$

The initial pesticide concentrations C_o were set to be close to the calculated solubilities of chlorpyrifos in methanol-water mixtures. The systems were treated as described in the determination of the amount adsorbed on the walls of the centrifuge tubes and sorption data fitted using the linear Freundlich isotherm model;

$$S_e = K_m C_e \tag{4}$$

K_m being the sorption co-efficient in mixed solvents at a given f_c. Using the Solvophobic model proposed by Rao and Nkedi-Kizza (Rao *et al.*, 1985, Nkedi-Kizza *et al.*, 1985) to describe the sorption of hydrophobic organic compounds in mixed solvent, graphs of sorption coefficients $\ln(K_m)$ against (f_c) were used to estimate sorption coefficients in aqueous system K_W at $f_c = 0$, by extrapolating the graph to zero. The sorption coefficient values in aqueous system (K_W) obtained were then normalized to sediment organic carbon content (OC) to obtain K_{oc}. Organic carbon content (OC) was determined by the Walkley - Black Method (Walkley and Black, 1934; Arnold, 1986) and sediment organic matter was calculated from organic carbon values as described by Arnold (Arnold, 1986). Sediment moisture content of fresh (not air dried) samples was calculated as described by Stewart (Stewart, 1989).

Degradation Experiments

Degradation experiments were conducted on sediment samples collected from two regions of the study area; Waiya bay with the highest organic carbon content and Thurston bay with the lowest organic carbon content. This is because, organic carbon is reported to affect pesticide mobility and microbial populations and hence, transformation rates are known to increase with organic carbon content (Jack *et al.*, 1997). Samples from these regions that did not contain chlorpyrifos were preserved for this study. Standard solutions for chlorpyrifos used for these experiments were obtained from Protocol Analytical Company.

Sediment/water systems were designed according to guidelines used in studies for the registration of pesticides as described by Bromilow *et al.* (2002). In 250 ml wide-necked conical flasks, wet sediment (200g giving a 4-cm deep sediment layer) and lake water (200ml) were introduced. The system was allowed to settle, and 200 μg of chlorpyrifos (164μl of a 1.22mg/ml solution) were introduced into flasks. The standard solutions used were prepared in methanol as the vehicle for the pure pesticide. The water in each flask was gently stirred by a glass rod and left to settle in laboratory conditions at an average temperature of 25°C. Three replicate flasks were set up and sub samples (10g) were then taken off from each flask at intervals of 3, 7, 14, 30, 60, and 90 days. The moisture content of the sediments was maintained by adding appropriate mass of water after each sampling date. The sub samples were extracted, cleaned and analyzed as described before. Similar experimental set ups were put in place but using sterilized sediment samples. Sediment samples were sterilized using an auto clave at 121°C for 4 hours. The purpose of using sterilized sediment samples was to assess the effect of microorganisms on the degradation of the pesticide.

Degradation of pesticide was defined in terms of the concentration C_t (μg/g) of the pesticide at a time (t), using exponential first order kinetics. That is to say;

$$C_t = C_o e^{-Kt} \tag{5}$$

where C_o (μg/g) is the initial concentration of the pesticide (at time t = 0) and K (1/day) is the first order rate constant. If C_t^* (μg/g) is the extracted concentration of the pesticide at time t (days), then;

$$C_t = C_t^*/f^c \tag{6}$$

f^c being the method extraction efficiency and is assumed a constant. Substituting equation [6] into equation [5] gives equation [7]. Note that $f^c < 1$. However, at time zero ($C_o f^c = C_o$, since $f^c = 1$)

$$C_t^* = Cof^c e^{-Kt} \tag{7}$$

Log transformation of equation [7] yields equation [8]

$$Ln(C_t^*) = Ln(C_o f^c) - Kt \tag{8}$$

A plot of $Ln(C_t^*)$ against t gave a straight line of slope –K and intercept $Ln(C_o f_c)$ and therefore, the decay constant K was calculated independent of the initial concentration C_o at time t = 0. The pesticide half-life ($t_{1/2}$) was estimated from the expression;

$$t_{1/2} = \ln(0.5)/K \tag{9}$$

RESULTS AND DISCUSSION

Chlorpyrifos Levels in Lake Victoria Sediments

The detection frequency of chlorpyrifos was 16.3 % with a mean value of 4.18 μg/kg. Chlorpyrifos is registered for use in the country. A survey on "non-POPs Strategies for Crop Protection in Uganda" lists chlorpyrifos as one of the pesticides being used in sugar plantations as well as in the forestry sector (IPEP, 2005). Indeed chlorpyrifos was mostly detected in Thurston bay, a region that receives storm water from sugar plantations in the vicinity.

A threshold effect concentration (TEC) is used to indicate contaminant concentration levels in sediments below which toxicity to benthic organisms is unlikely to occur is not available for chlorpyrifos (MacDonald et al., 2000; Burton, 2002). However, the mean concentration of chlorpyrifos found in this study (4.18 μg/kg) could have a harmful impact on aquatic invertebrates, as this level is greater than the calculated acute and chronic toxicity for Daphnia magna (Pfeuffer, 2010)

Table 3. Chlorpyrifos Recovery, Mean, Range and LLD in sediments in Lake Victoria, Uganda

Variable	Value
Average Percentage recovery	70.70±0.16
Range (μg/Kg)	0.59 – 10.52
Mean (μg/Kg)[a,b]	4.18±2.79
LLD = 3.29 x standard deviation (ng)	9.47±2.88
TEC	N/A

[a]Results for n = 5;
[b]Concentration based on oven dry weight of sediments; N/A = Not Available.

Sorption of Chlorpyrifos within Sediments from Different Regions of the Lake

Since sorption of organic compounds by soil or sediment has been shown to be highly correlated with soil organic carbon (OC) content (Briggs, 1981; Karickhoff, 1984; Alexandra, 2005), only sediment samples from the regions that recorded the highest (Waiya bay) and the lowest (Thurston bay) organic carbon content were selected for sorption studies. Samples from these two regions in which chlorpyrifos had not been detected were used for sorption studies.

Table 4. Selected Sediment Properties

Sampling zone	Mean moisture content (%)	Mean Organic Carbon (OC) (%)	Mean Organic matter (OM) = OC x1.724 (%)
Napoleon Gulf	72.66	25.27	43.56
Thurston Bay	21.04	7.13	12.30
Waiya Bay	78.67	27.65	47.66
Murchison Bay	68.56	20.24	34.89

Based on the ratios of concentrations of the pesticide in the solution before the sorption process, to the concentrations in solutions at equilibrium, results indicate that at $f_c = 0.1$ (10% methanol- 90% water), 71.1 % of chlorpyrifos remained in the aqueous system (Table 5).

Although working in mixed solvents was aimed at eliminating sorption on container walls, results in Table 5 indicate that even at volume fraction 0.4 (40% methanol- 60%) there was considerable pesticide loss. This suggests that in addition to sorption on walls of the Teflon lined centrifuge tubes that were used in sorption experiments, the chemicals were exposed to other surfaces onto which they sorbed. In case of HPLC analysis, the technique used in this study, the possible steps where the chemicals would adsorb before analysis include; preparation of standard solutions, transfer of solutions to centrifuge tubes, the centrifuge tubes themselves and HPLC syringe if one is using a manual injection as was the case in this study. In this study, standard solutions of chlorpyrifos were prepared in Teflon centrifuge tubes using pure methanol in order to eliminate sorption on the container walls.

However, account has not been made for the sorption of Chlorpyrifos by surfaces of the HPLC syringe which is composed of a graduated glass tube and a stainless steel needle.

Table 5. Sorption of Chlorpyrifos on Teflon-line centrifuge tubes at various volume fractions of methanol (f_c)

Chlorpyrifos (spiked concentration C_0 = 0.14 μg/ml)		
f_c (Methanol)	Concentration in solution (C_e μg/ml)	Percentage recovered
0.1	0.099	71.1
0.2	0.096	78.8
0.3	0.120	85.7
0.4	0.131	93.8
0.5	0.133	95.3
0.6	0.138	98.7
0.7	0.139	99.3
0.8	0.139	99.5
0.9	0.140	99.7

To estimate the percentage recoveries (%R) at various f_c, the Solvophobic model (Rao *et al.*, 1985; Nkedi-Kizza *et al.*, 1985) has been employed. This model has been applied to sorption of several organic compounds having moderate and intermediate hydrophobicity (Nkedi-Kizza *et al.* 1985). The theory of the model is based upon the relationship between solubility of a hydrophobic molecule in aqueous and in mixed solvents, and is expressed by the following equation (Nkedi-Kizza *et al.*, 1985).

$$\text{In}X_m = \text{In}X_w + \sigma_C f_C \tag{10}$$

where X_m is the mole fraction solubility in mixed solvents, X_w is the mole fraction solubility in water, and f_c is the volume fraction of the cosolvent. The term σ_c reflects the solute–liquid interactions and is represented by the expression;

$$\sigma_C = \left[\Delta\gamma^c \text{HSA} + \Delta\varepsilon^c \text{PSA}\right]/(kT) \tag{11}$$

In this case, γ^c is the interfacial free energy at the organic co-solvent interface with hydrocarboneous surface area (HSA) of the organic molecule, ε^c is the interfacial free energy at the organic co-solvent interface with polar surface area (PSA) of the organic molecule at ambient temperature T, and k is the Boltzmann constant.

On the other hand, the relationship between sorption coefficient K_w (in aqueous) normalized with organic carbon content (OC) of the sorbent and the mole fraction aqueous solubility (X_w) is expressed as;

$$\text{In}(K_w/OC) = -\alpha \text{In}X_w - \frac{\Delta S_f (T^* - T)}{RT} - \beta\,; \tag{12}$$

For binary solvent systems (water plus water-miscible organic co-solvent), the relation is expressed as;

$$\text{In}\left(K_m/OC\right) = -\alpha \text{In} X_m - \frac{\Delta S_f\left(T^* - T\right)}{RT} - \beta \tag{13}$$

α and β being empirical constants. Thus substituting for $\text{In}X_m$ from equation (10) into equation (13) gives equation (14).

$$\text{In}\left(K_m/K_w\right) = -\alpha\sigma_C f_C \tag{14}$$

The percentage recoveries R^M in Table 5 are expressed as;

$$R^M = C_e/C_o \tag{15}$$

where Ce = solution concentration (ug/ml) and Co = Initial concentration (ug/ml). If V = volume of solution (ml) and A = surface area of the centrifuge tubes covered by solution (cm^2), then the adsorbed concentration Se ($\mu g/cm^2$) is given as;

$$S_e = V/A(C_o\text{-}C_e) \tag{16}$$

At equilibrium, the Freundlich linear isotherm is such that;

$$S_e = K_m\, C_e \tag{17}$$

K_m being the sorption coefficient in mixed solvents (ml/g). Substituting equations (15) and (16) in equation (17) gives equation (18);

$$K_m = V/A(1/R^M\text{-}1) \tag{18}$$

Similarly, in aqueous solution, equation 19 is obtained.

$$K_w = V/A(1/R^W\text{-}1) \tag{19}$$

where K_W = Sorption coefficient in aqueous system (ml/g) and R^W = Fraction recovered in water (f_c =0).

Equation (19) above relates K_m to K_w by the relation $\text{In}\left(K_m/K_w\right) = -\alpha\sigma_C f_C$.

Substituting equations (17) and (18) in equation (19) produces equation (20)

$$V/A(1/R^M\text{-}1) = V/A\ (1/R^W\text{-}1)\ e\text{-}^{\alpha\sigma cfc} \tag{20}$$

Equation (20) simplifies further to equation (21),

$$(1/R^M-1) = (1/R^W-1)\ e^{-\alpha\sigma c f c} \quad (21)$$

Log transformation of equation [21] gives $\text{Ln}\,(1/R^M-1) = \text{Ln}\,(1/R^W-1) - \alpha\sigma_c f_c$. Therefore a plot of $\text{Ln}\,(1/R^M-1)$ against f_c is linear with intercept $\text{Ln}\,(1/R^W-1)$ which was used to estimate the percentage recovery in water R^W at volume fraction $f_c = 0$.

Percentage recoveries ($\%R^M$) at various volume fraction f_c can then be obtained from equation (21) as indicated in equation 22;

$$\%R^M = [1/\{1+(1/R^W-1)*\exp(-\sigma_c\alpha f_c)\}]\%100 \quad (22)$$

Equation (22) was used to simulate the data and is the model (line). Analysis of the data in Table 5 using the model described above indicates that the percentage recovery ($\%R^W$) of chlorpyrifos in water was 52.

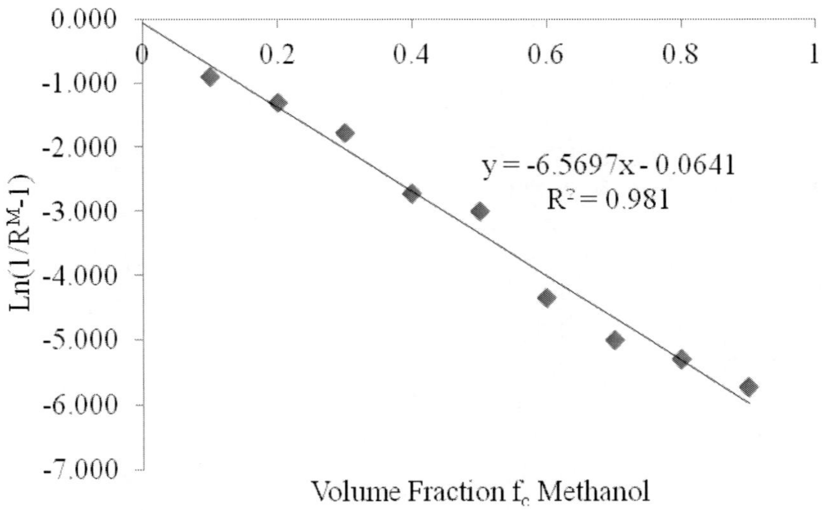

Figure 2. A Plot of $\text{Ln}(1/R^{m}-1)$ against f_c for Chlorpyrifos.

The model also indicates that at volume fraction of 0.5 and above ($f_c \geq 0.5$), there was essentially no sorption on the walls of the Teflon lined centrifuge tubes. This implies that sorption experiments conducted in aqueous solutions or at volume fractions $f_c < 0.5$ (methanol) may lead to variability in sorption data of chlorpyrifos. This may explains why there are wide variations in the K_{OC} values reported in literature; 3,715 to 30,903 for chlorpyrifos (Verschueren, 1983; Howard, 1991; Montgomery, 2000).

To account for these effects (low solubility in water and sorption on container walls) the sorption data for chlorpyrifos considered in this study were determined in methanol-water systems at volume fractions $0.5 \leq f_c < 0.8$. On examining the sorption isotherm data of chlorpyrifos on sediments at different volume fraction f_c values in methanol – water mixtures it was realized that the sorption coefficients K_m decreased with increasing f_c (Table 6). This trend was also observed in a study on sorption of anthracene, diuron and atrazine by soils in mixed solvents (Nkedi-Kizza *et al.*, 1985), as well as in a study on sorption of endosulfan on two Indian soils (Umar-Iraqi and Eram-Iraqi1; 2000).

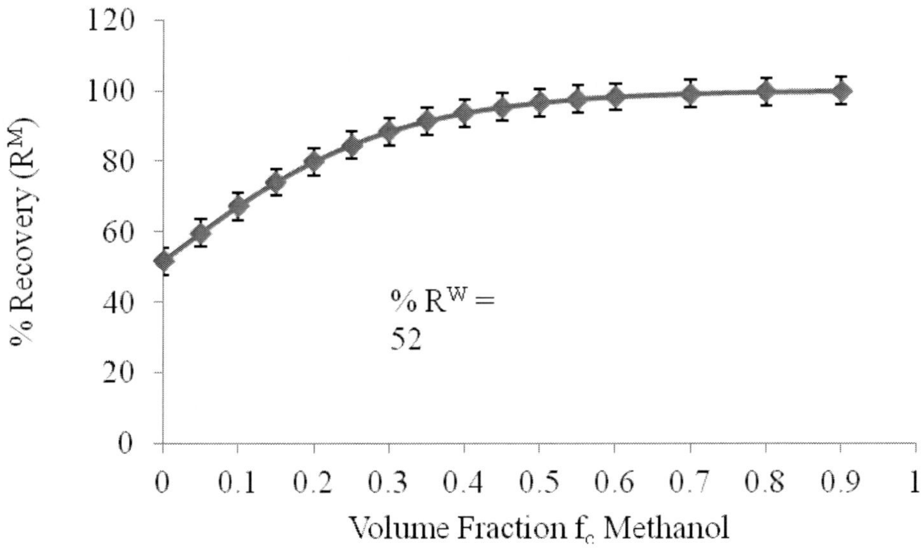

Figrue 3. % Recovery (R^M) against f_c for Chlorpyrifos.

The lower adsorption at high f_c values was attributed to the presence of co-solvents in the aqueous phase implying that the affinity of chlorpyrifos for sediments is low in presence of co-solvents and decreases with increasing amounts of co-solvent in mixtures.

Higher adsorption of chlorpyrifos was also observed in Waiya bay sediments than in Thurston bay sediments at all volume fractions as revealed by the K_m values (Table 6). The higher adsorption on Waiya bay sediments than on Thurston bay sediments is due to the larger amount of organic matter in Waiya bay sediments (Table 7). Organic matter is known to be the most important sediment or soil component affecting sorption of the pesticides because it provides the greatest number of binding sites and has an extremely large surface area. (Huddleston, 1996; Briggs, 1981; Karickhoff, 1984; Alexandra, 2005)

Table 6. Sorption coefficient (Km) values for chlorpyrifos at various fractions cosolvent (f_c)

	Chlorpyrifos	
	Waiya	Thurston
f_c	K_m	K_m
0.5	982.50	68.01
0.6	740.20	64.17
0.8	638.70	46.34

Table 7. Koc values for chlorpyrifos in Lake Victoria sediment

		K_w	OC	K_{OC}	Log K_{OC}
Chlorpyrifos	Waiya	1804.4	0.28	6444	3.81
	Thurston	136	0.07	1942	3.29

Applying the solvophobic model by plotting $\ln(K_m)$ against f_c, a linear plot with slope = $\alpha\sigma_c$ and intercept = $\ln(K_w)$ was obtained from which the sorption coefficients K_w (in aqueous systems) were calculated.

From the data in Figures 4 and 5, it is evident that sorption of chlorpyrifos was higher on Waiya bay sediment than on Thurston bay sediment with K_w values of 1804.4 and 136.0, respectively [ln(Kw) = intercept of the graph). Sorption of chlorpyrifos on soils and sediments has been investigated by a number of researchers (Felsot and Dahm, 1979; Macalady and Wolfe, 1985; Wu and Laird, 2004) although none of them considered sorption of chlorpyrifos on container walls, which may result in larger K_d and K_{OC} values. Results from these studies have reported moderately high partition coefficients for chlorpyrifos and this has been attributed to the non-polar nature of chlorpyrifos coupled with its tendency to associate strongly with organic matter (Wu and Laird, 2004). Waiya bay sediment recorded higher organic matter content than Thurston bay sediment (Table 7) and the variation in K_w values in the two regions could be attributed to the differences in the organic matter content.

In laboratory studies that involved use of batch equilibrium method, sorption K_d values for chlorpyrifos were reported to have varied between 13.4 to 1862.0 and the variations were attributed to the organic matter content (Racke 1993). In a study on the influence of lipophilicity and formulation on the distribution of pesticides in laboratory-scale sediment/water systems, a K_d value of 83 on sediments was reported (Bromilow *et al.*, 2002). Results from this study lie in the range reported by Racke (1993), although the organic carbon content at which the highest K_d value of 1862.0 was recorded, is not reported.

Sorption coefficients in aqueous system (K_w) were normalized to sediment organic carbon content (OC) to obtain K_{OC}, using the relationship $K_w = K_{OC} \times OC$. The results indicate that based on the Solvophobic theory approach, log K_{OC} values for chlorpyrifos on Lake Victoria sediment vary between 3.29 – 3.81 (Table 7). Note that the sorption coefficient (K_d) values vary by a factor of 13 and the K_{OC} values vary by a factor of 3 (Table 7). A variation in K_{OC} values of hydrophobic pesticides by a factor of about 2 across soils has been reported in the literature (Nkedi-Kizza *et al.*, 1985). It should also be noted that the slopes in Figures 4 and 5 are equal to [$\sigma_c\alpha = 1.33$] and are a property of the interaction of the solvents and HSA of chlorpyrifos [parameter σ_c], and the interaction of the solvents and soil organic matter [parameter α]. The slopes are essentially a constant (Figures 4 and 5).

Log K_{OC} values for chlorpyrifos were below the maximum of (4.49) reported for soils (Montgomery, 2000; Racke, 1993) possibly due to sorption on container walls that was not accounted for by the authors. The sorption data indicate that organic matter functionality might be slightly different in Waiya and Thurston sediments. In the Waiya sediment K_{OC} values were more that three times those obtained in the Thurston sediment.

Although sorption processes are known to be influenced by a number of factors (Schrap and Opperhuizen, 1989), the only variable considered in this chapter was organic carbon content. This is because, sorption of hydrophobic organic compounds is reported to be strongly controlled by the presence of soil organic material (Lion *et al.*, 1990). Thus the variations in K_{OC} values in Table 7 are attributed to differences in organic carbon content in the various regions of Lake Victoria. The reason for the strong influence of organic carbon content on sorption of hydrophobic organic compounds stems from the theory of sorption of chemicals to sediment from solution.

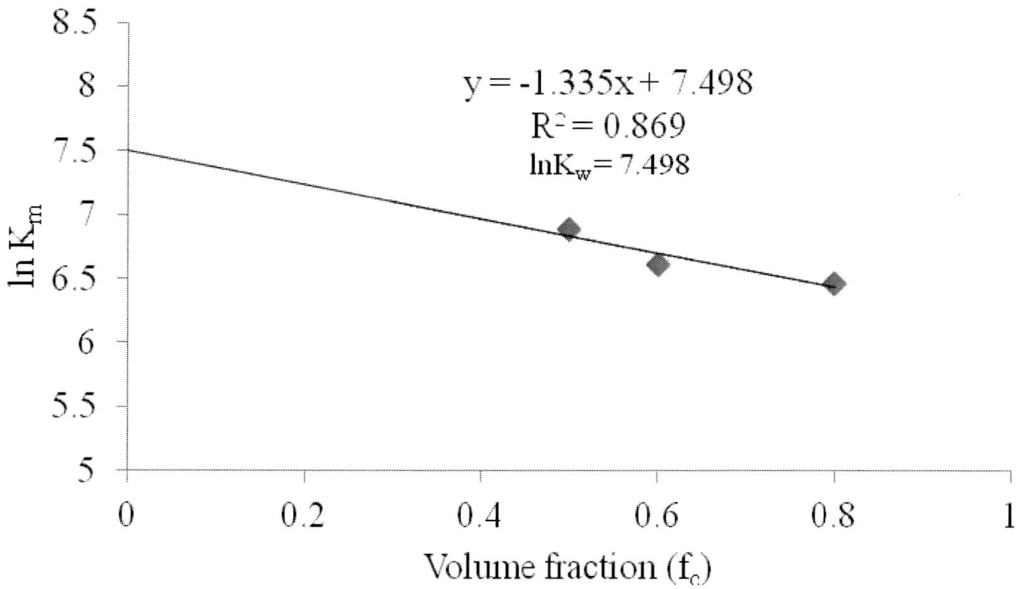

Figure 4. A plot of lnK_m Vs cosolvent fraction (f_c) for chlorpyrifos adsorption on Waiya bay sediment.

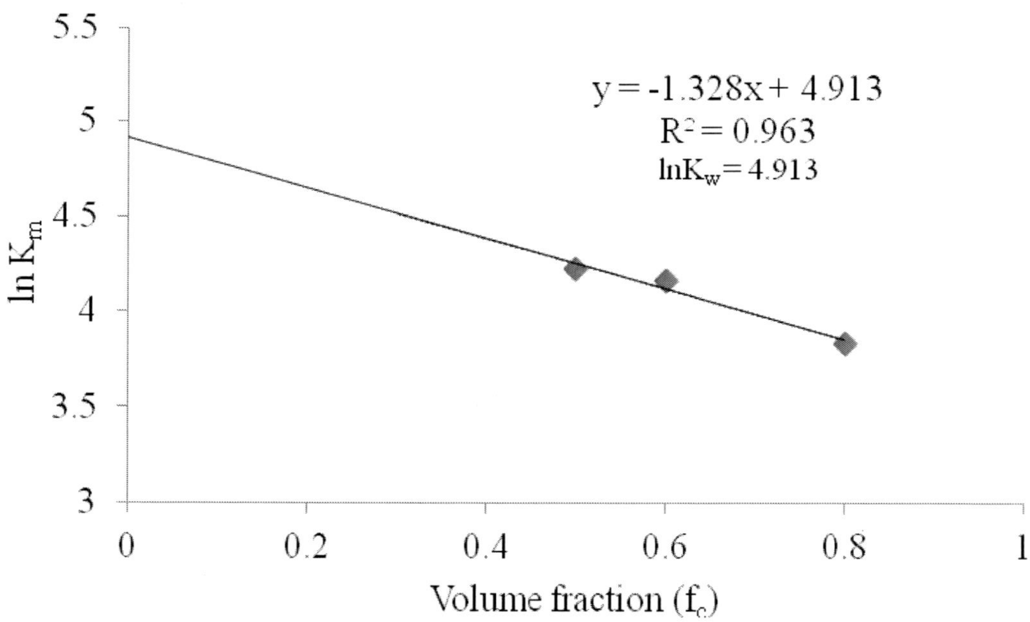

Figure 5. A plot of lnK_m Vs cosolvent fraction (f_c) for chlorpyrifos adsorption on Thurston bay sediment.

Unlike top soils which are in contact with the atmosphere and are aerobic, the environment around bottom sediments is usually anaerobic, so sediments are subjected to reducing conditions (Manahan, 2000; Mackay, 2001). In view of the generally anaerobic nature of bottom sediments, ionized species in aqueous systems adsorb water molecules at the sediment surface via strong electrostatic interactions, where the proton on water associates with the ion (Schrap and Opperhuizen, 1989). For a chemical to undergo chemisorption at the ion-aquated sediment surface, the sorbing molecule must displace one of these strongly bound waters, a difficult and energy-intensive interaction. For this reason, the K_{sorp} would be expected to be low for neutral organic compounds, higher for polar compounds that can compete for charged sites. When hydrophobic molecules associate with soil organic matter, however, there is no competition with water, and the interaction resembles a dissolution reaction into an organic solvent. This requires much less energy and the associated sorption co-efficient is much lower (Schrap and Opperhuizen, 1989).

Pesticide Transformation by Different Microbial Communities in Different Regions of the Lake and the Kinetics Involved

Degradation of chlorpyrifos in Lake Victoria sediments was defined in terms of the extractable concentration C_t^* (μg/g) of the pesticide at a time (t), using exponential first order kinetics. Plot of $\mathrm{Ln}(C_t^*)$ against t (days) for chlorpyrifos in unsterilized sediments from Waiya and Thurston bays of Lake Victoria are given in Figures 6 and 7 while those generated using sterilized sediments are given in Figures 8 and 9.

From the Figures above, the calculated half-lives based on Equation [9] for chlorpyrifos on both sterilized and un-sterilized sediment samples are given in Table 8.

The half–life values indicate that chlorpyrifos degrades faster in unsterilized than in sterilized sediment samples of Lake Victoria. This suggests that microorganisms (which are absent in sterilized samples) contribute to the degradation processes of chlorpyrifos.

Chlorpyrifos was more persistent in sterilized than in unsterilized sediments (Table 8). Major degradation pathways of chlorpyrifos are reported to be through; hydrolysis, microbial degradation, and volatility (Montgomery, 2000). In this case, the key contributing factors to the degradation of chlorpyrifos in unsterilized sediments could be hydrolysis and microbial degradation while in the sterilized sediment the key factor could be hydrolysis. Previous studies based on first-order biotic and abiotic degradation rates in estuarine water and sediment/water systems, have estimated biodegradation half-lives of chlorpyrifos to vary between 3.5-41 and 11.9-51.4 days, respectively (Walker *et al.*, 1988; Brajesh *et al.*, 2002).

In general, the chlorpyrifos was more persistent in Thurston bay sediment samples than in samples from Waiya bay. Thurston bay recorded lower mean organic matter content of 12.30 % than Waiya bay - 47.66 % (Table 4) and the discrepancies in degradation rates in the two regions may partly be attributed to the organic matter content. Organic matter content is known to affect the chemical, physical and biological properties of the sediment and its overall composition (Alexandra, 2005). In addition, composting (a process in which *microorganisms* rapidly consume organic matter) is reported to provide an optimal environment for pesticide destruction (Buyuksonmez *et al.,* 2000). Although many parameters are known to affect the degradation of chemicals in the environment (Howard, 1991), this chapter has established that microbial presence and organic matter content play key roles in the degradation of chlorpyrifos in Lake Victoria sediments.

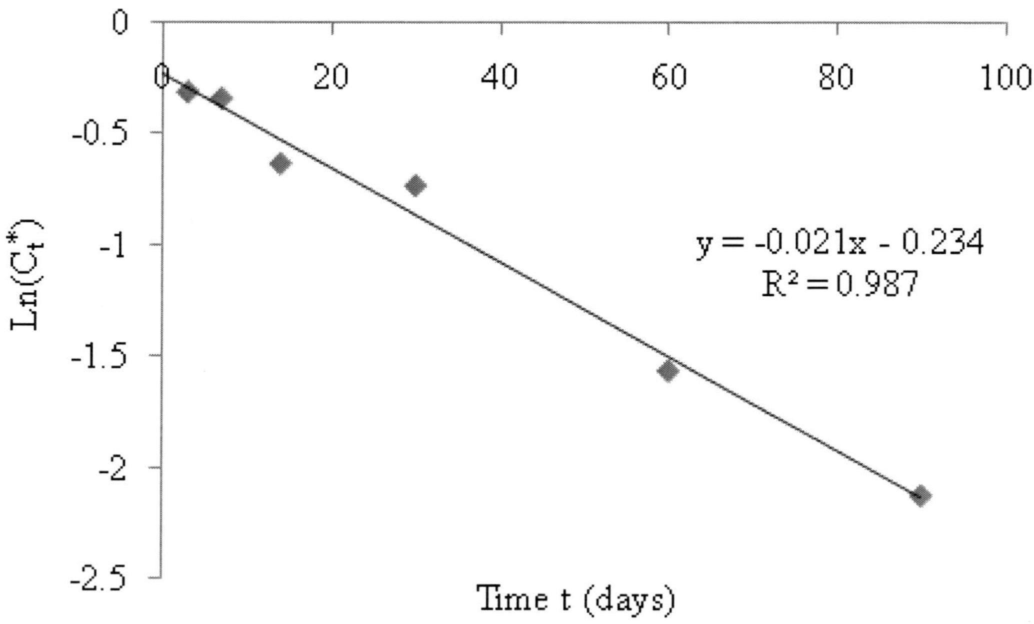

Figure 6. Degradation of Chlorpyrifos on unsterilized sediment from Waiya Bay – Lake Victoria.

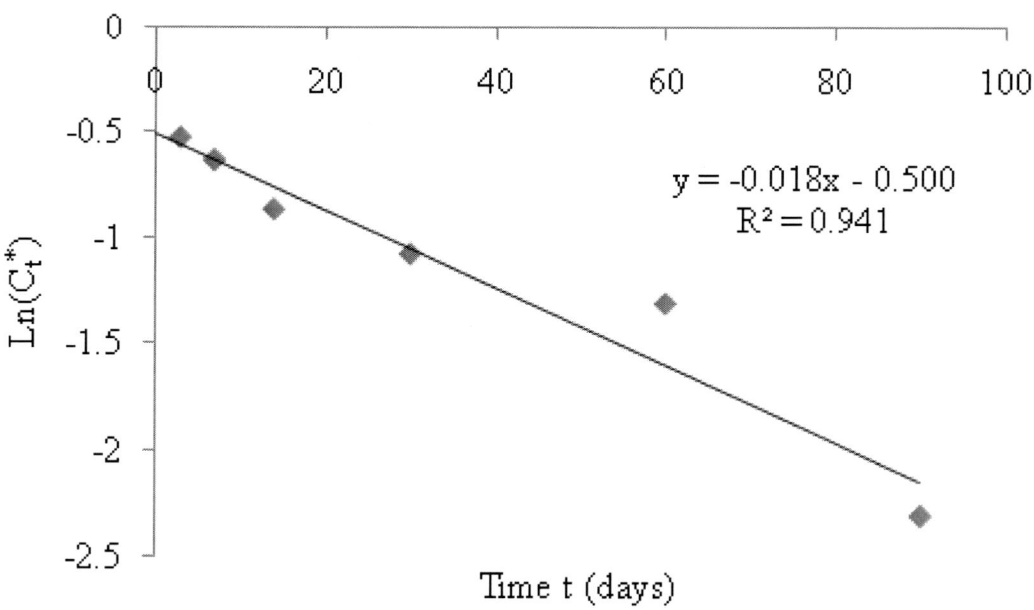

Figure 7. Degradation of Chlorpyrifos on unsterilized sediment from Thurston Bay – Lake Victoria.

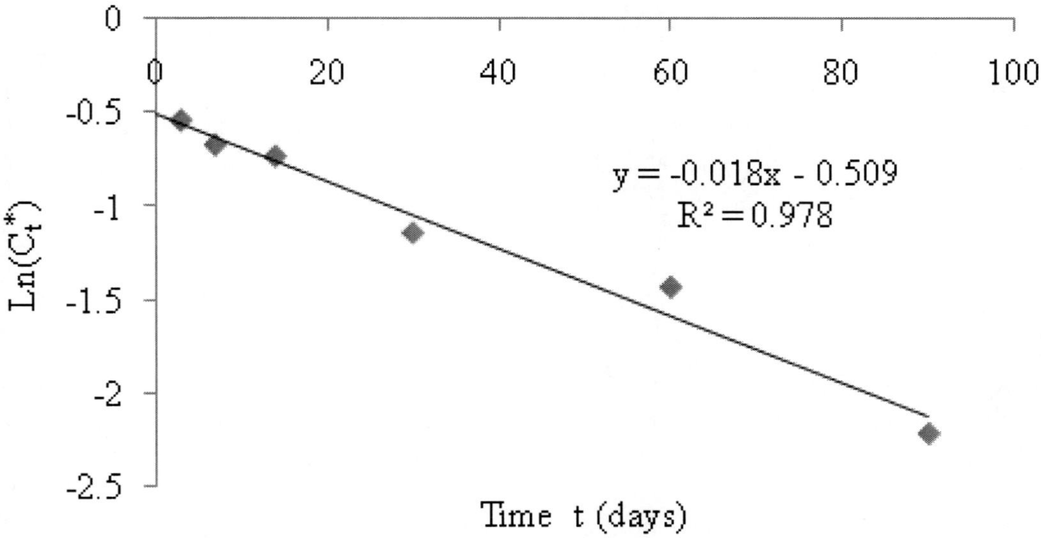

Figure 8. Degradation of Chlorpyrifos on unsterilized sediment from Waiya Bay – Lake Victoria.

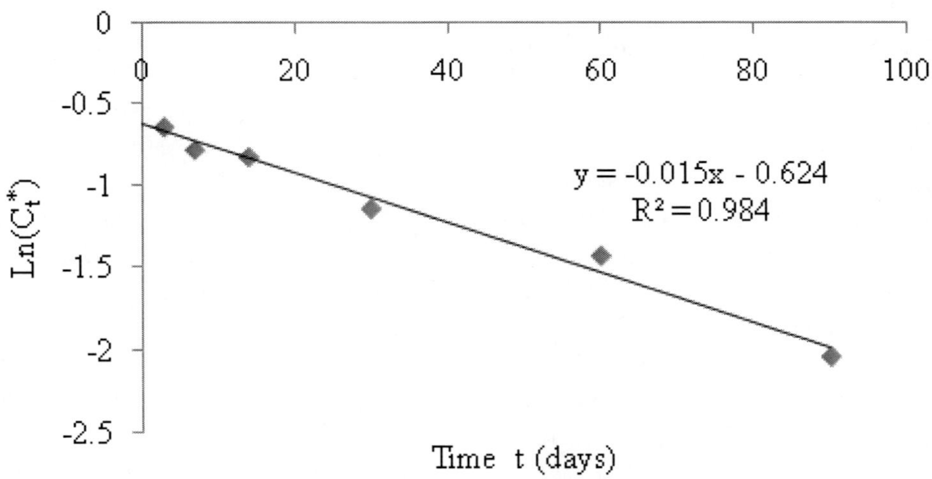

Figure 9. Degradation of Chlorpyrifos on unsterilized sediment from Thurston Bay – Lake Victoria.

Table 8. Half-lives of Chlorpyrifos in Lake Victoria Sediments

	Un-sterilized		Sterilized	
	Rate constant K_1	$t_{\frac{1}{2}}$ (days)	Rate constant K_1	$t_{\frac{1}{2}}$ (days)
Waiya bay	0.021	33.0	0.018	38.5
Thurston bay	0.018	38.5	0.015	46.2

ACKNOWLEDGMENTS

We are thankful to the International Programme in Chemical Sciences (IPICS) under the International Science Programme (ISP), Uppsala University, Sweden, as well as Makerere University, Kampala, Uganda, and the University of Florida, Gainesville, USA, for sponsoring this study.

REFERENCES

Abu-qare Aqel W and Abou-Donia Mohamed B (2001). Simultaneous determination of chlorpyrifos, permethrin, and their metabolites in rat plasma and urine by highperformance liquid chromatography. *Journal of Analytical Toxicology*. 25 (4) 275-279.

Åkerblom, M., 1995. Environmental monitoring of pesticide residues: Guideline for SADC Region, SADC/ELMS, monitoring techniques series, Sweden.

Alexandra Bot (2005). The importance of soil organic matter - key to drought-resistant soil and sustained food production. ISBN 92-5-105366-9 FAO, Rome.

Anankware P.J., Fatunbi A.O., Afreh -Nuamah K., Obeng-Ofori D. and Ansah A.F., 2012. Efficacy of the Multiple-Layer Hermetic Storage Bag for Biorational Management of primary Beetle Pests of Stored Maize. *Academic Journal of Entomology* 5 (1): 47-53.

Arnold, J. G., Williams, J. R., Srinivasan, R., and King, K. W. (1995). "SWAT: Soil and water assessment tool," Grassland, Soil and Water Research Laboratory, USDA-ARS, Temple, TX.

Brajesh Kumar Singh, Allan Walker, and Denis J. Wright (2002). Degradation of Chlorpyrifos, Fenamiphos and Chlorothalonil alone and in Combination and their effects on Soil and Microbial Activity. *Journal of Environmental Toxicology and Chemistry* 2 (12) 2600–2605

Briggs, G.G. 1981. Theoretical and experimental relationships between soil adsorption, octanol–water partition coefficients, water solubilities, bioconcentration factors, and the parachor. *J. Agric. Food Chem*. 29:1050–1059.

Bromilow H Richard, Evans A Avis, Nicholls H Peter, 2002. The influence of lipophilicity and formulation on the distribution of pesticides in laboratory-scale sediment/water systems. *Pest Management Science* 59 (2) 238 – 244

Buyuksonmez F, Rynk R, Hess T.F, and Bechinski E. (2000). Occurrence, degradation and fate of pesticides during composting. Part II: occurrence and fate of pesticides in compost and composting systems. *Compost Sci. Util*. 8,61–81.

Burton, A.G. Jr., 2002. Sediment quality criteria in use around the world. Limnol. 3, 65–75.

Calamari, D. and Barg, U. 1993. Hazard assessment of agricultural chemicals by simple simulation models. In: Prevention of Water Pollution by Agriculture and Related Activities. Proceedings of the FAO Expert Consultation, Santiago, Chile, 20-23 Oct. 1992. Water Report 1. FAO, Rome. pp. 207-222.

Cohen, A.S., Kaufman, L. and Ogutu-Ohwayo, R. 1996. Anthropogenic threats, impacts and conservation strategies in the African Great, Lakes: A review. In: *The Limnology,*

Climatology and Paleoclimatology of the East African Lakes. Johnson, T.C. and Odada, E. (eds). Gordon and Breach, Toronto, pp. 575-624.

Eaton, A.D., Clesceri, L.S., Greenberg, A.E., 1995. Standard Methods for the Examination of Water and Wastewater. 19th Ed. ISBN 0-87553-223-3 American Public Health Association.

Felsot, A., and P.A. Dahm. (1979). Sorption of organophosphorus and carbamate insecticides by soil. *J. Agric. Food Chem*. 27:557–563.

Hecky R.E. 1993. The eutrophication of Lake Victoria. *Verh. Internat. Verein. Limnol*.25, 39-48.

Howard, Philip H. (1991). Handbook of Environmental Fate and Exposure Data for Organic Chemicals. Volume III, Pesticides. Lewis Publishers, Chelsea, MI.

Huddleston J. H. 1996. How soil properties affect ground water vulnerability to pesticide contamination. Oregon State University Extension Service.

International POPs Elimination Project – IPEP (2005) Non-POPs Strategies for Crop Protection in Uganda Website- www.ipen.org.

Jack E. Barbash and Elizabeth A. Resek (1997) Pesticides in Ground Water, Distribution trends and governing factors Vol 2 pg 418 ISBN 1-57504-005-0 Lewis Publishers.

Johnson, T. C. and E. O. Odada, 1996. *The Limnology, Climatology and Paleoclimatology of the East African Lakes*. Gordon and Breach Publishers, Amsterdam, the Netherlands Pp 664.

Karickhoff, S.W. 1984. Organic pollutant sorption in aquatic systems. *J. Hydraulic Eng*. 110:707–735.

Kulkarni, A. R.; Soppimath, K. S; Dave, A. M.; Mehta, M. H.; Aminabhavi,T. M., 2000. Solubility study of hazardous pesticide (chlorpyrifos) by gas chromatography. *Journal of hazardous Materials*. A80, 9-13.

Lion L. W., Stauffer T. B., and MacIntyre W. G. (1990) Sorption of hydrophobic compounds on aquifer materials; analysis methods and the effect of organic carbon. *J. Contam. Hydrol*. 5, 215-234.

LVBC (Lake Victoria Basin Commission), 2006. Special Report on the Declining of Water Levels of Lake Victoria EAC Secretariat Arusha, Tanzania.

Macalady, D.L., and N.L. Wolfe. (1985). Effects of sediment sorption and abiotic hydrolysis. 1. Organophosphorothioate esters. *J. Agric. Food Chem*. 33:167–173.

MacDonald, D.D., Ingersoll, C.G., Berger, T.A., 2000. Development and evaluation of consensus-based sediment quality guidelines for freshwater ecosystems. *Archives of Environmental Contamination and Toxicology* 39, 20e31.

Mackay Donald (2001). Multimedia Environmental Models – The Fugacity Approach. Second Edition Lewis Publishers, Boca Raton.

Manahan, Stanley E. (2000) Environmental Science, Technology, and Chemistry *Environmental Chemistry* Boca Raton: CRC Press LLC

Montgomery, J.H., 2000. Groundwater Chemicals Desk Reference, third ed. Lewis Publishers, CRC Press, Boca Raton, FL.

Muwamba, A.; Nkedi-Kizza, P.; Rhue, D. R.; Keaffeber, J. J.. 2009. Use of Mixed Solvent Systems to Eliminate Sorption of Strongly Hydrophobic Organic Chemicals (SHOCs) on Container. *J. Environ, Qual*. 38, 1170-1176.

NEMA –National Environment Management Authority, Uganda, 2007. State of Environment Report for 2006/2007: NEMA Secretariat, Kampala.

Nkedi-Kizza, P., Rao P.S.C., Hornsby A.G.; 1985. Influence of organic cosolvents on sorption of hydrophobic organic chemicals by soils. *Environ. Sci. Technol.* 19:975–979.

Nowell, L.H., Capel, P.D., Dileanis, P.D., 1999. Pesticides in stream sediment and aquatic biota: distribution, trends, and governing factors. Lewis, New York.

Ogutu-Ohwayo, R., 2001. *Changes in the Life History Characteristics of Nile Perch, Lates Niloticus L. in Lake Victoria and their Implications on the Future in the Lake.* First National Conference, Lake Victoria Environment Management Project 17th – 21st September, Mukono-Uganda.

Payne, A.I., 1986. The Ecology of Tropical Lakes and Rivers. John Wiley and Sons Ltd., New York.

Peterson, S.M., and G.E. Batley. 1993. The fate of endosulfan in aquatic ecosystems. *Env. Pollut.* 82:143–152.

Pfeuffer Richard J., 2010. Pesticide Surface Water and Sediment Quality Report March 2010 Sampling Event. South Florida Water Management District MSC 7281 3301 Gun Club Road West Palm Beach, FL 33406.

Racke, K.D. 1993. Environmental fate of chlorpyrifos. *Rev. Environ. Contam. Toxicol.* 131:1–154.

Rao P.S.C., A.G. Hornsby, D.P. Kilcrease, P. Nkedi-Kizza; 1985. Sorption and transport of hydrophobic organic chemicals in aqueous and mixed solvent systems: Model development and preliminary evaluation. *J. Environ. Qual.* 14: 376-383.

Salyha Y., 2010. Biological Effects Assessment of Chlorpyrifos and Some Aspects of Its Neurotoxicity. *Biology series.* 54, 3-14.

Schrap S. M. and Opperhuizen A. 1989. Quantifying the sorption of organic chemicals on sediments. *Chemosphere* 18, 1883-1893.

Stewart E. Allen (1989) Chemical analysis of ecological materials. 2nd Edition, ISBN – 0632 - 01742-2 Blackwell Scientific Publications, London.

Stoytcheva Margarita, 2011. Pesticides in the Modern World – Pesticides Use and Management. Janeza Trdine 9, 51000 Rijeka, Croatia 978-953-307-459-7.

Tariq Siddique, Zahir A. Zahir, William T. Frankenberger, 2004. Reversed-Phase Liquid Chromatographic Method for Analysis of Endosulfan and Its Major Metabolites. *Journal of Liquid Chromatography & Related Technologies.*

Umar Iraqi S. M. and Eram Iraqi1; 2000. Sorption of the Pesticide Endosulphan on Two *Indian Soils Journal of Colloid and Interface Science* 224, 155–161.

Verschueren, K., 1983. Hand book of Environmental Data on Organic Chemicals. Van Nostrand Reinhold New York. ISBN 0-442-28802-6.

Walker, W.W., Cripe, C.R., Pritchard, P.H., and Bourquin, A.W. (1988). Biological and abiotic degradation of xenobiotic compounds in *in vitro* estuarine water and sediment/water systems, *Chemosphere*, 17(12):2255-2270.

Walkley A. and Black I. A (1934). An examination of Degtjareff method for determining soil organic matter and a proposed modification of the chromi acid titration method. *Soil Sci.* 37:29-38.

Wasswa, J., 2009. Characterisation of selected pesticide residues in sediments of the Ugandan side of Lake Victoria. PhD thesis, Department of Chemistry, Makerere University, Kampala, Uganda.

Wu, J., Laird, D.A. 2004. Interactions of chlorpyrifos with colloidal materials in aqueous systems. *J. Environ. Qual.* 33, 1765-1770.

In: Chlorpyrifos
Editor: Caitlin Mayes

ISBN: 978-1-63482-111-7

Chapter 9

CHLORPYRIFOS: ENVIRONMENTAL OCCURRENCE IN THE ATMOSPHERE AND WATER

Renata Raina-Fulton**, Liu Yang and Noof Alzahrani
Department of Chemistry and Biochemistry, University of Regina, Regina, SK, Canada

ABSTRACT

Chlorpyrifos is used worldwide in agricultural and urban environments for insect control. The major environmental studies on the occurrence of chlorpyrifos in the air and water have taken place in North America. Major areas of study in the United States have included California, Midwestern United States and along the Mississippi River. In Canada the major usage region for chlorpyrifos is the prairie agriculture region which has the highest grain and oil seed production in Canada. Environmental measurements of chlorpyrifos have also taken place in National Park sites in both Canada and the United States. The focus of this chapter is a review of atmospheric, surface water, or precipitation measurements that have taken place in North America. New measurements in two agricultural regions of Canada (Okanagan Valley and Prairie Agricultural Region) are also presented. Atmospheric concentrations of chlorpyrifos have been measured at Bratt's Lake, SK since 2003. Large annual shifts in atmospheric concentrations of chlorpyrifos have been observed in the Canadian prairies with a seasonal maximum concentration of chlorpyrifos in late July-early August. Evidence for potential for regional to long-range atmospheric transport of chlorpyrifos is presented from historical measurements at National Park or remote sites as well as new measurements in areas with little local/regional usage of chlorpyrifos. Both urban and agricultural sources have contributed to the occurrence of chlorpyrifos in the atmosphere or surface waters. In the United States environmental levels and usage of chlorpyrifos have declined since the 1990s. Usage and atmospheric concentrations of chlorpyrifos in the Canadian prairies has also declined since 2003 and attributed to less demand for chlorpyrifos for control of grasshoppers in the prairies.

* Corresponding author

INTRODUCTION

China is a major user and exporter of chlorpyrifos estimated to produce ~ 85% of global capacity. Studies in China have focused on dietary risk of chlorpyrifos residue in food commodities such as rice, maize and soybeans [1, 2]. Application of chlorpyrifos according to good agricultural practices (GAPs) at the recommended dose levels show that crops are unlikely to pose any public health issues from dietary exposure [1, 2]. Dermal exposure has also been identified as a major route of exposure for workers involved in applications of chlorpyrifos for maize cultivation in China [3]. Little environmental studies have been conducted in Asia or Europe. Emissions of chlorpyrifos in Europe in 2000 are relatively low as compared to usage in Canada and the United States [4]. The majority of information on environmental occurrence of chlorpyrifos is available from North America where it has been used for agriculture as well as residential applications.

North America continues to be a major usage region of chlorpyrifos. Emission of organophosphorus pesticides (OPs) has declined more than 60% from 1990 to 2006 in the United States [5]. Chlorpyrifos was the most widely used OP insecticide used in the United States in 2007 with usage range of 3.6-5 million kg of active ingredient, while in 2001 usage of chlorpyrifos was in the range 5-7.3 million kg and it was the 2nd most widely used OP (malathion most widely used) in the United States [5]. In Canada in 2008 chlorpyrifos was the top sold OP for use with usage > 500,000 kg and among the top insecticides sold in Canada (others include mineral oil, hydrogen peroxide, and sulfur) [6]. During 2009-2011 usage of chlorpyrifos has declined to the >100,000 kg category but chlorpyrifos is still one of the preferred insecticides for use in agriculture [6]. In the United States chlorpyrifos has been used on orchards and grapes, soybeans, and corn with higher usage within states where these crops are grown including California, Washington, North and South Dakota, and midwestern United States (*Indiana, Illinois, Iowa, Missouri,* eastern *Nebraska,* and eastern *Kansas*) [7]. In Canada the highest usage region of chlorpyrifos has been the prairie provinces (Alberta, Saskatchewan, and Manitoba) with > 95% of total usage in the early 2000s [8, 9]. The focus of studies on the occurrence of chlorpyrifos in the atmosphere has been on agricultural areas in high usage regions, areas with high agricultural pesticide application rates or urban areas where there is nonagricultural usage of chlorpyrifos. Chlorpyrifos has been detected in the atmosphere in a large range of sites across North America as shown in Table 1 and will be discussed further within this chapter. Few measurements are available in Europe and Asia.

Chlorpyrifos has been primarily detected in the atmosphere in the gas phase [11, 16]. Atmospheric sampling of the gas phase fraction has generally involved the use of polyurethane foam (PUF), XAD-2, XAD-4, or PUF/XAD-2 sorbent combination [25, 26]. For high-volume air sampling a higher collection efficiency for chlorpyrifos was observed when a PUF/sorbent combination was used as compared to PUF alone [25]. Sorbent tubes used for worker and community measurements were more prone to artifact transformation with chlorpyrifos degradation to chlorpyrifos oxon when XAD-2 was used in comparison to PUF [26]. Chlorpyrifos has relatively low water solubility and higher vapour pressure (2.19 X10^{-3} Pa) and Henry's law constant of 0.13-0.17mPa m^3/mol such that there should be fewer detection in rain relative to other pesticides with similar gas phase concentrations [11, 27].

Table 1. Occurrence of Chlorpyrifos in the Atmosphere

Sampling Location	Sampling Period	Atmospheric Concentration Chlorpyrifos (pg/m^3) Maximum (Medium or Average*)	Reference
		United States Urban Impacted	
Mississippi River	June 1994	1600 (290)	[10]
Misssissippi River Valley Urban Site Agriculture Site	April-Mid September 1995	 3500 (1500) 3100 (nd)	[11]
		United States : Agriculture and Urban Impacted	
Subestuary of Chesapeake Bay	April-July 1995	2000 (200 ± 390)*	[12]
Parlier, CA	Jan-Dec 2006	7000-150000** (23100)*	[13]
Tulare County, CA	Spring/summer 1996	16000-1340000**	[14]
		Canadian Agriculture Impacted	
Abbotsford, BC	July-Aug 2003	(281)***	[15]
Abbotsford, BC	2004 2005	348 (229)* 1100 (742)*	[16]
Abbotsford, BC	Jan-Dec 2005	3.9-86.7**	[17]
Osoyoos, BC	April-Oct 2011	2877 (56)	This study
Bratt's Lake, SK	2004 2005	170000 (3100)* 4910 (1590)*	[15]
Bratt's Lake, SK	April 2003-Dec 2003 Jan-Dec 2005	250000 0.6-1380**	[17]
Bratt's Lake, SK	Jan-Dec 2010	10255 (74)	This study
Egbert, ON	July-August 2003	(378)***	[15]
Egbert, ON	2004 2005	184 (62.6)* 246 (90.8)	[16]
Vineland, ON	2004 2005	167000 (21900)* 107000 (20600)*	[16]
Downsview, ON	2005	371 (190)*	[16]
St. Anicet, QC	July-August 2003	(768)***	[15]
St. Anicet, QC	2004 2005	478 (260)* 1260 (955)*	[16]
Baie St. Francois, QC	July-August, 2003	(742)***	[15]
Baie St. Francois, QC	2004 2005	2050 (891)* 6740 (3820)*	[16]
Kensington, PEI	July-August, 2003	(107)***	[15]
Kensington, PEI	2004 2005	618 (104)* 759 (284)*	[16]
South Tabacco Creek Watershed, Manitoba	1994-1996	10000-103000**	[18]
		Other Agriculture Impacted Sites	
Centre Region, France	2006-2008	110-97770 (2860)	[19]

Table 1. (Continued)

Sampling Location	Sampling Period	Atmospheric Concentration Chlorpyrifos (pg/m^3) Maximum (Medium or Average*)	Reference
		Remote or National Park Sites	
Mt. Bachelor Observatory, OR	April 2004 to May 2006	< MDL -6	[20]
Revelstoke National Park, BC	2003-2004	0.8-1.7** (1.1)	[21]
Yoho National Park, BC	2003-2004	0.6-4.6** (0.8)	[21]
Observation Peak, Banff National Park, AB	2003-2004	0.4-1.3** (0.7)	[21]
Ontario,10 remote inland lakes	2003-2005	60 (7)	[22]
New Zealand southern alpine site	Jan-Feb 2009	< MDL-12	[23]
Eastern Asia to High Arctic (highest East Japan Sea)	July-Sept 2010	0.5-146 (13.5)*	[24]

* Average; ** concentration range (min-max above MDL); *** pooled air samples.

ATMOSPHERIC CONCENTRATIONS IN URBAN IMPACTED SITES OF UNITED STATES

Early reports from the midwestern United States and along the Mississippi River in April-September 1995 found that the chlorpyrifos could be detected in >80% of air samples collected at urban sites in Iowa and Mississippi with a significant lower detection frequency at nearby agricultural sites [10]. Minnesota sites had a much lower detection frequency of chlorpyrifos in atmospheric samples (< 20%) [10]. The highest air concentrations of chlorpyrifos (~ 1000-3500 pg/m^3) in the Mississippi River Valley Region were observed in July-mid-September [11]. At Mississipppi sites methyl parathion was also more prevalent than chlorpyrifos in air and water, while for other sites in the Mississippi River Valley Region chlorpyrifos was the most frequently detected OP insecticide in the atmosphere [10, 28]. From air sampling in June 1994 along the Mississippi River the maximum and median air concentrations of chlorpyrifos were 1600pg/m^3 and 290pg/m^3, respectively [29]. There was 100% detection frequency and highest concentrations of chlorpyrifos were observed near urban areas of New Orleans, LA; Memphis, TN, St. Louis, MO or Minnesota, MN where they were used extensively at the time [29]. Chlorpyrifos was one of the insecticides that was more frequently detected in air and water at urban sites [29, 30] and these higher air concentrations of chlorpyrifos correlated to urban usage of chlorpyrifos such as for termite control, control of fire ants, and lawn applications [30]. Chlorpyrifos has also been used for flea control in United States with broadcast applications resulting in higher deposition on carpets than aerosol applications [31]. Chlorpyrifos is also a widely used turfgrass insecticide and golfers are exposed to chlorpyrifos with dermal exposure accounting for ~ 60% total loadings following applications to turfgrass [32].

ATMOSPHERIC CONCENTRATIONS IN AGRICULTURAL IMPACTED SITES

Chlorpyrifos has been and continues to be widely used in agriculture in the United States. Use of chlorpryrifos on corn has dropped significant in the United States since 2003 [7]. In California there are over 60 different crops including alfalfa, walnuts, oranges, cotton, and grapes where chlorpyrifos is applied for insect control. Sales of chlorpyrifos annually in California were highest during 1994-1997 and have declined since 2006 but were still between 632000 and 820000 kg active ingredient post 2006 as shown in Figure 1 [33]. In Sept 2014 the *California Department of Pesticide Regulation* (DPR) proposed to make all pesticide products containing the active ingredient chlorpyrifos, used to produce an agricultural commodity, a California restricted material and thus further reductions in chlorpyrifos agricultural usage are expected in the future [34].

In 2006 community studies in and around the San Joaquin agricultural valley, California found maximum atmospheric concentrations of chlorpyrifos ranging from 83, 150, and 1340ng/m^3 for Lompoc, Parlier, and Tulare County in California [13, 14, 35]. These have been the highest reported urban community measurements of chlorpyrifos in North America where it is known high agricultural application rates of chlorpyrifos occur such as within 5 miles of Parlier where 1424 kg of chlorpyrifos were used in 2006 [13]. It has also been detected in fog water collected near Monterey, CA where fog forms under stagnant inversion conditions in California's central valley [36]. Concentrations of chlorpyrifos in fog water ranged from < MDL to 0.19μg/L and higher levels were attributed to usage of chlorpyrifos on vegetables near the coast and dormant fruit orchards in the Central Valley of California [36].

The Chesapeake Bay region of the United States has been another area of air and surface water measurements of pesticides due to high pesticide application rates. In the Chesapeake Bay the atmospheric concentrations of chlorpyrifos during 1995 were lower than other US areas with maximum of 97pg/m^3 and highest levels observed during June in the southern portion of the Bay when atmospheric deposition into the Bay took place [37]. In a subestuary of Chesapeake Bay (mouth of Patuxent River in Solomons, MS) maximum air concentrations were higher at 2000 pg/m^3 which occurred during May 1995 [12] and chlorpyrifos was the predominate OP used in agriculture in this region of the Chesapeake Bay area (109000 kg a.i./yr) [38].

In Canada the Okanagan Valley of British Columbia is an intensive agricultural area for tree fruit crops and grapes. Based on usage in the southern interior of BC where the sampling site at Osoyoos, BC is located only 294 an 280 kg of chlorpyrifos were used in 2010 and 2003, respectively so low usage was also expected for 2011 when measurements were taken [39, 40]. Figure 2 shows that relatively high atmospheric concentrations were observed particularly during the spring although regional usage in the Okanagan Valley was expected to be low. Higher usage of chlorpyrifos has been reported in the lower mainland region of British Columbia with usage in 2010 and 2003 usage at 3941 and 3929 kg, respectively [39, 40]. Atmospheric concentrations of chlorpyrifos are lower in the Lower Fraser Valley (within lower mainland) agricultural region as compared to the Okanagan valley of British Columbia. At Abbotford, BC located in the Lower Fraser Valley maximum concentrations were between 87-1100pg/m^3 in 2003 -2005 [16, 17]. Diazinon and malathion are recommended for use on tree fruits, and malathion is also recommended for use on grapes in the Okanagan Valley, but chlorpyrifos is not recommended for these crops [41, 42]. Chlorpyrifos is recommended for tree fruit crops in Washington State [43] and elevated exposure has been associated with children in homes near farmland [5]. Sites in the Centre Region, France in a region impacted by vineyards, and orchards also observed higher atmospheric concentrations of chlorpyrifos reaching 97770pg/m^3 [19].

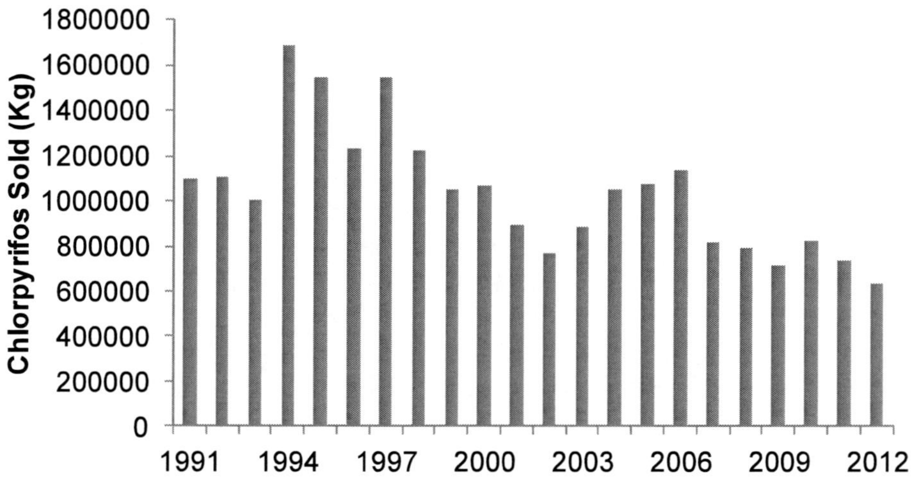

Figure 1. Active Ingredient Chlorpyrifos Sold in California. Adapted from pesticide sales reports [33].

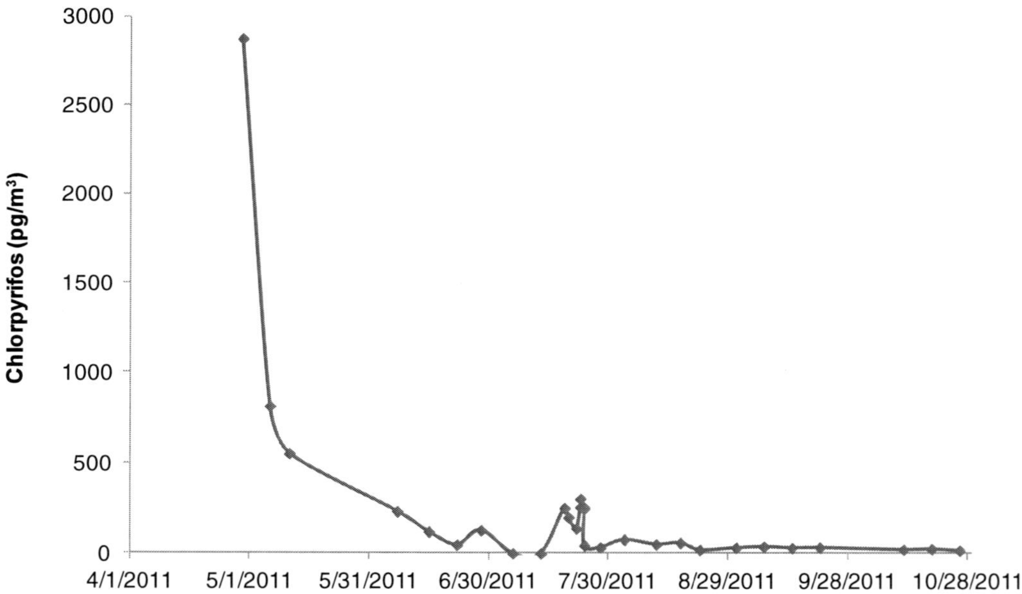

Figure 2. Gas Phase Atmospheric Concentrations of Chlorpyrifos in southern Okanagan Valley.

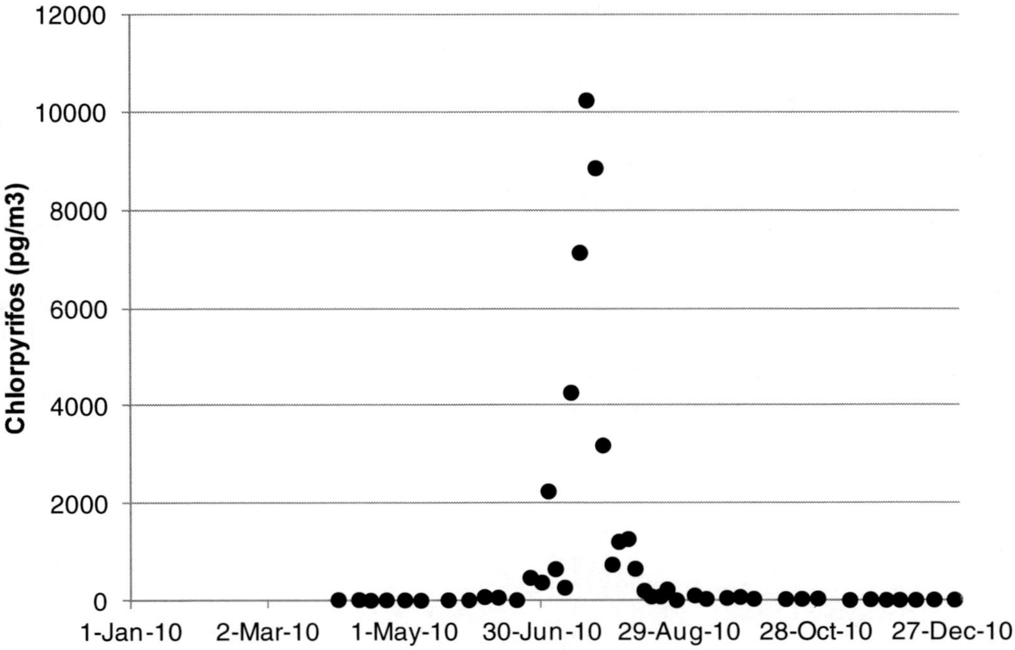

Figure 3. Gas Phase Atmospheric Concentrations of Chlorpyrifos at Bratt's Lake, SK during 2010.

In Canada the highest concentrations of chlorpyrifos in the atmosphere during the agricultural growing season in 2005 from 7 sites impacted by agricultural activity were found at Bratt's Lake, SK and Vineland, ON with maximum concentrations of chlorpyrifos of 170000 and 167000 pg/m^3, respectively [16]. These regions differ in agricultural crops from grains and oil seeds near Bratt's Lake, SK to fruit and vegetables near the Vineland site. Average concentrations at these sites were also high (2004: 31000pg/m^3 and 2005: 1590 pg/m^3 at Bratt's Lake, SK and 2004: 21900pg/m^3 and 2005:20600pg/m^3 at Vineland, ON) [16]. The highest reported concentration of chlorpyrifos at a agricultural site in North America was 250000pg/m^3 at Bratt's Lake in August, 2003 [17]. In 2003 there was a high density of grasshopper infestations in Saskatchewan [43]. Chlorpyrifos is commonly used in the prairie provinces for grasshopper eradication [17]. Although annual concentrations can vary significantly at Bratt's Lake, SK (see Table 1) the seasonal maximum in atmospheric concentration of chlorpyrifos occurs in late July-August. Figure 3 shows the atmospheric gas phase concentrations of chlorpyrifos at Bratt's Lake during 2010. Similar to our previous studies [17], this study shows high atmospheric concentrations of chlorpyrifos in late July-August. No usage data for chlorpyrifos is available currently for 2010 for Saskatchewan but Canadian sales of chlorpyrifos have declined throughout 2009-2011 [6]. Grasshopper infestations in 2010 were lower than 2003, but had increased relative to the prior years [43, 44]. In a small watershed in Manitoba (South Tabacco Creek watershed) high atmospheric concentrations of chlorpyrifos (maximum 103000 pg/m^3) were measured in 1996 which also coincided with local applications of chlorpyrifos between late July and mid August [18].

In Saskatchewan and the prairies, chlorpyrifos can be used for a wider range of control of other insects including armyworms, cutworms, and wheat midge on a wide variety of crops including canola, wheat, barley, oats, flax, lentils, sunflowers, and potatoes [44]. In 2003 when the highest atmospheric concentrations of chlorpyrifos occurred at Bratt's Lake, the total amount of chlorpyrifos usage in Saskatchewan was 205562 kg which was very high amount and occurred in a year with high grasshopper infestations [9, 17]. In comparison usage of chlorpyrifos was lower in prior years to 2003 (2001 and 2002 were 7267 kg, 121824 respectively) [9]. In 2003 approximately 45 % of the total usage of chlorpyrifos in Saskatchewan was applied in the dark brown soil zone where the sampling site at Bratt's Lake is located, while the remaining amount of usage (55 %) was applied in the dark grey soil zone (north of the dark brown soil zone) [9]. 2002 and 2003 were years of higher than normal grasshopper infestations in Saskatchewan [43]. The degradation products chlorpyrifos oxon and 3,5,6-trichloro-2-pyridinol (TCP) were detected in atmospheric samples at Bratt's Lake, SK with maximum concentrations of 31500 pg/m^3 and 10600 pg/m^3, respectively [17]. Chlorpyrifos oxon was more frequently detected in atmospheric samples than TCP and showed a similar seasonal maximum in late July-early August [17].

Atmospheric Concentrations in National Parks or Remote Sites-Impacted or Background Locations

Measurements of chlorpyrfios at more remote sites include Mt. Bachelor Observatory in Oregan's Cascade Range where atmospheric concentrations were < MDL to 6 pg/m^3 [20]. Generally concentrations of chlorpyrifos were low with sporadic increases in concentration and were associated with when air masses spent longer periods of time over highly populated areas of western United States where in urban areas one of the frequent uses of chlorpyrifos was for roach and termite control [20]. Measurements at an alpine site in New Zealand's southern alps also showed low levels of chlorpyrifos in the atmosphere with maximum of 12 pg/m^3 [23]. At remote lakes in Ontario where there was no local agricultural application of chlorpyrifos atmospheric concentrations reached a maximum of 60 pg/m^3 [22]. Although limited environmental measurements exist in Asia, air measurements from Eastern Arctic to the High Arctic, showed the highest concentrations of chlorpyrifos in the atmosphere in the Sea of Japan where concentrations of chlorpyrifos reached 146 pg/m^3 [24]. Higher concentrations were general reported near eastern Asia, whereas concentrations of chlorpyrifos in the Bering and Chukchi seas were < 1pg/m^3 and these low concentrations were attributed to phase-out of chlorpyrifos usage for outdoor urban applications such as termite control in United States [24].

Chlorpyrifos: Surface Water, Snowpack or Precipitation

In late 2001 a mandated phase out of chlorpyrifos and diazinon in the United States for outdoor urban settings resulted in a rapid decline in concentrations of these OPs in urban streams and rivers in the northeastern and Midwestern United States [45]. The decline of surface water concentrations of chlorpyrifos began as early as 2000 at some sites with detections frequency ranging from 23 to 45% pre 2000 to 2 to 8% after October 2001 [45]. Similar declines in surface water concentrations of chlorpyrifos were also observed in Texas [46]. Some agricultural uses of chlorpyrifos such as on tomatoes and post-bloom apples also stopped in December 2000 [47]. In the United States, 10 of 11 sites used for monitoring stream surface water concentrations showed significant decreases in concentrations of chlorpyrifos between 1996-2004, however only 2 of 5 sites had significant downward trend from 2000-2008 and was attributed to fewer sites having adequate detection frequency with higher concentrations of chlorpyrifos observed in the south [47]. Great than 40% of water samples collected in Iowa City and Cedar Rapids, IA, and Jackson, MS had detection of chlorpyrifos with lower detection frequency in Minneapolis and Princeton, MN and Rolling Forks, MS [28]. Maximum concentrations in water were 9ng/L and 40ng/L for urban and rural sites, respectively [11].

Chlorpyrifos was also one of the OPs that were widely used in agriculture, urban, and residential application in the Chesapeake Bay region [48]. It has been detected in air, rain, and in surface waters from the Chesapeake Bay and its tributaries [12, 37, 38, 49-51]. This region has been a focus of surface water measurements in the United States due to high agricultural pesticide application rates. Chlorpyrifos has been used on corn, orchard fruits, and strawberries in Maryland with declining agricultural usage from 1997 to 2000 in this region [48]. In 1995 prior to phase-out of chlorpyrifos, the maximum concentration of chlorpyrifos was 1.67ng/L and occurred in March [37]. Pastuxent and Choptank River are located in two tributaries of the Chesapeake Bay and discharge into opposite shores of the Bay. In Pastuxent river concentrations of chlorpyrifos in water were 1.5 ± 0.8ng/L with a maximum concentration of 3.1ng/L in April 1997 [49]. At Choptank river site concentrations of chlorpyrifos were 0.5 ± 0.3ng/L with maximum in late May 1997 of 1.2ng/L, but no measurements were taken in April [49]. Susquehanna River watershed drains into the Chesapeake Bay and is in a region of heavy agricultural usage of pesticides. Maximum concentration of chlorpyrifos at the river site was 2ng/L (0.5ng/L average) for water samples collected from February 1997 to March 1998 [50]. The average surface water concentrations of chlorpyrifos for 2000, 2001, 2002, and 2003 at the Choptank River site were1.0, 1.2, 4.5, and 2.3ng/L with maximum concentration of 180ng/L observed in 2002 [51]. Chlorpyrifos in surface waters is also prone to degradation with the major degradation products 3,5,6-trichloro-2-pyridinol and chlorpyrifos oxon [49, 51].

Figure 1 shows that there was a decline of chlorpyrifos usage in California after 1997. In the 1970s and through the 1980s OPs such as chlorpyrifos were recommended for use for dormant spray of almond and stone fruit orchards (almonds, peaches, nectarines, plum, prunes) in the San Joaquin valley and Sacramento valley region [52]. Due to concerns in the 1990s from surface water concentrations of chlorpyrifos and runoff after irrigation or storm water, these applications of OPs during the dormant season were replaced in the late 1990s with applications of Bacillus thiringienis (BT), spinosad, pyrethroids and carbamates [52]. Dormant sprayed orchards were a major source of inputs during storm water events in the San Joaquin River [53, 54]. Use of chlorpyrifos on almonds has declined since 1992 but OPs were still in higher use than pyrethroids, BT, and carbamates on almonds in California and OPs (chlorpyrifos and diazinon) have been shown to contribute to load of pesticides in the San Joanquin River watershed [54]. Reductions in agricultural use of chlorpyrifos were recommended to meet CDFA (15 ng/L) or EPA (41 ng/L) water quality standard for chlorpyrifos [55].

Chlorpyrifos has also been the most widely used OP insecticide in the San Joaquin Valley with greatest usage from May through August [56]. It has been detected in 100% of snow and rain samples collected in Yosemite National Park ~50 km east of the San Joaquin Valley at concentrations ranging from 0.14 to 4.9ng/L in 2008 to 2009 [56]. Concentrations of chlorpyrifos in snow were higher than rain samples and were also higher than reported at the Rocky mountains or Sequoia National Park [57-59]. Chlorpyrifos has been detected in snowpack from Sequoia, Rocky Mountains, Rainier, Glacier National Park (NP) sites, and two Alaskan Parks (Denali NP, and Noatak National Preserve and Gates of the Arctic NP) [60]. Sequioa, Glacier, and N. Cascades had the highest maximum concentrations of chlorpyrifos in snowpack during 2003-2005 at 2.9, 1.9, and 1.0 ng/L, respectively as compared to background impacted sites Denali and Noatak/Gates at 0.10 and 0.059 ng/L, respectively [60]. From snowpack collected in spring 2003 concentrations of chlorpyrifos were highest at Sequoia (2.3 ng/L) as compared to 0.069, 0.033, 0.052ng/L at Rocky, Glacier, and Rainier NPs [61]. Regional atmospheric transport of chlorpyrifos usage within western region of the United States was a major contribution to snowpack concentrations of chlorpyrifos at Rainier, Glacier, N. Cascades, Olympic, and Sequoia NPs [60]. Glacier NP may also be influenced by air masses with origin in British Columbia and to smaller extent prairies [60]. Chlorpyrifos showed a positive correlation coefficient to pesticide usage (from available 1997 pesticide inventory) within 150 km of the parks [61]. Agricultural activity in California's Central Valley has been identified as an important source of OP insecticides to the atmosphere with the highest atmospheric concentrations of OPs observed during the summertime agricultural use [62].

Chlorpyrifos and chlorpyrifos oxon have also been detected in fog water collected in the coastal region of California with concentrations of chlorpyrifos and chlorpyrifos oxon reaching 190ng/L and 230ng/L near Monterey, CA [63]. Thion to oxon conversion occurs during atmospheric transport from agricultural areas to more urban sites with higher ratio to chlorpyrifos/chlorpyrifos oxon measured at nonagricultural sites [63]. Studies of storm and irrigation runoff from residential and agricultural land in a coastal southern California watershed have shown that loads of OPs from residential land may exceed those from agricultural areas for a given rainfall [64]. Comparison of coastal and Central Valley of California where the importance of residential and agricultural sources differ showed highest concentrations of chlorpyrifos in fog water occurred in the Central Valley (up to 7700ng/L) suggesting the agricultural sources remain a main source of chlorpyrifos in California [63].

Surface water concentrations of chlorpyrifos showed high variability (more than 10-fold) with irrigation and storm runoff effects at agricultural sites in southern California, while residential areas showed less variability [64]. Irrigation run-off from an agricultural site was positively correlated with the amount of chlorpyrifos applied during the preceding 45 days, however the loading at the sites in Calleguas Creek watershed in coastal southern California were lower than in the San Joaquin Valley [64]. In runoff concentrations of chlorpyrifos were < 30ng/L for an irrigation event in Oct, 1999 and April 2000 [64]. Northern California is also impacted by agricultural activity as well as use of chlorpyrifos for structural pest control prior to 2001. Surface water concentrations of chlorpyrifos in urban streams in Stockton and Sacramento between November 1994 and May 1995 reached 190ng/L with median of 50ng/L and ~80 % of sampling periods exceeding California water-quality criteria in 1995 of 20ng/L [36].

In the Canadian prairies in a watershed impacted by usage of chlorpyrifos for agriculture maximum concentrations of chlorpyrifos in precipitation were 38, 16, and 32ng/L in 1994, 1995, and 1996 [18]. In the Red River and its tributaries in Manitoba maximum concentrations of chlorpyrifos in surface waters were 11.8ng/L (average 0.57 ± 2.11ng/L; median < 0.082ng/L) for the period June-October 1993 and March-October 1994-1995 [65].

At remote lakes in Ontario not impacted by local agricultural sources, concentrations of chlorpyrifos in precipitation reached 43ng/L (median 0.76ng/L) with much lower surface water concentrations of chlorpyrifos at maximum of 0.5ng/L (median 0.02 ng/L) [22].

Cruise measurements taken from eastern China to high Arctic showed that the most frequently detected current use pesticide in seawater was chlorpyrifos with the highest concentration at 111pg/L [24]. Similar to atmospheric measurements the highest concentrations of chlorpyrifos in seawater were found near eastern Asia [24].

REFERENCES

[1] Li, R., He, L., Hao, L., Ji, X., Zhou, Y., Wang, Q. Chlorpyrifos residue levels on field crops (rice, maize, and soybean) in China and their dietary risks to consumers. *Food Control,* 2015, 51, 212.

[2] Chen, C., Qian, Y., Liu, X., Tao, C., Liang, Y., Li, Y. Risk assessment of chlorpyrifos on rice and cabbage in China. *Regulatory Toxicology and Pharmacology,* 2012, 62, 125.

[3] An, X., Ji, X., Wu, M., Hu, X., Hu, X., Yu, R., Zhao, X., Cai, L. Risk assessment of applicators to chlorpyrifos through dermal contact and inhalation at different maize plant heights in China. *J. Agric. Food Chem.* 2014, 62, 7072.

[4] Sarigiannis, D. A., Kontoroupis, P., Solomou, E. S., Nikolaki, S., Karabelas, A. J. Inventory of pesticide emissions into the air in Europe. *Atmos. Environ.* 2013, 75, 6.

[5] Grube, A., Donaldson, D., Kiely, T., Wu, L. *Pesticide sales and usage, 2006 and 2007 market estimates.* Biological and Economic Analysis Division, Office of Pesticide Programs, Office of Chemical Safety and Pollution Prevention, United States Environmental Protection Agency, Washington, DC, 2011.

[6] *Health Canada, Pest Control Product Sales Report for 2007 and 2008; 2009; 2010; and 2011.* Available at http://www.hc-sc.gc.ca/cps-spc/pubs/pest/_corp-plan/sales-2011-ventes/index-eng.php.

[7] United States Geological Survey, USGS, *National Water-quality Assessment (NAWQA) Program, Pesticide use maps 1992 to 2011 for chlorpyrifos.* Available at http://water.usgs.gov/nawqa/pnsp/usage/maps/.

[8] Brimble, S., Bacchus, P., Caux; P.-Y. *Pesticide Utilization in Canada: A Compilation of Current Sales and Use Data;* Environment Canada, 2005.

[9] Saskatchewan Government, Saskatchewan Ministry of Agriculture, *Saskatchewan usage inventory 2003;* 2005 (unpublished data).

[10] Foreman, W. T., Majewski, M. S., Goolsby, D. A., Wiebe, F. W., Coupe, R. H. Pesticides in the atmosphere of the Mississippi River Valley, part II-air. *Sc. Total Environ.* 2000, 248, 213.

[11] Coupe, R. H., Manning, M. A., Foreman, W. T., Goolsby, D. A., and Majewski, M. S. Occurrence of pesticides in rain and air in urban and agricultural areas of Mississippi, April-September 1995: *Sci. Total Environ.* 2000, 248, 227.

[12] Harman-Fetcho, H. A., McConnell, L. L., Rice, C. F., Baker, J. E. Wet deposition and air-water gas exchange of currently used pesticides to a subestuary of the Chesapeake Bay. *Environ. Sci. Technol.* 2000, 34, 1462.

[13] California Department of Pesticide Regulation, December 2009. *Pesticide air monitoring in Parlier, CA (Final report),* California Department of Pesticide Regulation, Sacramento, CA.

[14] CARB, April 7 1998. *Report for the application and ambient air monitoring of chlorpryfios (and oxon analogue) in Tulare County during spring/summer 1996.* California Department of Pesticide Regulation, Sacramento, CA.

[15] Yao, Y., Tuduri, L., Harner, T., Blanchard, P., Waite, D., Poissant, L., Murphy, C., Belzer, W., Aulagnier, F., Li, Y.-F., Sverko, E. Spatial and temporal distribution of pesticide air concentrations in Canadian agricultural regions. *Atmos. Environ.* 2006, 40, 4339.

[16] Yao, Y., Harner, T., Blanchard, P., Tuduri, L., Waite, D., Poissant, L., Murphy, C., Belzer, W., Aulagnier, F., Sverko, E. Pesticides in the atmosphere across Canadian agricultural regions. *Environ. Sci. Technol.* 2008, 42, 5931.

[17] Raina, R., Hall, P., Sun, L. Occurrence and relationship of organophosphorus insecticides and their degradation products in the atmosphere in western Canada agricultural regions. *Environ. Sci. Technol.* 2010, 44, 8541.

[18] Rawn, D. F. K., Muir, D. C. G. Sources of chlopyrifos and dacthal to a small Canadian Prairie watershed. *Environ. Sci. Technol.* 1999, 33, 3317.

[19] Coscolla, C., Colin, P., Yahyaoui, A., Petrique, O., Yusa, V., Mellouki, A., Pastor, A. Occurrence of currently used pesticides in ambient air of Centre Region (France), *Atmos. Environ.* 2010, 44, 3915.

[20] Primbs, T., Wiulson, G., Schmedding, D., Higginbotham, C., Simoch, S. M. Influence of asian and western United States agricultural areas and fires on the atmospheric transport of pesticides in the western United States. *Environ. Sci. Technol.* 2008, 42, 6519.

[21] Daly, G., Lei, Y., Teixeira, C., Muir, D. C. G., Wania, F. Pesticides in Western Canadian Mountain air and soil. *Environ. Sci. Technol.* 2007, 41, 6020.

[22] Kurt-Karakus, P. B., Teixeira, C., Small, J., Muir, D., Bidleman, T. F. Current-use pesticides in inland lake waters, precipitation, and air from Ontario, Canada. *Environ. Toxicol. Chem.* 2011, 30, 1539.

[23] Lavin, K. S., Hageman, K. J., Marx, S. K., Dillingham, P. W., Kamber, B. S. Using trace elements in particulate matter to identify sources of semivolatile organic contaminants in air at an alpine site. *Environ. Sci. Technol.* 2012, 46, 268.

[24] Zhong, G., Xie, Z., Cai, M., Moller, A., Sturm, R., Tang, J., Zhang, G., He, J., Ebinghaus, R. Distribution and air-sea exchange of current-use pesticides (CUPs) from East Asia to the High Arctic Ocean. *Environ. Sci. Technol.* 2012, 46, 259.

[25] Raina, R., Hall, P. Field Evaluation of solid sorbents for ambient air sampling of pesticides. *Air, Soil, Water Res.* 2010, 3, 57.

[26] Armstrong, J. L., Fenske, R. A., Yost, M. G., Tchong-French, M., Yu, J. Comparison of polyurethane foam and XAD-2 sampling matrices to measure airborne organophosphorus pesticides and their oxygen analogs in an agricultural community. *Chemosphere,* 2013, 92, 451.

[27] Rice, C. P., Chernyak, S. M., McConnell, L. L. Henry's law constants for pesticides measured as a function of temperature and salinity. *J. Agric. Food Chem.* 1997, 45, 2291-2298.

[28] Majewski, M. S., Foreman, W. T., Goolsby, D. A. Pesticides in the atmosphere of the Mississippi River Valley, part I- rain. *Sc. Total Environ.* 2000, 248, 201.
[29] Majewski, M. S., Foreman, W. T., Goolsby, D. A., Nakagaki, N. Airborne pesticide residues along the Mississippi river. *Environ. Sci. Technol.* 1998, 32, 3689.
[30] Whitemore, R. W., Kelly, J. E., Reading, P. L. *National home and garden pesticide use survey.* Research Triangle Institute RTI/5100/17-015F; 1992.
[31] Lu, C., Fenske, R. A. Air and surface chlorpyrifos residues following residential broadcast and aerosol pesticide applications. *Environ. Sci. technol.* 1998, 32, 1386.
[32] Putman, R. A., Doherty, J. J., Clark, J. M. Golfer exposure to chlorpyrifos and carbary following application to turfgrass. *J. Agric. Food Chem.* 2008, 56, 6616.
[33] California Department of Pesticide Regulation, *Reports of pesticides sold in California, 1998- 2012.* Sacramento, CA. available at http://www.cdpr.ca.gov/docs/mill/nopdsold.htm.
[34] California Department of Pesticide Regulation, DPR 14-002, *Designating chlorpyrifos as a restricted material,* Notice 9/26/14, Comment period closed 11/12/14.
[35] California Department of Pesticide Regulation, March 2003. *Report of ambient air monitoring for pesticides in Lompoc,* California Department of Pesticide Regulation, Sacramento, CA.
[36] Bailey, H. C., Deanovic, L., Reyes, E., Kimball, T., Larson, K., Cortright, K., Connor, V., Hinton, D. E. Diazinon and chlorpyrifos in urban waterways in Northern California, USA. *Environ. Toxicol. Chem.* 19, 2000, 82.
[37] McConnell, L. L., Nelson, E., Rice, C. P., Baker, J. E., Johnson, W. E., Harman, J. A., Biaklek, K. Chlorpyrifos in the air and surface water of Chesapeake Bay: predictions of atmospheric deposition fluxes. *Environ. Sci. Technol.* 31, 1997, 1390-1398.
[38] Maryland Department of Agriculture, United States Department of Agriculture, Maryland Agricultural Statistics Service, *Maryland Pesticide Statistics for 1994:* Maryland Department of Agriculture: Annapolis, MD, 1996; MDA-265-95.
[39] Enkon Environmental Limited, *Survey of pesticide use in British Columbia: 2003, prepared for Environment Canada,* Project No. 1004-011, October 2005, Vancouver, BC.
[40] Wins-Purdy, A. *2010 Pesticide sales in British Columbia, Integrated Pest Management Program,* Ministry of Environment British Columbia, 2013.
[41] *2010 Integrated fruit production guide for commercial tree fruit growers,* Interior of British Columbia, 2010, BC Fruit Growers' Association, Kelowna BC.
[42] *2010 Best practices guide for grapes for British Columbia growers,* 2010, BC Wine Grape Council, Peachland, BC.
[43] Olfert, O., Giffen, D., Hartley, S. Analysis of grasshopper (Orthoptera: Acrdidae) population surveys in Saskatchewan: 1972-2004. *Can. Entomol.* 2006, 138, 875.
[44] *Government of Saskatchewan, Ministry of Agriculture,* 2010 Grasshopper Forecast, Grasshopper Count-Saskatchewan Crop Insurance Corporation, 2009.
[45] Phillips, P. J., Ator, S. W., Nystrom, E. A. Temporal changes in surface-water insecticide concentrations after the phaseout of diazinon and chlorpyrifos. *Environ. Sci. Technol.* 2007, 41, 4246.
[46] Banks, K. E., Hunter, D. H., Wachal, D. J. Chlorpyrifos in surface waters before and after federally mandated ban. *Environ. Int.* 2005, 31, 351.

[47] Ryberg, K. R., Vecchia, A. V., Martin, J. D., Gilliom, R. J. *National Water-Quality Assessment Program, Trends in pesticide concentrations in urban streams in the United States,* 1992-2008. Scientific Investigations Report 2010-5139. U.S. Department of the Interior, U. S. Geological Survey. U. S. Geological Survey, Reston, Virginia: 2011.
[48] *Maryland Pesticide Statistics for 1997,* MDA-265-99; Maryland Department of Agriculture: Annapolis, MD, 1999.; and Maryland Pesticide Statistics for 2000, MDA-265-02, Maryland Department of Agriculture: Annapolis, MD, 2002D.
[49] Lehotay, S. J., Harman-Fetcho, J. A., McConnell, L. L. Agricultural pesticides in oysters and water from two Chesapeake Bay tributaries. *Marine Pollut. Bull.* 1999, 37, 32.
[50] Liu, B., McConnell, L. L., Torrents, A. Herbicide and insecticide loadings from the Susquehanna River to the North Chesapeake Bay. *J. Agric. Food Chem.* 2002, 50, 4385.
[51] Goel, A., McConnell, L. L., Torrents, A. Wet Deposition of current use pesticides at a rural location on the Delmarva Peninsula: Impact of rainfall patterns and agricultural activity. *J. Agric. Food Chem.* 2005, 53, 7915.
[52] Epstein, L., Bassein, S., Zalom, F. G. Almond and stone fruit growers reduce OP, increase pyrethroid use in dormant sprays. *California Agriculture* 2000, 54, 14.
[53] Domagalski, J. L., Dubrovsky, N. M., Kratzer, C. R. Pesticides in the San Joaquin River, California: inputs from dormant sprayed orchards. *J. Environ. Quality.* 1997, 26, 454.
[54] Zhang, X., Liu, X., Luo, Y., Zhang, M. Evaluation of water quality in an agriculutral watershed as affected by almond pest management practices. *Water Research,* 2008, 42, 3685.
[55] California Regional Water Quality Control Board Central Valley Region, 2006. *Amendments to the water quality control plan for the Sacramento River and San Joanquin River Basins for the control of diazinon and chlorpyrifos runoff into the Sacramento-San Joaquin Delta,* June 2006. Final Staff Report. Resolution No. R5-2006-0061, Sacramento, CA, USA.
[56] *2014 Crop protection guide for fruit trees in Washington,* Washington State University Extension EB0419 Online available at http://www.tfrec.wsu.edu/pages/cpg/Home; Wenatchee WA, copyright 2014.
[57] Sparling, D. W., Fellers, G. M., McConnel, L. L. Pesticides and amphibian declines in California, USA. *Environ. Toxicol. Chem.* 2001, 20, 1591.
[58] Mast, M. A., Foreman, W. T., Skaates, S. V. Current-use pesticides and organochlorine compounds in precipitation and lake sediment from two high-elevation National Parks in the Western United States. *Arch. Environ. Contam. Toxicol.* 2007, 52, 294.
[59] Hageman, K. J., Hafner; W. D., Campbel, D. H., Jaffe, D. A., Simonich, S. L. M. Variability in pesticide deposition and source contributions to snowpack in western US National Parks. *Environ. Sci. Technol.* 2010, 44, 4452.
[60] Fenske, R. A., Lu, C., Barr, D., Needham, L. Children's Exposure to chlorpyrifos and parathion in an agricultural community in central Washington State. *Environ. Health Perspectives,* 2002, 110, 549.
[61] Hageman, K. J., Simonich, S. L., Campbell, D. H., Wilson, G. R., Landers, D. H. Atmospheric deposition of current-use and historic-use pesticides in snow at National Parks in the Western United States. *Environ. Sci. Technol.* 2006, 40, 3174.

[62] Le Noir, J. S., McConnel, L. L., Fellers, G. M., Cahill, T. M., Sieber, J. N. Summertime transport of current-use pesticides from California's Central Valley to the Sierra Nevada Mountain Range, USA. *Environ. Toxicol. Chem.* 1999, 18, 2715.

[63] Schomburg, C. J., Glotfelty, D. E., Seiber, J. N. Pesticide occurrence and distribution in fog collected near Monterey, California. *Environ. Sci. Technol.* 1991, 25, 155-160.

[64] Pedersen, J. A., Yeager, M. A., Suffet, I. H. Organophosphorus insecticides in agricultural and residential runoff: field observations and implications for total maximum daily load development. *Environ. Sci. Technol.* 2006, 40, 2120.

[65] Rawn, D. F. K., Halldorson, T. H. J., Woychuk, R. N., Muir, D. C. G. Pesticides in the Red River and its tributaries in southern Manitoba: 1993-95. *Water Qual. Res. J. Canada,* 1999, 34, 183.

In: Chlorpyrifos
Editor: Caitlin Mayes

ISBN: 978-1-63482-111-7

Chapter 10

ORGANIC PHASE THICK FILM BIOSENSOR FOR CHLORPYRIFOS PESTICIDE DETECTION

V. Somerset[1*], P. Baker[2] and E. Iwuoha[2]

[1]NRE, CSIR, Stellenbosch, South Africa, South Africa.
[2]SensorLab, Department of Chemistry, University of the Western Cape, Bellville, South Africa

ABSTRACT

Biosensors represent alternative electro-analytical techniques that can be used to complement or replace classical analytical methods for pesticide detection. Biosensors have also emerged as simple, rapid and sensitive tools that can be applied for environmental monitoring, food safety and quality control. This electrochemical technique further allows for simpler sample preparation and efficient on site analysis, thereby decreasing the cost per sample analysis. This chapter reports the work done in the construction of a thick-film acetylcholinesterase biosensor for the determination of chlorpyrifos pesticide concentrations in selected aqueous organic solvent solutions. The constructed biosensor was designed by encapsulating acetylcholinesterase (AChE) as enzyme in a polyaniline (PANI) polymer composite, followed by the drop coating of poly(vinyl acetate) (PVAc) on top to secure the biosensor film from disintegration in organic solvents. Biosensor construction also involved the optimisation of various performance and stability parameters to determine its influence on the amperometric behaviour or the biosensor. These factors included optimal enzyme loading, effect of pH, long-term stability of the biosensor, and the effect of both polar and non-polar organic solvents on enzyme behaviour. The constructed AuE/MBT/PANI/AChE/PVAc thick-film biosensor offered a reasonable linear range between 0.24 – 0.70 μg/L for chlorpyrifos. The biosensor also exhibited a detection limit of 0.012 μg/for chlorpyrifos. It was further determined that the detection limit was below the allowable concentration of 1 μg/L set by the WHO guideline value for drinking water.

Keywords: Organic phase biosensor; polyaniline; mercaptobenzothiazole; acetylcholinesterase; chlorpyrifos

---

* Email: vsomerset@csir.co.za; vsomerset@gmail.com

1. INTRODUCTION

Traditionally, pesticides have been developed to protect agricultural crops and to control public health pests. However, decades of extensive use of pesticides especially in agriculture have seen these compounds accumulate in the aquatic environment, with sediment serving as a sink. Therefore, the threat of pesticide residues in the environment has become a growing threat as it can enter the food chain and pose a serious human health risk (Noyrod et al., 2014; Viswanathan et al.,2009).

One group of pesticides of particular interest is the organophosphates (OPs). This group of pesticides have been widely used for both agricultural and domestic purposes as insecticides. The danger of these pesticides lies in the fact that they can inhibit the functioning of acetylcholinesterase (AChE), an essential enzyme in most animals involved in nerve transmission. For this reason, OPs have also been used as nerve agents in chemical warfare (Istamboulie et al., 2009; Somerset et al., 2009).

The focus of the work reported in this study is on chlorpyrifos, which is known to be extensively used in agriculture, household and urban applications. The urban use of this pesticide has also been associated with its presence in urban aquatic environments. Chlorpyrifos is moderately toxic and studies have shown that exposure to this pesticide leads to neurological defects and auto-immune disorders (Banks et al., 2005; Eaton et al., 2008; Uygun and Dilgin, 2013).

In European Union countries the use of set targets to limit the use and discharge of these pesticides has been set, while other countries are slower to adopt these mitigation strategies. In order to protect our water resources and to limit the threat to human health, due to the presence and adverse effects of the OPs, early detection and monitoring has become important. Traditional methods that have been widely used for the analysis and quantification of OPs include gas chromatography (GC), hyphenated GC techniques combined with mass spectroscopy (GC-MS), capillary electrophoresis, high performance liquid chromatography (HPLC), and HPLC coupled with mass spectroscopy (HPLC-MS) or UV detection (HPLC-UV) (Somerset et al., 2009; Noyrod et al., 2014). These established techniques are reliable and efficient, but also have some disadvantages. Some of the limitations include the requirement of a long assay time, skilled operating technicians, the use of toxic organic solvents, and non-portability of instruments for on-site detection (Somerset et al., 2009; Istamboulie et al., 2009; Noyrod et al., 2014). Therefore, biosensors have been identified as the alternative analytical devices to compliment the traditional techniques. Biosensors offer to provide real-time qualitative information, good selectivity, fast response, lower operating costs, and on-site analysis with minimum preparation required (Istamboulie et al., 2009; Somerset et al., 2011; Noyrod et al., 2014).

The focus of the work reported in this study is to show that a thick film AChE biosensor can be successfully applied in the determination of chlorpyrifos as pecticide compound in organic solvent mixtures. The thick film AChE biosensor was constructed that consisted of a gold disc electrode surface, followed by a mercaptobenzothiazole (MBT) selfassembled monolayer (SAM), onto which a polyaniline (PANI) polymer film was electrosynthesised. Consequently, AChE as enzyme was immobilised onto the PANI film followed by drop-coating of a poly(vinylacetate) (PVAc) layer. The biosensor was abbreviated as AuE/MBT/PANI/AChE/PVAc for this publication.

2. MATERIALS AND METHODS

2.1. Apparatus

A BAS-50/W electrochemical workstation with BAS-50/W software (Bioanalytical Systems, Lafayette, IN, USA) was used to perform all of the electrochemical protocols. The different modes of voltammetry utilised included cyclic voltammetry (CV), Oysteryoung square wave voltammetry (OSWV or SWV), differential pulse voltammetry (DPV) or time-based amperometric modes. The conventional three electrode set-up was used consisting of a gold disc electrode (diameter: 1.6 mm; area: 2.01 × 10−2 cm2), a silver/silver chloride (Ag/AgCl – 3MNaCl type) reference electrode and a platinum wire as auxiliary electrode (Somerset et al., 2009; Silwana et al., 2014.)

2.2. Reagents and Materials

Disodium hydrogen phosphate (98%), aniline (99%), potassium dihydrogen phosphate (99%), diethyl ether (99.9%), acetylthiocholine chloride (99%), mercaptobenzothiazole (MBT), acetylcholinesterase (AChE, from Electrophorus electricus, EC 3.1.1.7;~850 U/mg), acetylcholine chloride (99%), acetone (99.8%, pestanal), hydrogen peroxide (30%), ethanol (99.9%, absolute grade), acetonitrile (99.9%, pestanal grade), and chlorpyrifos (pesticide standard) were purchased from Sigma-Aldrich, South Africa. All solutions were prepared using double-distilled water.

2.3. Electrode Surface Preparation

Prior to use, the gold electrodes were first etched for approximately 5 to 10 minutes in a hot 'Piranha' solution {1:3 (v/v) 30% H_2O_2 and concentrated H_2SO_4} and rinsed thoroughly with double-distilled water. This was followed by polishing of the electrode surface on alumina micropolish and polishing pads (Buehler, IL, USA), using aqueous slurries of 1 μm, 0.3 μm and 0.05 μm alumina powder. Afterwards, the gold electrode surface was cleaned electrochemically by cycling the potential scan between -200 and + 1500 mV (vs. Ag/AgCl) in 0.05 M H_2SO_4 at the scan rate of 40 mV/s for approximately 10 min or until the CV characteristics for a clean gold electrode were obtained. The platinum auxiliary electrode was cleaned in the flame of a Bunsen burner and rinsing in double-distilled water (Somerset et al., 2009).

2.5. Thick Film AChE Biosensor Construction

The thick film biosensor construction involved several steps and layers, in order to ensure that the biosensor was operational in organic solvents. The thick film strategy adopted in this study was to first coat the gold (Au) electrode surface with a self-assembled monolayer of mercaptobenzothiazole (MBT). This was followed by coating the Au/MBT electrode surface with a film of polyaniline (PANI), which also served as the immobilisation matrix for the AChE enzyme.

The last layer consisted of a film of poly(vinylacetate) (PVAc) that acted as a binder, to hold the immobilised AChE attached to the polymer film when in contact with organic solvents (Somerset, 2007). The mercaptobenzothiazole (MBT) film was coated by immersing the cleaned Au electrode in an ethanol solution containing 10 mM of MBT for a period of 2 hours. The Au/MBT electrode was then rinsed with doubled distilled water and placed into a freshly prepared 10 mL solution of 0.1 M aniline in aqueous 1 M HCl for the polyaniline film coating. The Au/MBT/PANI electrode was then prepared for the AChE enzyme immobilisation (Somerset et al., 2006). For the bio-electrode construction, the AuE/MBT/PANI electrode was transferred to a batch cell, containing a 1 mL solution of argon degassed 0.1 M phosphate buffer (pH 7.2). This was done to reduce the PANI polymer film at a potential of – 500 mV (vs. Ag/AgCl) until a steady current was achieved (approximately 30 minutes). The next step involved transferring the AuE/MBT/PANI electrode to a second batch cell containing 60 μL of AChE in a 0.1 M phosphate buffer (pH 7.2) solution. This solution was also argon degassed, before enzyme immobilization was performed by oxidation of the PANI film in the presence of AChE at a potential of +400 mV (vs. Ag/AgCl) until a steady current was achieved (approximately 40 minutes). This step ensured that the AChE enzyme was electrostatically attached to the PANI polymer film (Somerset et al., 2006). The constructed Au/MBT/PANI/AChE biosensor was thoroughly rinsed with double-distilled water, in order to remove any unbound enzyme followed by storage in the working 0.1 M phosphate buffer (pH 7.2) solution at 4^0C. The final step in the biosensor construction involved the drop coating of the AuE/MBT/PANI/AChE bioelectrode (arranged vertically) with a 2 μL drop of poly(vinyl acetate) (PVAc) solution (0.3 M) prepared in acetone. This was followed by air drying for approximately 2 minutes. The thick film biosensor was abbreviated as AuE/MBT/PANI/AChE/PVAc (Somerset et al., 2006; 2009).

2.6. Electrochemical Detection of Chlorpyrifos

The detection of chlorpyrifos was performed using the percentage inhibition method. This involved placing the AuE/MBT/PANI/AChE/PVAc biosensor in a stirred 1 mL of 0.1 M phosphate (0.1 M KCl, pH 7.2) solution (anaerobic conditions), followed by multiple additions of a standard acetylthiocholine (ATCh) substrate solution until a stable current response was obtained.

The biosensor was thereafter rinsed with double-distilled water and incubated in a chlorpyrifos solution of specific concentration and the AChE senzyme was exposed for 20 minutes to the pesticide concentration in the aqueous-organic solvent mixture. After pesticide exposure, the biosensor was thoroughly rinsed with double-distilled water and placed into a freshly stirred 1 mL of 0.1 M phosphate (0.1 M KCl, pH 7.2) solution (anaerobic conditions), followed by multiple additions of a standard acetylthiocholine (ATCh) substrate solution until a stable current response was obtained. This exposure of the AChE enzyme was repeated for several different concentrations of chlorpyrifos, to obtain a standard calibration curve. The maximum current responses before and after pesticide exposure were further used to calculate the percentage inhibition ($I\%$) for each concentration of chlorpyrifos that the enzyme was exposed to. The formula used is:

$$I\% = \frac{(I_1 - I_2)}{I_1} \times 100 \quad \text{(Eqn. 1)}$$

where, I% is the degree of inhibition, I_1 is the steady-state current obtained in buffer solution, I_2 is the steady-state current obtained in buffer solution after the biosensor was exposed to chlorpyrifos for a 20 minute period in a phosphate buffer-organic solvent mixture (Somerset et al., 2006; 2009).

The electrochemical responses were examined with the anodic difference square wave voltammogram (SWV) responses collected in an oxidation direction by applying a linear potential scan between – 400 mV and + 1800 mV (vs. Ag/AgCl).

For the SWV responses a step potential of 4mV, a frequency of 5 Hz, and a square amplitude of 50 mV was used. Similarly, the anodic difference differential pulse voltammogram (DPV) responses were collected in an oxidation direction also, by applying a linear potential scan between – 400mVand+1800 mV (vs. Ag/AgCl).

The DPV responses were performed at a scan rate of 10 mV/s1 and a pulse amplitude of 50 mV. The sample width, pulse width and pulse period were 17 ms, 50 ms and 200 ms, respectively (Somerset et al., 2009).

3. Results and Discussion

3.1. Voltammetric Characterisation of AChE Biosensor

The functioning of the AuE/MBT/PANI/AChE/PVAc biosensor was evaluated by measuring the voltammetric responses of the constructed biosensor in a 0.1 M phosphate buffer (pH = 7.2) solution.

Anodic difference square wave voltammogram (SWV) measurements were conducted for successive addition of the substrate, acetylthiocholine (ATCh), to the AChE biosensor.

The SWV measurements were performed at a scan rate of 10 mV/s and in a shorter potential window between +200 to +850 mV (vs. Ag/AgCl), as shown in Figure 1 (a), for the increasing concentrations of ATCh as substrate between 0.2 – 0.6 mM (Somerset et al., 2009).

Analysis of the SWV responses have shown that it was very sensitive (compared to cyclic voltammetry) towards the performance of the AChE biosensor in the reduction of ATCh (Figure 1).

The results in Figure 1 further shows that well stable cathodic peaks for the reduction of ATCh were obtained, for the AChE biosensor under optimised conditions and between 0.2 – 1.6 mM of ATCh.

For concentrations higher than 1.6 mM, the graph started to level out as the current response remained relatively similar. The linear regression between current and concentration of ATCh for the linear section of the calibration plot is given by:

$$I\,(A) = -2 \times 10^{-6} C_{ATCh} - 8 \times 10^{-7}; R^2 = 0.974 \quad \text{(Eqn. 2)}$$

These results have thus indicated that the AChE biosensor is effective in ATCh reduction, especially up to an ATCh concentration of 1.6 mM, for which a relatively linear response was obtained. The AChE biosensor was thus successful in hydrolysing the substrate ATCh, with the increase in oxidative current as the preferred route of reaction, and the peak potentials shifting negatively in sequence (Somerset et al., 2009; 2011).

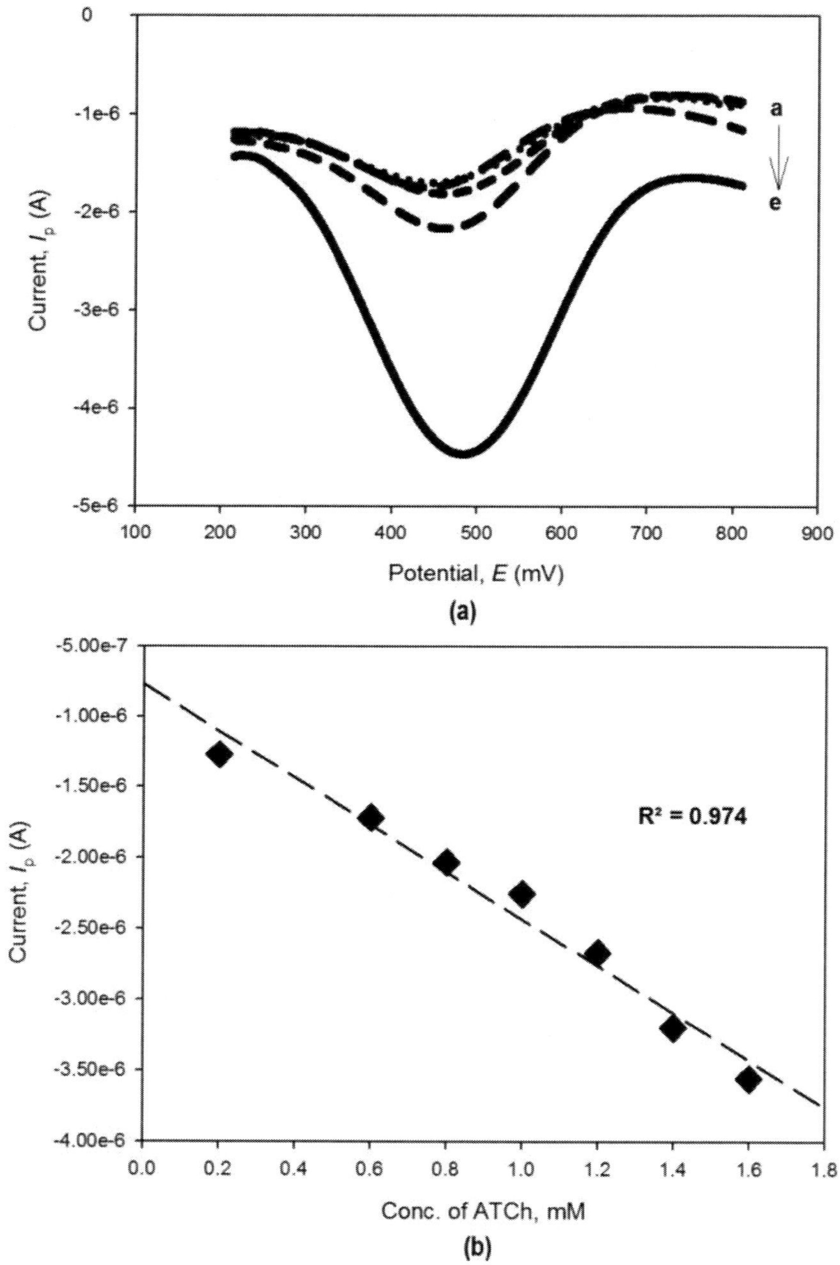

Figure 1. Square wave voltammetric responses for successive ATCh additions at the AuE/MBT/PANI/AChE/PVAc biosensor in (a); where a-e corresponds to 0.2 to 1.6 mM of ATCh in 0.1 M phosphate buffer (pH = 7.2; 0.1 M KCl) solution. In (b) the calibration plot for the ATCh measurement is shown.

3.2. Optimisation of Biosensor Performance Parameters

To ensure the high performance of the AChE biosensor, certain experimental parameters were optimised. These parameters included the pH conditions, temperature, AChE enzyme amount, and storage stability.

The relationship between the activity of the AChE enzyme and various pH conditions was evaluated in 0.1 M phosphate (pH = 7.2; 0.1 M KCl) solution, with the results obtained shown in Figure 2.

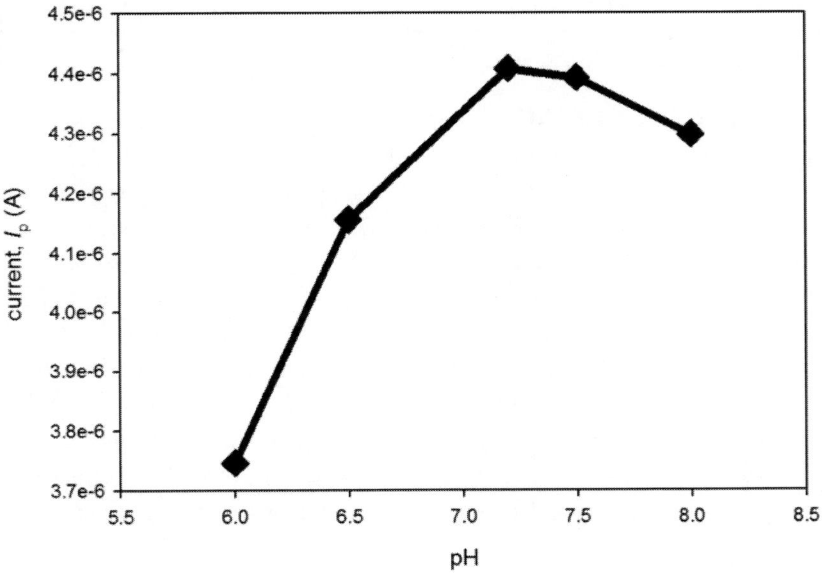

Figure 2. Results obtained for the optimisation of the pH of the AuE/MBT/PANI/AChE/PVAc thick-film biosensor in 0.1 M phosphate buffer, (pH = 7.2; 0.1 M KCl) solution, with 2 mM of acetylthiocholine chloride (ATCh) as substrate added. The absolute values of the peak current are shown for graphing purposes.

The results in Figure 2 indicate that the amperometric response of the AChE biosensor increased with the increase of pH and reached a maximum at 7.2. A decrease was observed at higher pH values, therefore a pH 7.2 was used in the detection solution.

The second parameter optimised was the influence of the AChE amount in the biosensor operation, with the results obtained shown in Table 1.

Analysis of the results shown in Table 1 has indicated that for the different amounts of AChE evaluated, e.g. 40 to 80 U/1 mL, different analytical responses were obtained. The graph of current response vs. successive ATCh substrate addition, not shown here, showed the best response for an AChE enzyme concentration of 60 U/1 mL used (Somerset et al., 2009). The amperometric responses for this amount of enzyme evaluated in 0.1 M phosphate buffer (pH = 7.2; 0.1 M KCl) solution, showed an increase in current response to an ATCh concentration of 1.6 mM, followed by a plateau in the current response. The results for the linear range and sensitivity was also the best for the concentrations evaluated, therefore a 60 U/1 mL of AChE was used in pesticide detection.

Table 1. Response characteristics of AuE/MBT/PANI/AChE/PVAc biosensor for different amounts of enzyme incorporated into the biosensor. Responses measured in a 0.1 M phosphate buffer, KCl (pH = 7.2) solution at 25°C

AChE (U/1 mL)	Liner range (mM)	Sensitivity (μA/mM)	Correlation (R^2)
40	0.2 – 0.5	1.0×10^{-6}	0.943
50	0.2 – 0.9	1.5×10^{-6}	0.948
60	0.2 – 1.0	3.8×10^{-6}	0.989
70	0.2 – 0.5	1.8×10^{-6}	0.952
80	0.2 – 0.9	2.4×10^{-6}	0.975

Stability testing of the AuE/MBT/PANI/AChE/PVAc biosensor was performed by storing a freshly constructed biosensor at 4◦C for a length of approximately30 days (Figure 3). Testing of the AChE biosensor was performed every 7 days, using a freshly prepared 0.1 M phosphate buffer (pH = 7.2; 0.1 M KCl) solution. To 1 mL of this buffer solution, aliquots of the ATCh substrate were sequentially added and the redox response of the biosensor was measured. The AChE biosensor response was recorded on 0, 7, 14, 21 and 28 days with the results obtained displayed in Figure 3 (Somerset et al., 2009). Analysis of the results in Figure 3 have shown that for each of the time interval evaluated, the AChE biosensor reached a maximum current response when 0.4 to 1.0 mM of ATCh substrate was added. Within the first 21 days, the current response showed minimum changes after reaching a plateau of 1.1 mM of the substrate added. However, after the 21 days of exposure, the results of the 28th day evaluation clearly indicated a much reduced maximum current response compared to the first 21 days of operation.

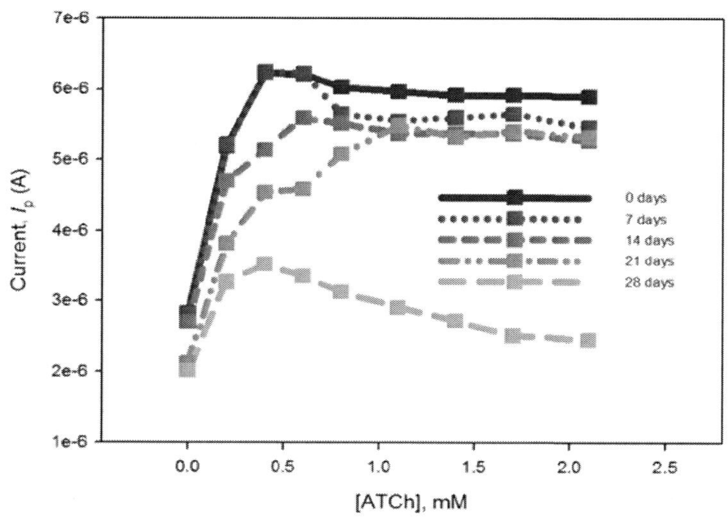

Figure 3. Effect of time (0, 7, 14, 21, 28 days) on the peak current response of the AuE/MBT/PANI/AChE/PVAc thick-film biosensor in 0.1 M phosphate buffer, KCl (pH = 7.2) solution, with 2 mM of acetylthiocholine chloride (ATCh) as substrate added. The absolute values of the peak current are shown for graphing purposes.

The above results compared favourable to that obtained by Sen et al. (2004), in which an AChE biosensor gave favourable redox responses up to 25 days (Somerset et al., 2009).

The results obtained for the influence of the temperature of the solution environment to which the AChE enzyme was exposed was also studied. The amperometric responses of the Au/MBT/PANI/AChE/PVAc biosensor to successive additions of the substrate ATCh in a 1 mL of 0.1 M phosphate buffer (pH = 7.2; 0.1 M KCl) solution, was determined at different temperatures ranging between 10 to 35 °C (283 – 308 K) (Somerset et al., 2009).

A summary of the peak current results obtained for each temperature is shown in Table 2.

Table 2. Summary of the peak current responses obtained for the AuE/MBT/PANI/AChE/PVAc thick-film biosensor evaluated at different temperature values in 0.1 M phosphate buffer, KCl (pH = 7.2) solution, with 2 mM of acetylthiocholine chloride (ATCh) as substrate added

T (°C)	T (K)	Ip (A)
10	283	-3.01×10^{-6}
15	288	-3.80×10^{-6}
20	293	-5.23×10^{-6}
25	298	-6.30×10^{-6}
30	303	-3.99×10^{-6}
35	308	-4.79×10^{-6}

The dependence of the peak current response on temperature was further evaluated, using the electrochemical version of the Arrhenius equation, shown in Eqn. 3 (Nandini et al., 2013):

$$I\,(T) = i_0 \exp(\frac{-E_a}{RT}) \qquad \text{(Eqn. 3)}$$

where i_0 is the peak current corresponding to each temperature, R is the gas constant, T is the temperature in Kelvin degrees, and E_a is the activation energy. The data in Table 2 was then used to plot a graph of log I vs. $\frac{1}{T}$ for the temperature range, 283 – 308 K. From the slope of this plot the activation energy was calculated to be 1.84 kJ/mol. This small E_a value was an indication that the thick-film AChE biosensor displayed a high enzymatic activity and affinity to ATCh as substrate.

3.3. Analytical Performance of the AChE Biosensor

The data plotted in Figure 1 was further examined and analysed to summarise the analytical characteristics of the AuE/MBT/PANI/AChE/PVAc biosensor, obtained under optimised conditions using acetythiocholine chloride as substrate.

The apparent Michaelis-Menten constant, K_M^{app}, which describes the enzyme affinity, was calculated using the electrochemical version of the Lineweaver-Burk equation, given below in Eqn. 4 (Nandini et al., 2013):

$$\frac{1}{I_{ss}} = \frac{1}{I_{max}} + \frac{K_M^{app}}{I_{max}} \cdot \frac{1}{C} \qquad \text{(Eqn. 4)}$$

where I_{ss} is the steady state current measured after the addition of ATCh as substrate; I_{max} represents the maximum current measured under saturated ATCh substrate concentration; and C is the ATCh substrate concentration at saturation. A graph of $\frac{1}{I_{ss}}$ vs. $\frac{1}{C}$ was then constructed and from the slope of the graph, the value of K_M^{app} was estimated to be 1.55×10^{-3} M (Table 3).

Table 3. Summary of the results for the analytical characteristics of the AuE/MBT/PANI/AChE/PVAc biosensor obtained from the graphs displayed in Figure 1

Linear range (mM)	Sensitivity (μA/mM.cm²)	R^2	I_{max}	Michaelis-Menten constant (K_M^{app}; M)
0.2 – 1.6	26.68	0.974	-6.30×10^{-6}	1.55×10^{-3}

Analysis of the results in Table 3 has shown that the thick-film AChE biosensor is very sensitive with a favourable linear range towards ATCh as substrate.

3.4. Detection of Chlorpyrifos Pesticide

The detection of chlorpyrifos as pesticide was achieved using the percentage inhibition method. This method involved placing the biosensor first in a stirred 1 mL of 0.1 M phosphate (pH = 7.2; 0.1 M KCl) solution (anaerobic conditions), to which multiple additions of a standard acetylthiocholine (ATCh) substrate solution was added until a stable current was obtained. This steady state current obtained was related to the activity of the enzyme in the biosensor when no chlorpyrifos was present (Somerset et al., 2009).

Immediately after a steady current was obtained, the biosensor was incubated in anaerobic conditions for 20 min with a standard chlorpyrifos phosphate buffer-organic solvent mixture. The activity of the AChE biosensor was then again evaluated by multiple additions of a standard ATCh substrate solution (anaerobic conditions) in a fresh 1 mL of 0.1 M phosphate (pH = 7.2; 0.1 M KCl) solution (anaerobic conditions), until a stable current was obtained. This procedure was repeated for several concentrations of chlorpyrifos. The percentage inhibition for each chlorpyrifos concentration was then calculated using the formula, shown in Eqn. 5 (Somerset et al., 2009; Zamfir et al., 2011; Raghu et al., 2012; Zhai et al., 2014):

$$I\% = \frac{(I_1 - I_2)}{I_1} \times 100 \qquad \text{(Eqn. 5)}$$

where $I\%$ is the degree of inhibition, I_1 is the steady-state current obtained in buffer solution before exposure to chlorpyrifos, I_2 is the steady-state current obtained after the biosensor was incubated for 20 min in a chlorpyrifos phosphate buffer-organic solvent mixture (Somerset et al., 2009).

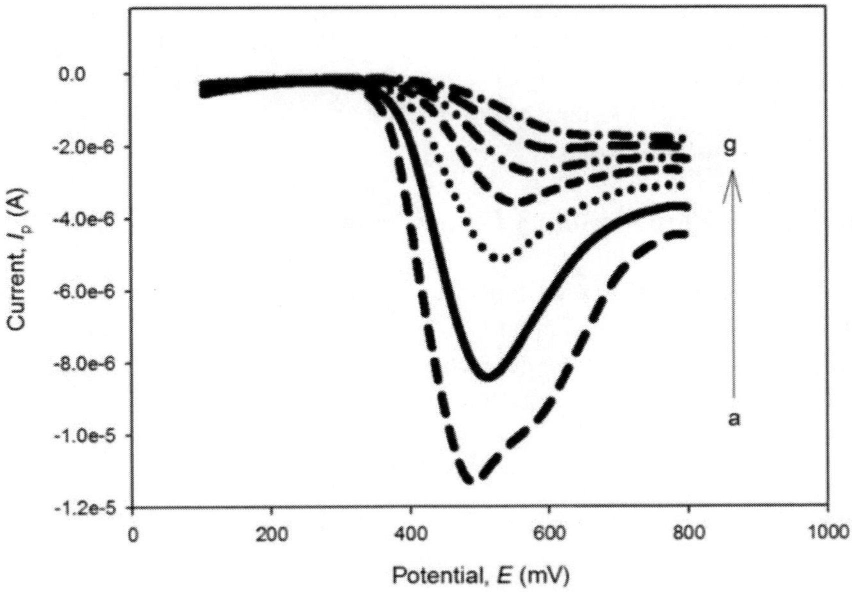

Figure 4. Results obtained for the anodic difference SWV responses for ATCh additions to the AuE/MBT/PANI/AChE/PVAc biosensor in 0.1 M phosphate buffer (pH = 7.2; 0.1 M KCl) solution, after inhibition and exposure for 20 min in a chlorpyrifos phosphate buffer-organic solvent mixture. The a-g corresponds to 0.2 to 1.6 mM of ATCh in 0.1 M phosphate buffer (pH = 7.2; 0.1 M KCl) solution.

The anodic difference square wave voltammograms (SWV) responses was collected in an oxidation direction only by applying a linear potential scan in a shorter potential window between +100 mV and +800 mV (vs. Ag/AgCl), at a scan rate of 10 mV/s. For the SWV responses a step potential of 4 mV, a frequency of 5 Hz, and square amplitude of 50 mV was used (Somerset et al., 2009). Typical results obtained are shown in Figure 4.

The performance of the AuE/MBT/PANI/AChE/PVAc biosensor was further compared to other reported AChE biosensor, with the results summarised in Table 4. The limit of detection (LOD) was determined using percentage inhibition data for chlorpyrifos detection, and using the equation of $LOD = \frac{3 \times SD}{m}$, with a signal-to-noise (S/N) ratio of 3 and where SD is the standard deviation of the blank signal (*n* = *10*), *m* is the slope of the calibration plot (Silwana et al., 2013). The results obtained are listed in Table 4:

Table 4. Results for the comparison of the performance of the AuE/MBT/PANI/AChE/PVAc biosensor for the detection of chlorpyrifos with other AChE biosensors from literature

Biosensor	Linear range	LOD	Reference
AuE/MBT/PANI/AChE/PVAc	0.24 – 0.70 μg/L	0.012 μg/L	This work
GCE/GN-AuNPs/CLDH-AChE	0.05 – 150 μg/L	0.05 μg/L	Zhai et al., 2014
GCE/CMCS-AuNPs/AChE	0.1 – 100 μg/L	0.007 μg/L	Sun et al., 2013

The results in Table 4 have shown that the AChE biosensor constructed in this study exhibited a more sensitive linear range between 0.24 – 0.70 μg/L and the detection limit was 0.012 μg/L, more sensitive than the AChE biosensor constructed by Zhai et al. (2014) and relatively similar in sensitivity to the sensor constructed by Sun et al. (2013).

3.5. Inhibition Plots for Standard Chlorpyrifos Samples

Following the results obtained for the detection of chlorpyrifos with the AChE biosensor (Figure 4; Table 4), the percentage inhibition results were calculated.

In Figure 5(a) the results for the percentage inhibition vs. chlorpyrifos concentration is plotted, while in Figure 5(b) the results for the percentage inhibition vs. – log chlorpyrifos concentration is shown.

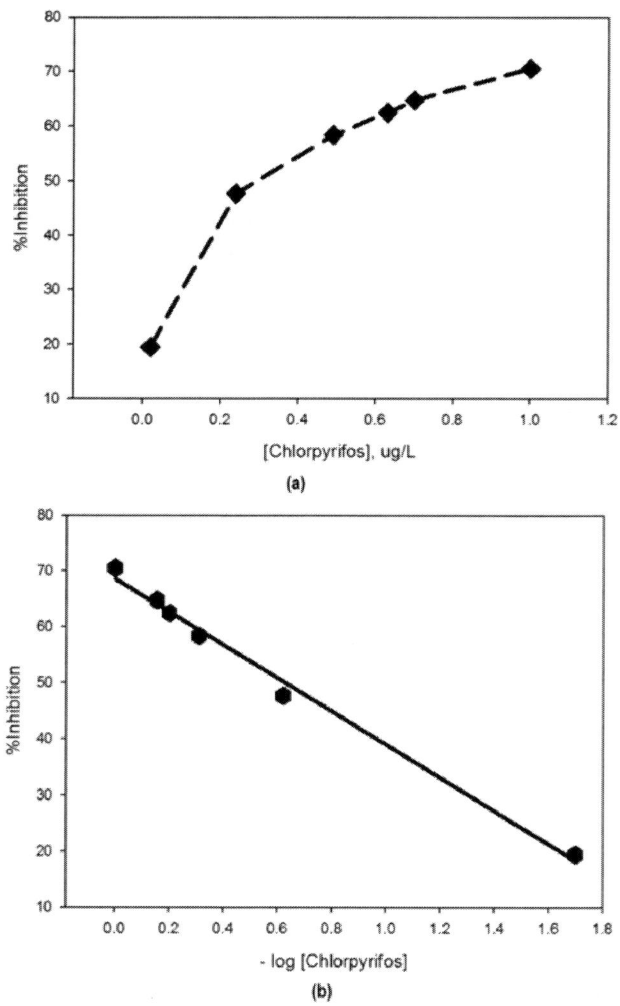

Figure 5. Graphs of percentage inhibition vs. [Chlorpyrifos] results in (a) for concentrations between 0.02 – 1 μg/L; and in (b) the graph of percentage inhibition vs. – log [Chlorpyrifos]. Results were obtained with the use of the AuE/MBT/PANI/AChE/PVAc biosensor.

Analysis of the results in Figure 5 has shown a steady increase in the percentage inhibition was obtained as the chlorpyrifos concentration was evaluated between 0.02 – 1 μg/L. The results further indicated that a very good inhibition of approximately 70% can be achieved with the AChE biosensor constructed in study, when the biosensor is applied to chlorpyrifos concentration less than 1 μg/L.

3.6. Biosensor Behaviour in Aqueous Organic Solvent Solutions

Another important aspect of the AChE biosensor, its behaviour in organic solvents, was also investigated in this study. The effect of organic solvents on the activity of the constructed AuE/MBT/PANI/AChE/PVAc biosensor was evaluated in the presence of various organic solvents containing a 0 – 10% aqueous solution of distilled water. The organic solvents evaluated in this study included acetonitrile, acetone, ethanol and hexane.

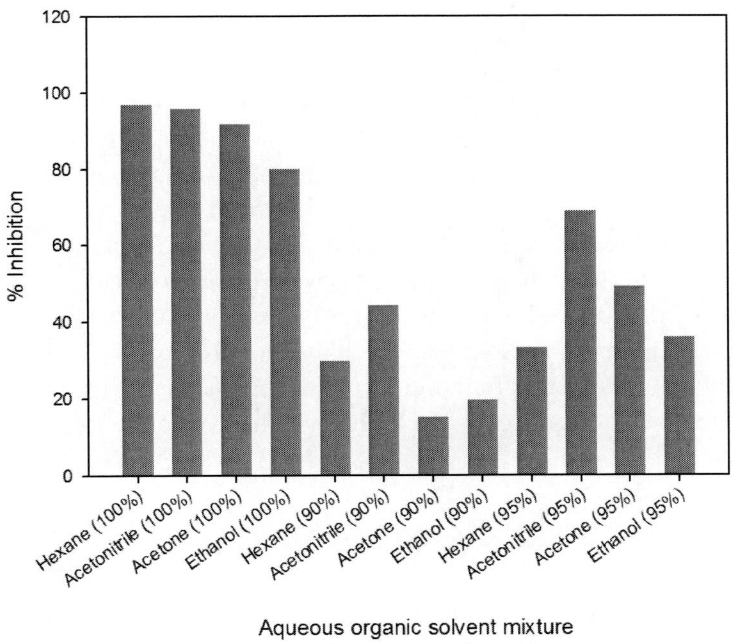

Figure 6. Results obtained for the inhibition behaviour of the AChE enzyme in the Au/MBT/PANI/AChE/PVAc thick-film biosensor evaluated in: (a) 5% water-organic solvent mixture; (b) 10% water-organic solvent mixtures; and (c) pure organic solvent (Somerset et al., 2009).

In Figure 6, the results obtained for the behaviour of the AChE biosensor in the different aqueous-solvent mixtures are displayed.

Analysis of the results in Figure 6 has shown that the behaviour of the AuE/MBT/PANI/AChE/PVAc biosensor was different in each of the organic solvents evaluated. In each of the cases where 100% organic solvent was used, the AChE biosensor results showed the highest percentage inhibition, with the inhibition the highest in hexane > acetonitrile > acetone > ethanol.

The results for the 10% water-organic solvent mixture have shown that the highest decrease in the catalytic activity of the AChE biosensor was observed in acetonitrile > hexane

> ethanol > acetone. Analysis of the results for the 5% water-organic solvent mixture showed that the highest inhibition was experienced by the AChE biosensor in acetonitrile > acetone > ethanol > hexane.

The above results have illustrated that the enzyme, AChE, needs to stay hydrated when in contact with polar organic solvents in order to optimise the functioning of the biosensor. The selection of organic solvent should be determined from the best results obtained for the AChE biosensor performance in the aqueous-organic solvent mixtures.

These results are therefore in line with the investigations reported by Evtugyn et al. (1998), highlighting the presence of water in the enzyme's active centre during biosensor functioning (Andreescu et al., 2002; Somerset et al., 2009; 2011).

Conclusion

This study described the construction and application of the AuE/MBT/PANI/AChE/PVAc thick-film biosensor for the detection of chlorpyrifos pesticide in various organic solvents of acetonitrile, acetone, ethanol, and hexane. The biosensor results have shown that a self-assembled monolayer can be applied in thick film biosensor construction and that the poly(vinyl acetate) film does not interfere with the PANI-AChE electrocatalytic activity towards acetylthiocholine chloride, as substrate. Application of the constructed biosensor to aqueous-organic solvent mixtures illustrated the effect of water on the functioning of the AChE enzyme activity, showing that the presence of water has a considerable effect on the functioning of the AChE-biosensor. The constructed AuE/MBT/PANI/AChE/PVAc thick-film biosensor offered a reasonable linear range between 0.24 – 0.70 μg/L for chlorpyrifos detection. The biosensor also exhibited a detection limit of 0.012 μg/L for chlorpyrifos. It was further determined that the detection limit was below the allowable concentration of 1 μg/L set by the WHO guideline value for drinking water.

Acknowledgments

The authors express their gratitude to the National Research Foundation (NRF) of South Africa for their financial support. The work reported was conducted at the SensorLab, Chemistry Department, University of the Western Cape, Bellville, South Africa.

References

Andreescu, S., Noguer, T., Magearu, V. and Marty, J-L. (2002). Screen-printed electrode based on AChE for the detection of pesticides in presence of organic solvents. *Talanta*, 57: 169–176.

Banks, K.E., Hunter, D.H. and Wachal, D.J. (2005). Chlorpyrifos in surface waters before andafter a federally mandated ban. *Environment International*, 31: 351–356.

Eaton, D.L., Daroff, R.B., Autrup, H., Bridges, J., Buffler, P., Costa, L.G., Coyle, J., McKhann, G., Mobley, W.C., Nadel, L., Neubert, D., Schulte-Hermann, R. and Spencer,

P.S. (2008). Review of the toxicology of chlorpyrifos with an emphasis on human exposure and neurodevelopment. *Critical Reviews in Toxicology*, 38: 1–125.

Evtugyn, G.A., Budnikov, H.C. and Nikolskaya, E.B. (1998). Sensitivity and selectivity of electrochemical sensors for inhibitor determination. *Talanta*, 46, 465–484.

Istamboulie, G., Fournier, D., Marty, J-L. and Noguer, T. (2009). Phosphotriesterase: A complementary tool for the selective detection of two organophosphate insecticides: Chlorpyrifos and chlorfenvinfos. *Talanta*, 77: 1627–1631.

Kesik, M., Kanik, F.E., Turan, J., Kolb, M., Timur, S., Bahadir, M. and Toppare, L. (2014). An acetylcholinesterase biosensor based on a conducting polymer using multi-walled carbon nanotubes for amperometric detection of organophosphorous pesticides. *Sensors and Actuators B*, 205: 39–49.

Nandini, S., Nalini, S., Manjunatha, R., Shanmugam, S., Melo, J.S. and Suresh, G.S. (2013). Electrochemical biosensor for the selective determination of hydrogen peroxide based on the co-deposition of palladium, horseradish peroxidase on functionalized-graphene modified graphite electrode as composite. *Journal of Electroanalytical Chemistry*, 689: 233–242.

Noyrod, P., Chailapakul, O., Wonsawat, W. and Chuanuwatanakul, S. (2014). The simultaneous determination of isoproturon and carbendazim pesticides by single drop analysis using a graphene-based electrochemical sensor. *Journal of Electroanalytical Chemistry*, 719: 54–59.

Raghu, P., Swamy, B.E.K., Reddy, T.M., Chandrashekar, B.N. and Reddaiah, K. (2012). Sol–gel immobilized biosensor for the detection of organophosphorous pesticides: A voltammetric method. *Bioelectrochemistry*, 83: 19–24.

Sen, S., Gülce, A. and Gülce, H. (2004). Polyvinylferrocenium modified Pt electrode for the design of amperometric choline and acetylcholine enzyme electrodes. *Biosensors and Bioelectronics*, 19: 1261–1268.

Silwana, B., van der Horst, C., Iwuoha, E. and Somerset, V. (2013). Inhibitive determination of metal ions using a horseradish peroxidase amperometric biosensor. In T. Rinken (Ed.), *State of the Art in Biosensors - Book 2*. (pp 105-120). INTECH, Croatia. ISBN 980-953-307-691-6.

Silwana, B., Van der Horst, C., Iwuoha, E. and Somerset, V. (2014). Amperometric determination of cadmium, lead, and mercury metal ions using a novel polymer immobilised horseradish peroxidase biosensor system. *Journal of Environmental Science and Health*, A49: 1501–1511.

Somerset, V.S., Klink, M.J., Sekota, M.M.C., Baker, P.G.L. and Iwuoha, E.I. (2006). Polyaniline-Mercaptobenzothiazole Biosensor for Organophosphate and Carbamate Pesticides. *Analytical Letters*, 39, 1683–1698.

Somerset, V.S. (2007). *Mercaptobenzothiazole-on-gold biosensor systems for organophosphate and carbamate pesticide sompounds*. Unpublished PhD Thesis. University of the Western Cape, Bellville, South Africa, 1–531.

Somerset, V., Baker, P. and Iwuoha, E. (2009). Mercaptobenzothiazole-on-gold organic phase biosensor systems: 1. Enhanced organosphosphate pesticide determination. *Journal of Environmental Science and Health,* B44: 164–178.

Somerset, V., Baker, P. and Iwuoha, E. (2011). Mercaptobenzothiazole-on-Gold Organic Phase Biosensor Systems: 4. Effect of Organic Solvents on Organophosphate and Carbamate Pesticide Determination. In M. Stoytcheva (Ed.), Pesticides - Strategies for

Pesticides Analysis. (pp 373-386). INTECH, Croatia (January 2011). ISBN: 978-953-307-460-3.

Sun, X., Zhai, C., Wang, X.Y. (2013). Amperometric Acetylcholinesterase Biosensor Based on O-Carboxymethyl Chitosan-Gold Nanoparticle Nanocomposite. *IEEE Sensors Journal*, 13(1): 172-179.

Uygun, Z.O. and Dilgin, Y. (2013). A novel impedimetric sensor based on molecularly imprintedpolypyrrole modified pencil graphite electrode for trace leveldetermination of chlorpyrifos. *Sensors and Actuators B*, 188: 78– 84.

Viswanathan, S., Radecka, H. and Radecki, J. (2009). Electrochemical biosensor for pesticides based on acetylcholinesterase immobilized on polyaniline deposited on vertically assembled carbon nanotubes wrapped with ssDNA. *Biosensors and Bioelectronics*, 24: 2772–2777.

Zhai, C., Guo, Y., Sun, X., Zheng, Y. and Wang, X. (2014). An acetylcholinesterase biosensor based on graphene–gold nanocomposite and calcined layered double hydroxide. *Enzyme and Microbial Technology*, 58–59: 8–13.

INDEX

#

A

B

C

D

E

F

G

H

I

J

K

L

M

N

O

P

Q

R

S

T

U